W0091730

Modern Genetic Analysis

Modern Genetic Analysis

Contributors

Weidong Mao & Jeonghwa Le et al.

AURIS
Reference

www.aurisreference.com

Modern Genetic Analysis

Contributors: Weidong Mao & Jeonghwa Le et al.

Published by Auris Reference Limited

www.aurisreference.com

United Kingdom

Copyright 2016
Printed in 2017 for Sale in the Indian Subcontinent

The information in this book has been obtained from highly regarded resources. The copyrights for individual articles remain with the authors, as indicated. All chapters are distributed under the terms of the Creative Commons Attribution License, which permit unrestricted use, distribution, and reproduction in any medium, provided the original author and source are credited.

Notice

Contributors, whose names have been given on the book cover, are not associated with the Publisher. The editors and the Publisher have attempted to trace the copyright holders of all material reproduced in this publication and apologise to copyright holders if permission has not been obtained. If any copyright holder has not been acknowledged, please write to us so we may rectify.

Reasonable efforts have been made to publish reliable data. The views articulated in the chapters are those of the individual contributors, and not necessarily those of the editors or the Publisher. Editors and/or the Publisher are not responsible for the accuracy of the information in the published chapters or consequences from their use. The Publisher accepts no responsibility for any damage or grievance to individual(s) or property arising out of the use of any material(s), instruction(s), methods or thoughts in the book.

Modern Genetic Analysis

ISBN: 978-1-78154-774-8

British Library Cataloguing in Publication Data
A CIP record for this book is available from the British Library

Printed in the United Kingdom

Exclusively distributed by CBS Publishers & Distributors Pvt. Ltd.

Sales & Distribution Rights only for India, Pakistan, Bangladesh, Sri Lanka, Nepal and Bhutan. This book is not to be sold outside these territories.

Contents

List of Abbreviations

AJCPD	Abnormal junction of the choledochopancreatic ducts
APC	Adenomatous polyposis coli
ASO	Allele-specific oligonucleotide
ABC	Approximate Bayesian Computation
BMI	Body mass index
CNS-GCTs	Central nervous system germ cell tumors
CGH	Comparative genomic hybridization
CMLM	Compressed mixed linear model
ERV	Endogenous retrovirus
EGDC	Equine Genetic Diversity Consortium
GIF	Genealogical Index of Familiality
GRF	General random forest
GWAS	Genome-wide association studies
GBS	Genotyping-by-sequencing
HVR-I	Hypervariable region I
IBD	Inflammatory bowel disease
IACUC	Institutional Animal Care and Use Committee
IALHA I	nternational Andalusian and Lusitano Horse Association Registry
LD	Linkage disequilibria
LTR	Long terminal repeat
MSY	Male specific euchromatic
MTA	Material Transfer Agreement
MPWD	Mean number of pairwise differences
MSI	Microsatellite instability
MAF	Minor allele frequency
mtDNA	Mitochondrial DNA
MAGIC	Multiparent advanced generation intercrosses
MSC	Multi-SNPs combination
NHANES	National Health and Nutrition Examination Survey
NAM	Nested association mapping
ORF	Optimum random forest
PCA	Principal components analysis
PAR	Pseudoautosomal region
PODs	Pseudo-observed datasets
QTLs	Quantitative trait loci
ROC	Receiver operating characteristic
RIL	Recombinant inbred line
RGE	Relative genotypic effect
rCRS	Revised Cambridge Reference Sequence
SSvR	Seedling survival rates

SSLP	Simple sequence length polymorphism
SSR	Simple-sequence repeat
SNP	Single nucleotide polymorphism
SNACs	Sinonasal adenocarcinomas
SDS	Sodium dodecyl sulfate
SAP	Sorghum association panel
SVM	Support Vector Machine Algorithm
TWHF	Tennessee Walking Horse Foundation
TRH	Thyrotropin-releasing hormone
UPDB	Utah Population Data Base
WGS	Whole Genome Shotgun

List of Contributors

Weidong Mao
Department of Mathematics & Computer Science Virginia State University Petersburg, VA 23806, USA

Jeonghwa Lee
Department of Computer Science Shippensburg University Shippensburg, PA 17257, USA

Anil V Parwani M.D., Ph.D.
Department of Pathology, The Johns Hopkins Hospital, Baltimore, Maryland

Joseph Geradts M.D.
Department of Pathology and Laboratory Medicine, Roswell Park Cancer Institute, Buffalo, New York

Eric Caspers M.D.
The Academic Medical Center, Amsterdam, The Netherlands

G Johan Offerhaus M.D.
The Academic Medical Center, Amsterdam, The Netherlands

Charles J Yeo M.D.
Department of Surgery, The Johns Hopkins Hospital, Baltimore, Maryland
Department of Oncology, The Johns Hopkins Hospital, Baltimore, Maryland

John L Cameron M.D.
Department of Surgery, The Johns Hopkins Hospital, Baltimore, Maryland

David S Klimstra M.D.
Department of Pathology, Memorial Sloan-Kettering Cancer Center, New York

Anirban Maitra M.D.
Department of Pathology, The Johns Hopkins Hospital, Baltimore, Maryland

Ralph H Hruban M.D.
Department of Pathology, The Johns Hopkins Hospital, Baltimore, Maryland
Department of Oncology, The Johns Hopkins Hospital, Baltimore, Maryland

Pedram Argani M.D.
Department of Pathology, The Johns Hopkins Hospital, Baltimore, Maryland

Takashi Kuramoto
Institute of Laboratory Animals, Graduate School of Medicine, Kyoto University, Sakyo-ku, Kyoto, Japan

Satoshi Nakanishi
Institute of Laboratory Animals, Graduate School of Medicine, Kyoto University, Sakyo-ku, Kyoto, Japan

Masako Ochiai
National Cancer Center Research Institute, Chuo-ku, Tokyo, Japan

Hitoshi Nakagama
National Cancer Center Research Institute, Chuo-ku, Tokyo, Japan

Birger Voigt
Institute of Laboratory Animals, Graduate School of Medicine, Kyoto University, Sakyo-ku, Kyoto, Japan

Tadao Serikawa
Institute of Laboratory Animals, Graduate School of Medicine, Kyoto University, Sakyo-ku, Kyoto, Japan

Dominik T Schneider
Clinic of Paediatric Oncology, Haematology and Immunology, Heinrich-Heine-University, Düsseldorf, Germany

Susanne Zahn
Clinic of Paediatric Oncology, Haematology and Immunology, Heinrich-Heine-University, Düsseldorf, Germany

Sonja Sievers
Clinic of Paediatric Oncology, Haematology and Immunology, Heinrich-Heine-University, Düsseldorf, Germany
Max Planck Institute of Molecular Physiology, Dortmund, Germany

Katayoun Alemazkour
Clinic of Paediatric Oncology, Haematology and Immunology, Heinrich-Heine-University, Düsseldorf, Germany

Guido Reifenberger
Department of Neuropathology, Heinrich-Heine-University, Düsseldorf, Germany

Otmar D Wiestler
German Brain Tumor Reference Center Institute of Neuropathology, University of Bonn, Bonn, Germany

Gabriele Calaminus
Clinic of Paediatric Oncology, Haematology and Immunology, Heinrich-Heine-University, Düsseldorf, Germany

Ulrich Göbel
Clinic of Paediatric Oncology, Haematology and Immunology, Heinrich-Heine-University, Düsseldorf, Germany

Elizabeth J Perlman
Department of Pathology, Children's Memorial Hospital, Chicago, IL, USA

Sue S Yom
Department of Radiation Oncology, The University of Texas MD Anderson Cancer Center, Houston, TX, USA

Asif Rashid
Department of Pathology, The University of Texas MD Anderson Cancer Center, Houston, TX, USA

David I Rosenthal
Department of Radiation Oncology, The University of Texas MD Anderson Cancer Center, Houston, TX, USA

Danielle D Elliott
Department of Pathology, The University of Texas MD Anderson Cancer Center, Houston, TX, USA

Ehab Y Hanna
Department of Head and Neck Surgery, The University of Texas MD Anderson Cancer Center, Houston, TX, USA

Randal S Weber
Department of Head and Neck Surgery, The University of Texas MD Anderson Cancer Center, Houston, TX, USA

Adel K El-Naggar
Department of Pathology, The University of Texas MD Anderson Cancer Center, Houston, TX, USA
Department of Head and Neck Surgery, The University of Texas MD Anderson Cancer Center, Houston, TX, USA

Rui Sousa-Neves
Department of Biology, Case Western Reserve University, Cleveland, Ohio, United States of America

Alexandre Rosas
Departamento de Fı́sica, CCEN, Universidade Federal da Paraı́ba, Joãᵒo Pessoa, Paraı́ba, Brazil

Qiuying Huang
Hubei Insect Resources Utilization and Sustainable Pest Management Key Laboratory, Huazhong Agricultural University, Wuhan, China

Ganghua Li
Hubei Insect Resources Utilization and Sustainable Pest Management Key Laboratory, Huazhong Agricultural University, Wuhan, China

Claudia Husseneder
Department of Entomology, Louisiana State University Agricultural Center, Baton Rouge, Louisiana, United States of America

Chaoliang Lei
Hubei Insect Resources Utilization and Sustainable Pest Management Key Laboratory, Huazhong Agricultural University, Wuhan, China

Yinghua Pan
Key Laboratory of Crop Heterosis and Utilization, Ministry of Education, Beijing Key Laboratory of Crop Genetic Improvement, China Agricultural University, Beijing, 100193, China
Rice Research Institute, Guangxi Academy of Agricultural Sciences, Nanning, Guangxi, 530005, China

Hongliang Zhang
Key Laboratory of Crop Heterosis and Utilization, Ministry of Education, Beijing Key Laboratory of Crop Genetic Improvement, China Agricultural University, Beijing, 100193, China

Dongling Zhang
Institute of Crop Science, Chinese Academy of Agricultural Sciences, the National Key Facility for Crop Gene Resources and Genetic Improvement, Key Laboratory of Crop Germplasm Resources and Biotechnology, Ministry of Agriculture, Beijing, 100081, China

Jinjie Li
Key Laboratory of Crop Heterosis and Utilization, Ministry of Education, Beijing Key Laboratory of Crop Genetic Improvement, China Agricultural University, Beijing, 100193, China

Haiyan Xiong
Key Laboratory of Crop Heterosis and Utilization, Ministry of Education, Beijing Key Laboratory of Crop Genetic Improvement, China Agricultural University, Beijing, 100193, China

Jianping Yu
Key Laboratory of Crop Heterosis and Utilization, Ministry of Education, Beijing Key Laboratory of Crop Genetic Improvement, China Agricultural University, Beijing, 100193, China

Jilong Li
Key Laboratory of Crop Heterosis and Utilization, Ministry of Education, Beijing Key Laboratory of Crop Genetic Improvement, China Agricultural University, Beijing, 100193, China

Muhammad Abdul Rehman Rashid
Key Laboratory of Crop Heterosis and Utilization, Ministry of Education, Beijing Key Laboratory of Crop Genetic Improvement, China Agricultural University, Beijing, 100193, China

Gangling Li
Key Laboratory of Crop Heterosis and Utilization, Ministry of Education, Beijing Key Laboratory of Crop Genetic Improvement, China Agricultural University, Beijing, 100193, China

Xiaoding Ma
Institute of Crop Science, Chinese Academy of Agricultural Sciences, the National Key Facility for Crop Gene Resources and Genetic Improvement, Key Laboratory of Crop Germplasm Resources and Biotechnology, Ministry of Agriculture, Beijing, 100081, China

Guilan Cao
Institute of Crop Science, Chinese Academy of Agricultural Sciences, the National Key Facility for Crop Gene Resources and Genetic Improvement, Key Laboratory of Crop Germplasm Resources and Biotechnology, Ministry of Agriculture, Beijing, 100081, China

Longzhi Han
Institute of Crop Science, Chinese Academy of Agricultural Sciences, the National Key Facility for Crop Gene Resources and Genetic Improvement, Key Laboratory of Crop Germplasm Resources and Biotechnology, Ministry of Agriculture, Beijing, 100081, China

Zichao Li
Key Laboratory of Crop Heterosis and Utilization, Ministry of Education, Beijing Key Laboratory of Crop Genetic Improvement, China Agricultural University, Beijing, 100193, China

William R. Yates
Laureate Institute for Brain Research, Tulsa, Oklahoma, United States of America
University of Oklahoma College of Medicine, Tulsa, Tulsa, Oklahoma, United States of America

Craig Johnson
Eating Recovery Center, Denver, Colorado, United States of America

Patrick McKee
Laureate Institute for Brain Research, Tulsa, Oklahoma, United States of America
University of Oklahoma College of Medicine, Oklahoma City, Oklahoma, United States of America

Lisa A. Cannon-Albright
Department of Medicine (Genetic Epidemiology), University of Utah School of Medicine, Salt Lake City, Utah, United States of America
George E. Wahlen Department of Veterans Affairs Medical Center, Salt Lake City, Utah, United States of America

Dong Zhang
Plant Genome Mapping Laboratory, University of Georgia, Athens, GA 30602, USA
Institute of Bioinformatics, University of Georgia, Athens, GA 30602, USA

Wenqian Kong
Plant Genome Mapping Laboratory, University of Georgia, Athens, GA 30602, USA
Department of Crop and Soil Sciences, University of Georgia

Jon Robertson
Plant Genome Mapping Laboratory, University of Georgia, Athens, GA 30602, USA

Valorie H Goff
Plant Genome Mapping Laboratory, University of Georgia, Athens, GA 30602, USA

Ethan Epps
Plant Genome Mapping Laboratory, University of Georgia, Athens, GA 30602, USA

Alexandra Kerr
Plant Genome Mapping Laboratory, University of Georgia, Athens, GA 30602, USA

Gabriel Mills
Plant Genome Mapping Laboratory, University of Georgia, Athens, GA 30602, USA

Jay Cromwell
Plant Genome Mapping Laboratory, University of Georgia, Athens, GA 30602, USA

Yelena Lugin
Plant Genome Mapping Laboratory, University of Georgia, Athens, GA 30602, USA

Christine Phillips
Plant Genome Mapping Laboratory, University of Georgia, Athens, GA 30602, USA

Andrew H Paterson
Plant Genome Mapping Laboratory, University of Georgia, Athens, GA 30602, USA
Institute of Bioinformatics, University of Georgia, Athens, GA 30602, USA
Department of Crop and Soil Sciences, University of Georgia
Department of Plant Biology, University of Georgia
Department of Genetics, University of Georgia

Jessica L. Petersen
University of Minnesota, College of Veterinary Medicine, St Paul, Minnesota, United States of America

James R. Mickelson
University of Minnesota, College of Veterinary Medicine, St Paul, Minnesota, United States of America

E. Gus Cothran
Texas A&M University, College of Veterinary Medicine and Biomedical Science, College Station, Texas, United States of America

Lisa S. Andersson
Swedish University of Agricultural Sciences, Department of Animal Breeding and Genetics, Uppsala, Sweden

Jeanette Axelsson
Swedish University of Agricultural Sciences, Department of Animal Breeding and Genetics, Uppsala, Sweden

Ernie Bailey
University of Kentucky, Department of Veterinary Science, Lexington, Kentucky, United States of America

Danika Bannasch
University of California Davis, School of Veterinary Medicine, Davis, California, United States of America

Matthew M. Binns
Equine Analysis, Midway, Kentucky, United States of America

Alexandre S. Borges
University Estadual Paulista, Department of Veterinary Clinical Science, Botucatu-SP, Brazil

Pieter Brama
University College Dublin, School of Veterinary Medicine, Dublin, Ireland

Artur da Caˆmara Machado
University of Azores, Institute for Biotechnology and Bioengineering, Biotechnology Centre of Azores, Angra do Heroı́smo, Portugal

Ottmar Distl
University of Veterinary Medicine Hannover, Institute for Animal Breeding and Genetics, Hannover, Germany

Michela Felicetti
University of Perugia, Faculty of Veterinary Medicine, Perugia, Italy

Laura Fox-Clipsham
Animal Health Trust, Lanwades Park, Newmarket, Suffolk, United Kingdom

Kathryn T. Graves
University of Kentucky, Department of Veterinary Science, Lexington, Kentucky, United States of America

Ge´ rard Gue´ rin
French National Institute for Agricultural Research-Animal Genetics and Integrative Biology Unit, Jouy en Josas, France

Bianca Haase
University of Sydney, Veterinary Science, New South Wales, Australia

Telhisa Hasegawa
Nihon Bioresource College, Koga, Ibaraki, Japan

Karin Hemmann
University of Helsinki, Faculty of Veterinary Medicine, Helsinki, Finland

Emmeline W. Hill
University College Dublin, College of Agriculture, Food Science and Veterinary Medicine, Belfield, Dublin, Ireland

Tosso Leeb
University of Bern, Institute of Genetics, Bern, Switzerland

Gabriella Lindgren
Swedish University of Agricultural Sciences, Department of Animal Breeding and Genetics, Uppsala, Sweden

Hannes Lohi
University of Helsinki, Faculty of Veterinary Medicine, Helsinki, Finland

Maria Susana Lopes
University of Azores, Institute for Biotechnology and Bioengineering, Biotechnology Centre of Azores, Angra do Heroı´smo, Portugal

Beatrice A. McGivney
University College Dublin, College of Agriculture, Food Science and Veterinary Medicine, Belfield, Dublin, Ireland

Sofia Mikko
Swedish University of Agricultural Sciences, Department of Animal Breeding and Genetics, Uppsala, Sweden

Nicholas Orr
Institute of Cancer Research, Breakthrough Breast Cancer Research Centre, London, United Kingdom

M. Cecilia T Penedo
University of California Davis, School of Veterinary Medicine, Davis, California, United States of America

Richard J. Piercy
Royal Veterinary College, Comparative Neuromuscular Diseases Laboratory, London, United Kingdom

Marja Raekallio
University of Helsinki, Faculty of Veterinary Medicine, Helsinki, Finland

Stefan Rieder
Swiss National Stud Farm, Agroscope Liebefeld-Posieux Research Station, Avenches, Switzerland

Knut H. Røed
Norwegian School of Veterinary Science, Department of Basic Sciences and Aquatic Medicine, Oslo, Norway

Maurizio Silvestrelli
University of Perugia, Faculty of Veterinary Medicine, Perugia, Italy

June Swinburne
Animal Health Trust, Lanwades Park, Newmarket, Suffolk, United Kingdom
Animal DNA Diagnostics Ltd, Cambridge, United Kingdom

Teruaki Tozaki
Laboratory of Racing Chemistry, Department of Molecular Genetics, Utsunomiya, Tochigi, Japan

Mark Vaudin
Animal Health Trust, Lanwades Park, Newmarket, Suffolk, United Kingdom

Claire M. Wade
University of Sydney, Veterinary Science, New South Wales, Australia

Molly E. McCue
University of Minnesota, College of Veterinary Medicine, St Paul, Minnesota, United States of America

Danielle A. Badro
The Lebanese American University, Chouran, Beirut, Lebanon

Bouchra Douaihy
The Lebanese American University, Chouran, Beirut, Lebanon

Marc Haber
The Lebanese American University, Chouran, Beirut, Lebanon
Institut de Biologia Evolutiva (CSIC-UPF), Departament de Cie`ncies de la Salut i de la Vida, Universitat Pompeu Fabra, Barcelona, Spain

Sonia C. Youhanna
The Lebanese American University, Chouran, Beirut, Lebanon

Ange´lique Salloum
The Lebanese American University, Chouran, Beirut, Lebanon

Michella Ghassibe-Sabbagh
The Lebanese American University, Chouran, Beirut, Lebanon

Brian Johnsrud
Modern Thought and Literature, Stanford University, Stanford, California, United States of America

Georges Khazen
The Lebanese American University, Chouran, Beirut, Lebanon

Elizabeth Matisoo-Smith
Allan Wilson Centre for Molecular Ecology and Evolution, University of Otago, Dunedin, New Zealand

David F. Soria-Hernanz
Institut de Biologia Evolutiva (CSIC-UPF), Departament de Cie`ncies de la Salut i de la Vida, Universitat Pompeu Fabra, Barcelona, Spain
The Genographic Project, National Geographic Society, Washington, DC, United States of America

R. Spencer Wells
The Genographic Project, National Geographic Society, Washington, DC, United States of America

Chris Tyler-Smith
The Wellcome Trust Sanger Institute, Wellcome Trust Genome Campus, Hinxton, United Kingdom

Daniel E. Platt
Computational Biology Centre, IBM TJ Watson Research Centre, Yorktown Heights, New York, United States of America

Pierre A. Zalloua
The Lebanese American University, Chouran, Beirut, Lebanon
Harvard School of Public Health, Boston, Massachusetts, United States of America

Barbara Wallner
Institute of Animal Breeding and Genetics, Department of Biomedical Sciences, University of Veterinary Medicine Vienna, Vienna, Austria

Claus Vogl
Institute of Animal Breeding and Genetics, Department of Biomedical Sciences, University of Veterinary Medicine Vienna, Vienna, Austria

Priyank Shukla
Institute of Animal Breeding and Genetics, Department of Biomedical Sciences, University of Veterinary Medicine Vienna, Vienna, Austria

Joerg P. Burgstaller
Institute of Animal Breeding and Genetics, Department of Biomedical Sciences, University of Veterinary Medicine Vienna, Vienna, Austria

Thomas Druml
Institute of Animal Breeding and Genetics, Department of Biomedical Sciences, University of Veterinary Medicine Vienna, Vienna, Austria

Gottfried Brem
Institute of Animal Breeding and Genetics, Department of Biomedical Sciences, University of Veterinary Medicine Vienna, Vienna, Austria

Stefania Vai
Dipartimento di Biologia Evoluzionistica, Università di Firenze, 50122 Florence, Italy

Silvia Ghirotto
Dipartimento di Scienze della Vita e Biotecnologie, Università di Ferrara, 44121 Ferrara, Italy

Elena Pilli
Dipartimento di Biologia Evoluzionistica, Università di Firenze, 50122 Florence, Italy

Francesca Tassi
Dipartimento di Scienze della Vita e Biotecnologie, Università di Ferrara, 44121 Ferrara, Italy

Martina Lari
Dipartimento di Biologia Evoluzionistica, Università di Firenze, 50122 Florence, Italy

Ermanno Rizzi
Institute for Biomedical Technologies, National Research Council, 20090 Segrate, Milan, Italy

Laura Matas-Lalueza
Institut de Biologia Evolutiva, CSIC-UPF, Barcelona 08003, Spain

Oscar Ramirez
Institut de Biologia Evolutiva, CSIC-UPF, Barcelona 08003, Spain

Carles Lalueza-Fox
Institut de Biologia Evolutiva, CSIC-UPF, Barcelona 08003, Spain

Alessandro Achilli
Dipartimento di Chimica, Biologia e Biotecnologie, Università di Perugia, 06123 Perugia, Italy

Anna Olivieri
Dipartimento di Biologia e Biotecnologie "L. Spallanzani", Università di Pavia, 27100,Pavia,Italy

Antonio Torroni
Dipartimento di Biologia e Biotecnologie "L. Spallanzani", Università di Pavia, 27100,Pavia,Italy

Hovirag Lancioni
Dipartimento di Chimica, Biologia e Biotecnologie, Università di Perugia, 06123 Perugia, Italy

Caterina Giostra
Dipartimento di Storia, Archeologia e Storia dell'arte, Università Cattolica del Sacro Cuore, 20123 Milano, Italy

Elena Bedini
Anthropozoologica L.B.A. s.n.c., 57123 Livorno, Italy

Luisella Pejrani Baricco
Soprintendenza per i Beni Archeologici del Piemonte, 10122 Turin, Italy

Giuseppe Matullo
Human Genetics Foundation, 10125 Turin, Italy

Cornelia Di Gaetano
Human Genetics Foundation, 10125 Turin, Italy

Alberto Piazza
Human Genetics Foundation, 10125 Turin, Italy

Krishna Veeramah
Department of Ecology and Evolution, State University of New York, Stony Brook, New York 11794– 5245, United States of America

Patrick Geary
School of Historical Studies, Institute for Advanced Study, Princeton, New Jersey 08540, United States of America

David Caramelli
Dipartimento di Biologia Evoluzionistica, Università di Firenze, 50122 Florence, Italy

Guido Barbujani
Dipartimento di Scienze della Vita e Biotecnologie, Università di Ferrara, 44121 Ferrara, Italy

Preface

Genetic analysis is the overall process of studying and researching in fields of science that involve genetics and molecular biology. The text *Modern Genetic Analysis* focuses on key advances in genetics and cutting-edge experiments and techniques. First chapter uses a combinatorial method to analyze the genetic data for Crohn's disease and search disease-associated factors for given case/control samples. In second chapter, we analyze large series of resected non–small cell and small cell gallbladder carcinomas for Dpc4, pRB, p16, and p53 protein expression and for K-ras gene mutation. The aim of third chapter is to clarify and characterize the origin and nature of albino and hooded mutations using molecular genetic approaches and combine the findings with historical records. Molecular genetic analysis of central nervous system germ cell tumors with comparative genomic hybridization has been presented in fourth chapter. Fifth chapter focuses on genetic analysis of sinonasal adenocarcinoma phenotypes. In sixth chapter, we investigate the genetic changes in three closely related species of *Drosophila* by making pair-wise comparisons of their genomes. Seventh chapter contributes to better understanding of the variance of genetic structure and reproductive mode in the genus *Reticulitermes*. Genetic analysis of cold tolerance at the germination and booting stages in rice by association mapping has been revealed in eighth chapter.Ninth chapter supports a specific genetic contribution in the risk for the low body mass index (BMI) phenotype in the Utah population database. In tenth chapter, we perform genome-wide association studies (GWAS) to investigate six components of sorghum inflorescence morphology and 6 traits related to plant height, then compare GWAS-based associations to positional evidence from meta-analysis of QTL likelihood intervals. Eleventh chapter describes the use of a genome-wide set of autosomal SNPs and 814 horses from 36 breeds to provide the first detailed description of equine breed diversity. Twelfth chapter highlights that Y-chromosome and MTDNA genetics reveal significant contrasts in affinities of modern Middle Eastern populations with European and African populations. Last chapter explores on the identification of genetic variation on the horse Y chromosome and the tracing of male founder lineages in modern breeds.

Chapter 1

A COMBINATORIAL ANALYSIS OF GENETIC DATA FOR CROHN'S DISEASE

Weidong Mao[1] & Jeonghwa Lee[2]

[1]Department of Mathematics & Computer Science Virginia State University Petersburg, VA 23806, USA.
[2]Department of Computer Science Shippensburg University Shippensburg, PA 17257, USA

ABSTRACT

The both environmental and genetic factors have roles in the development of some diseases. Complex diseases, suchas Crohn's disease or Typell diabetes, are caused by a combination of environmental factors and mutations in multiple genes. Patients who have been diagnosed with such diseases cannot easily be treated. However, many diseases can be avoided if people at high risk change their living style, one example being their diet. But how can we tell their susceptibility to diseases before symptoms are found and help them make informed decisions about their health? With the development of DNA microarray technique, it is possible to access the human genetic information related to specific diseases. This paper uses a combinatorial method to analyze the genetic data for Crohn's disease and search disease-associated factors for given case/control samples. An optimum random forest based method has been applied to publicly available genotype data on Crohn's disease for association study and achieved a promising result.

INTRODUCTION

Crohn's disease (also known as regional enteritis) is a chronic, episodic, inflammatory condition of the gastrointestinal tract characterized by transmural inflammation (affecting the entire wall of the involved bowel) and skip lesions (areas of inflammation with areas of normal lining in between). Crohn's disease is a type of inflammatory bowel disease (IBD) and can affect any part of the gastrointestinal tract from mouth to anus. As a result, the symptoms of Crohn's disease can vary among affected individuals. The exact cause of Crohn's disease is unknown. However, research shows that the inflammation seen in the people with Crohn's disease involves several factors: the genes the patient has inherited, the immune system itself, and the environment [1]. In other words, genetic factor has been invoked in the pathogenesis of the disease.

Although the Crohn's disease cannot easily be treated, it can be avoided if people at high risk change their living style, such as their diet. But how can we tell the susceptibility of people to the disease before symptoms are found and help them make informed decisions about their health? With the development of DNA microarray technique, it is possible to access the human genetic information related to specific diseases. Assessing the association between DNA variants and disease has been used widely to identify regions of the genome and candidate genes that contribute to disease [2].

99.9% of one individual's DNA sequences are identical to that of another person. Over 80% of this 0.1% difference will be Single Nucleotide Polymorphisms (SNP) and they promise to significantly advance our ability to understand and treat human disease. A SNP is a single base substitution of one nucleotide with another. Each individual has many single nucleotide polymorphisms that together create a unique DNA pattern for that person. It is important to study SNPs because they represent genetic differences among human beings. Genome-wide association studies require knowledge about common genetic variations and the ability to genotype a sufficiently comprehensive set of variants in a large patient sample [3]. High-throughput SNP genotyping technologies make massive genotype data, with a large number of individuals, publicly available. Accessibility of genetic data makes genome-wide association studies for complex diseases possible.

Success stories when dealing with diseases caused by a single SNP or gene, sometimes called monogenic diseases have been reported [4]. However, most complex diseases, such as psychiatric disorders, are characterized by a non-mendelian, multifactorial genetic contribution with a number of susceptible genes interacting with each other [5]. A fundamental issue in the analysis of SNP data is to define the unit of genetic function that influences disease risk. Is it a single SNP, a regulatory motif, an encoded protein subunit, a combination of SNPs in a combination of genes, an interacting protein complex, a metabolic or a physiological pathway [6]? In general, it may be impossible to associate a single SNP or gene with a disease because a disease may be caused by completely different modifications of alternative pathways, and each gene only makes a small contribution. This makes the identification of genetic factors difficult. Multi-SNP interaction analysis is more reliable but it is computationally infeasible. An exhaustive search among multi-SNP combination is computationally infeasible even for a small number of SNPs. Furthermore, there are no reliable tools applicable to large genome ranges that could rule out or confirm association with a disease.

It is important to search for informative SNPs among a huge number of SNPs. These informative SNPs are assumed to be associated with genetic diseases. Tag SNPs generated by the multiple linear regression based method [7] are good informative SNPs, but they are reconstruction-oriented instead of disease-oriented. Although the combinatorial search method [8] for finding disease-associated multi-SNP combinations has a better result, the exhaustive search is still very slow.

Multivariate adaptive regression spline models [9, 10] are used to detect associations between diseases and SNPs with some degree of success. However, the number of selected predictors is limited, and the type of possible interactions must be specified in advance. Multifactor dimensionality reduction methods [11, 12] are developed specifically to find gene-gene interactions among SNPs, but they are not applicable to a large set of SNPs.

Random forest model has been explored in disease association studies [13], but it was applied on simulated case-control data in which the interacting model among SNPs and the number of associated SNPs are specified, thus making the association model simple and the association is relatively easier to detect. For real data, such as Crohn's disease [14], multi-SNP interaction is much more complex , which involves more SNPs.

In Section 2 of this paper, we propose an optimum random forest model for searching the disease-associated multi-SNP combination for given case-control data. In the optimum random forest model, we generate a forest for each variable (e.g. SNP) instead of randomly selecting some variables to grow the classification tree. We can find the best classifier (a combination of SNPs which includes the SNP) for each SNP, and then we may have M classifiers if the length of the genotype is M. We rank classifiers according to their prediction rate, and the SNP with a higher prediction rates is more disease-associated.

The association of multi-SNP combination can be measured by the disease susceptibility prediction rate. In Section 3 we address the disease susceptibility prediction problem [15, 16, 17, 18]. The goal of disease susceptibility prediction is to assess accumulated information targeted to predicting susceptibility to complex diseases with significantly high accuracy and statistical power. The problem is based on the association study we described above. The Disease-associated multi-SNP combination found in association studies can be used to predict the susceptibility to diseases. On the other side, the prediction results can be used to evaluate the accuracy of the association studies. A higher prediction rate means the higher reliability of the association studies.

The proposed method is applied to analyze the genetic data of the Crohn's disease. We find the disease-associated multi-SNP combination and apply it to predict the susceptibility. The accuracy of the prediction is higher than that of all previously known methods. It can be also applied in disease prevention and control in the near future. For example, after training the available case-control genome data, we can find those significant SNPs which are well associated with the disease. When a patient comes, and we obtain his/her genetic data , we don't need to check the who le sequence, but only disease-associated SNP s ins tea d. This will save much money and time for diagnosis and can be done before the onset of diseases. Therefore, treatment could start earlier to prevent or delay the occurrence of the disease.

DISEASE ASSOCIATION SEARCH FOR CROHN'S DISEASE

In this section we first give an overview of the random forest tree and classification tree, then we will describe the genetic model. Next we propose the optimum random forest algorithm to search Tag SNPs.

Classification Trees and Random Forest

In machine learning, a Random Forest is a classifier that consists of many classification trees. Each tree is grown as follows:

1. If the number of cases in the training set is N, sample N cases at random - but with replacement, from the original data. This sample will be the training set for growing the tree.
2. If there are M input variables, a number $m \ll M$ is specified such that at each node, m variables are selected randomly out of the M and the best split on these m is used to split the node. The value of m is held constant during the forest growing.
3. Each tree is grown to the largest extent possible. There is no pruning [19].

   A different bootstrap sample from the original data is used to construct a tree. Therefore, about one-third of the cases are left out of the bootstrap sample and not used in the construction of the tree. Cross-validation is not required because the one-third oob (out-of-bag) data is used to get an unbiased estimate of the classification error as trees are added to the forest. It is also used to get estimates of variable importance. After each tree is built, we compute the proximities of each terminal node.

In every classification tree in the forest, put down the oob samples and make prediction the classification of the oob samples. In such way we can compute the importance score for variables in each tree based on the number of votes cast for the correct class. All variables can be ranked and those important variables can be found in this way.

Random forest is a sophisticated method in data mining to solve classification problems, and it can be used efficiently in disease association studies to find most disease-associated variables such as SNPs that may be responsible for diseases.

Genetic Model

Recent work has suggested that SNPs in human population are not inherited independently; rather, sets of adjacent SNPs are present on alleles in a block pattern, so called haplotype. Many haplotype blocks in human have been transmitted through many generations without recombination. This means although a block may contain many SNPs, it takes only a few SNPs to identify or to tag each haplotype in the block. A genome-wide haplotype would comprise half of a diploid genome, including one allele from each allelic gene pair. The genotype is the descriptor of the genome which is the set of physical DNA molecules inherited from the organism's parents. A pair of haplotype consists of a genotype.

SNPs are bi-allelic and can be referred as 0 for majority allele and 1, otherwise. If alleles on both haplotypes are the same, then the corresponding genotype is homogeneous, and can be represented as 0 or 1. If the two alleles on the two haplotypes are different, the genotype is heterozygous, represented as 2.

In Figure 1, there are four chromosomes, we assume the first two chromosomes belong to one person and the other two chromosomes belong to another person. We can find on most sites the four chromosomes are identical, but on some sites they are different, nucleotides on these sites are SNP. The haplotype is the concatenation of SNPs and a genotype is composed of two haplotypes.

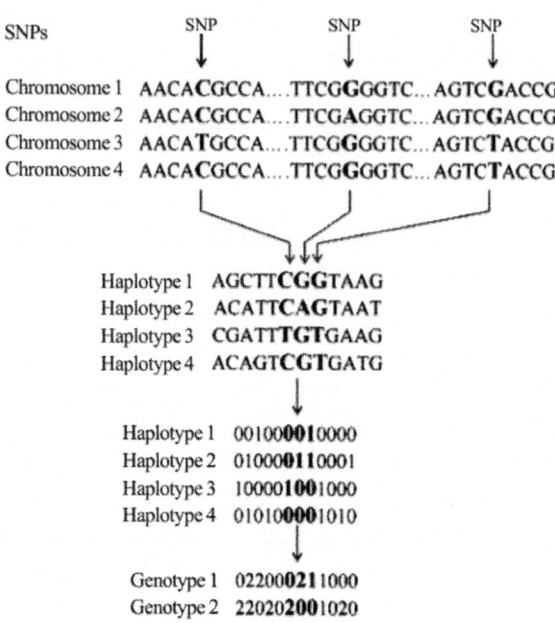

Figure 1: SNP, haplotype and genotype.

The case-control sample populations consist of N individuals who are represented in genotype with M SNPs. Each SNP attains one of the three values 0, 1, or 2. The sample G is an (0, 1, 2)-valued N x M matrix, where each row corresponds to an individual, each column corresponds to a SNP.

The sample G has 2 classes, case and control, and M variables, and each of them represents a SNP. To construct a classification tree, we split the sample S into 3 child sub-samples, depending on the value (0, 1, 2) of the variable (SNP) on the splitting site (loci). In fact we can construct a binary tree (split sample according to homozygous or heterozygous), but there is no way to tell the difference between major allele (1) and minor allele (0). In order to distinguish them we split the sample into 3 sub-samples instead of 2. We grow the tree to the largest possible extent. The construction of the classification tree for case-control sample is illustrated in Figure 2. In the first level, we split the sample (30 genotypes, 14 cases and 16 controls) into 3 sub-samples (17, 8, 5) at loci 5 (the 5th SNP). In the second level, the first sub-sample splits at loci 9 and the second sub-sample splits at loci 7. No splitting is required for the third sub-sample because it is a terminal node with only one class. In the third level, the only split node splits at loci 3. The relationship of a leaf to the tree on which it grows can be described by the hierarchy of splits of branches (starting from the trunk) leading to the last branch from which the leaf hangs. The collection of split site is a Multi-SNPs combination (MSC), which can be viewed as a classification tree. In this example, MSC = {5, 9, 7, 3} and m = 4, which is a collection of 4 SNPs, represented as their loci.

Searching for Disease Associated MultiSNPs

To fully understand the basis of complex diseases, it is important to identify the critical genetic factors involved, which is a combination of multiple SNPs. For a given sample G, S is the set of all SNPs (denoted by loci) for the sample, and a multi-SNPs combination (MSC) is a subset of S. In disease associations, we need to find a MSC which consists of a combination of SNPs that are well associated with the disease. To find such MSC, we need first rank all SNPs according to their association degree (measured as weight) with diseases. Based on the sorting, we can find the n most disease associated SNPs for a given threshold n.

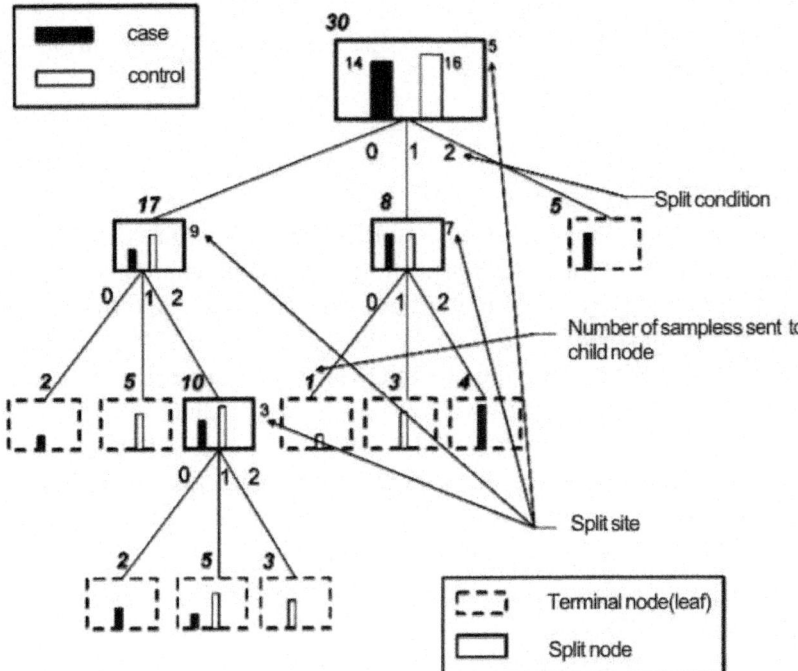

Figure 2: Classification tree for case-control sample.

Although there are many statistical methods to detect the most disease associated SNPs, such as odds ratio or risk rates, the result is not satisfactory. We decide to use the random forest to find them.

Optimum Random Forest

We randomly generate a group of MSCs for each SNP. The size of the MSC should be much less than the size of set S (m << M). Each MSC can be represented as a tree and all trees make the forest F. All trees (or MSCs) of the forest F_i(i=1, 2, ..., M) must include the ith SNP and the other (m-1) SNPs can be randomly chosen from S except the ith SNP. In this way, the M forests cover allSNPs in S.

We grow a classification tree for every MSC in each forest F_i. We run all the testing samples down these trees to get the classifier for each sample in the training set, then we can get a classification rate for each tree in F_i. The MSC_i is the representative for the forest F_i and the MSC_i has the highest classification rate among all trees in F_i. Each member (SNP) of the MSC_i] is assigned a weight w_{ij} (j ∈ MSC) based on the classification rate. The weights for SNPs in the same MSC are the same. We can find M MSCs for the M forests. If a SNP is not a member of MSC_i, then $w_{ij} = 0$.

The weight for each SNP W_j (j = 1, 2, ..., M) in M is the sum of weights from all MSCs.

$$W_j = \sum_{i=1}^{M} w_{i,j}$$

(1)

In the general random forest (GRF) algorithm, the MSC is selected completely randomly and m << M. It may miss some important SNPs if they are not chosen for any MSC. In our optimum random forest (ORF) algorithm, this scenario is avoided because we generate at least one MSC for each SNP. On the other hand, in GRF, the classifier (forest) consists of trees where there is a correlation between any two trees in the forest, and the correlation will decrease the rate of the classifier. But in ORF, we generate a forest by randomly choosing MSC and samples for each tree and the prediction for testing samples is in this forest only, which is completely independent from the other trees. In this way, we extinguish the correlation among trees.

All SNPs are sorted according to their cumulative weights. The most disease-associated SNP is the one with the highest weight. The contribution to diseases of each SNP is quantified by its weight, but in GRF there is no way tell the difference of contribution among SNPs. The GRF can only tell the difference among classifiers (trees).

DISEASE SUSCEPTIBILITY PREDICTION

In this section we first describe the input and the output of prediction algorithms and then show how to apply the optimum random forest to the disease susceptibility prediction.

Data sets have n genotypes and each has m SNPs. The input for a prediction algorithm includes:

(G1) Training genotype set $g_i = (g_{ij})$, i = 0, 1, ..., n, j = 1, ... m, $g_{i,j} \in \{0,1,2\}$

(G2) Disease status $s(g_i) \in \{0,1\}$, indicating if g_i, i = 0, 1, ..., n, is in case (1) or in control (0) , and

(G3) Testing genotype g_t without any disease status.

We will refer to the parts (G1-G2) of the input as the training set and to the part (G3) as the test set. The output of prediction algorithms is the disease status of the genotype $s(g_t)$.

We use leave-one-out cross-validation to measure the quality of the algorithm. In the leave-one-out cross-validation, the disease status of each genotype in the data set is predicted while the rest of the data is regarded as the training set.

We describe several universal prediction methods below. These methods are adaptations of general computer-intelligence classifying techniques.

Closest Genotype Neighbor (CN): For the test genotype g_t, find the closest (with respect to Hamming distance) genotype g_i in the training set, and set the status $s(g_t)$ equals to $s(g_i)$.

Support Vector Machine Algorithm (SVM) Support Vector Machine (SVM) is a generation learning system based on recent advances in statistical learning theory. SVMs deliver a state-of-the-art performance in real-world applications and have been used in case/control studies [18, 20]. There are some SVM softwares available and we decide to use libsvm-2.71 [19] with the following radial basis function:

$$exp(- \tau^* \mid u\text{-}v \mid^2)$$

General Random Forest (GRF): We use Leo Breiman and Adele Cutler's original implementation of RF version [19]. This version of RF handles unbalanced data to predict accurately. RF tries to perform a regression on the specified variables to produce the suitable model. RF uses bootstrapping to produce random trees and it has its own cross-validation technique to validate the model for prediction/classification.

Most Reliable 2 SNP Prediction (MR2) [17]: This method chooses a pair of adjacent SNPs (site of s_i and $s_i + 1$) to predict the disease status of the test genotype [g_t] by voting among genotypes from the training set which have the same SNP values as [g.sub.t] at the chosen sites s_i and s_{i+1} . They choose the 2 adjacent SNPs with the highest prediction rate in the training set.

LP-based Prediction Algorithm (LP): This method assumes that certain haplotypes are susceptible to the disease while others are resistant to the disease. The genotype susceptibility is then assumed to be a sum of susceptibilities of its two haplotypes.

We want to assign a positive weight to susceptible haplotypes and a negative weight to resistant haplotypes such that for any control genotype the sum of weights of its haplotypes is negative and for any case genotype it is positive. We would also like to maximize the confidence of our weight assignment which can be measured by the absolute values of the genotype weights. In other words, we would like to maximize the sum of absolute values of weights over all genotypes.

This method is based on a graph $X = \{H, G\}$, where the vertices H correspond to distinct haplotypes and the edges G correspond to genotypes connecting its two haplotypes. The density of X is increased by dropping SNPs which do not collapse edges with an opposite status. The linear program assigns weights to haplotypes that, for any non-diseased genotype, the sum of weights of its haplotypes is less than 0.5 and greater than 0.5 otherwise. We maximize the sum of absolute values of weights over all genotypes. The status of the testing genotype is predicted as sum of its endpoints [15].

Optimum Random Forest (ORF): In the training set, the optimum random forest algorithm we described above is used to sort all SNPs, and find out the m most disease associated SNPs for a given threshold m. The m most disease associated SNPs (Tag SNPs) are used to build the optimum random forest to test the left-out sample. In leave-one-out test, since the training set is different after leaving one sample out, we may have different Tag SNPs for different training sets. The m variables (SNPs) are used to grow many different classification trees by permuting the order of the splitting site (Note that the tree $\{3, 9, 5\}$ is different from the tree $\{5, 9, 3\}$). We may use the m Tag SNPs to grow many (say, 500) trees and choose the best tree (classifier) to predict the disease status of the testing genotype. The best tree has the highest average prediction rate (over 1000 trials) in the training set. Then we run the testing genotype down the best tree to get its disease status. The Optimum Random Forest algorithm is illustrated in Figure 3.

RESULTS & DISCUSSION

In this section we first describe the genetic data of the Crohn's disease and then discuss our experimental results.

Data Set

The genetic data is derived from the 616 kilobase region of human Chromosome 5q31 that may contain a genetic variant responsible for Crohn's disease by genotyping 103 SNPs for 129 trios [14]. All offspring belong to the case population, while almost all parents belong to the control population. In the entire data, there are 144 case and 243 control individuals. The missing genotype data and haplotypes have been inferred using the 2SNP phasing method [21].

Measures of Prediction Quality

To measure the quality of prediction methods, we need to measure the deviation between the true disease status and the result of predicted susceptibility, which can be regarded as measurement error. We will present the basic measures used in epidemiology to quantify the accuracy of our methods.

The basic measures are:

Sensitivity: the proportion of persons who have the disease and who are correctly identified as cases.

Specificity: the proportion of people who do not have the disease and who are correctly classified as controls.

Input:	Training genotype set $G^{N,M}$, N: the number of samples, M: the number of SNPs Disease status of $G^{N,M}$, $s^{N,M}$, The threshold m, Testing genotype g_t.

	Sorting the M SNPs, find the MSC with the m most disease-associated SNPs
	For $i = 1$ to 500,
	    Permute the order of MSC, generate a tree T_i,
	        For $j = 1$ to 1000,
	        Randomly generate a bootstrapped sample S_j from G,
	        Run S_j down the tree T_i to get the classification tree,
	        Predict testing sample G'_j ($G'_j = G - S_j$) to get the prediction rate $p_{i,j}$,
	    Compute the average prediction rate $\bar{p}_i$ for T_i,
	Find the best tree T_b which has the highest $\bar{p}$,
	Run g_t down the best tree T_b to get the disease status.

Output:	Disease status of the test genotype $s(g_t)$.

Figure 3: Optimum Random Forest Algorithm.

The definitions of these two measures of validity are illustrated in Table 1.

In this table:

a = True positive, people with the disease who test positive b = False positive, people without the disease who test positive

c = False negative, people with the disease who test negative

d = True negative, people without the disease who test negative

From Table 1, we can compute Sensitivity (accuracy in classification of cases, Specificity (accuracy in classification of controls) and accuracy:

$$Sensitivity = \frac{a}{a+c}$$

(2)

$$Specificity = \frac{d}{b+d}$$

(3)

$$Accuracy = \frac{a+d}{a+b+c+d}$$

(4)

Sensitivity is the ability to correctly detect a disease. Specificity is the ability to avoid calling normal as disease. Accuracy is the percent of the population that are correctly predicted.

Results and Discussion

The normalized weights of 103 SNPs are shown in Figure 4. SNPs with higher weights are more associated with the disease.

In Table 2 we compare the optimum random forest (ORF) method with the other 5 methods we described in Section 3. The best accuracy is achieved by ORF 74.4%. From the results we can find that the ORF has the best result since we select the most disease-associated multi-SNPs to build the random forest for prediction. Because these SNPs are well associated with the disease, the random forest may produce a good classifier to reflect the association.

Table1: Classification contingency table

		True Status	
		+	-
Classified	+	a	b
Status	-	c	d

Table 2: The comparison of the prediction rates of 6 prediction methods

Measures	Prediction Methods					
	CN	SVM	GRF	MR2	LP	ORF
Sensitivity	45.5	20.8	34.0	30.6	37.5	70.1
Specificity	63.3	88.8	85.2	85.2	88.5	76.9
Accuracy	54.6	63.6	66.1	65.5	69.5	74.4

Figure 5 shows the receiver operating characteristics (ROC) curve for 6 methods. A ROC curve represents the tradeoffs between sensitivity and specificity. The ROC curve also illustrates the advantage of ORF over all previous methods.

If the size of MSC is m, and the total number of SNPs is M, to get a good classifier, then m should be much less than M. The prediction rate depends on the size of MSC, as shown in Figure 6. In our experiment, we found that the best size of MSC is 19.

CONCLUSION

In this paper, we discuss the potential of applying random forest on disease association studies. The proposed genetic susceptibility prediction method based on the optimum random forest is shown to have a high prediction rate and the multi-SNPs being selected to build the random forest are well associated with diseases. Actually the cause of complex diseases is the combination of the environmental, genetic factors and some other factors such as infection and races. In our future work we are going to analyze the interactive contribution of these factors for the development of complex diseases. Our next project is going to find the relationship between the genetic factor and race in the development of Type 2 Diabetes. The integrated software will be available soon for public use.

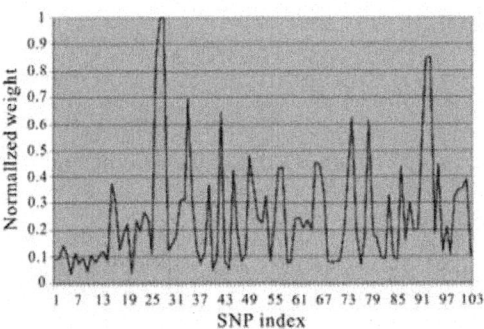

Figure 4: Normalized weights for 103 SNPs.

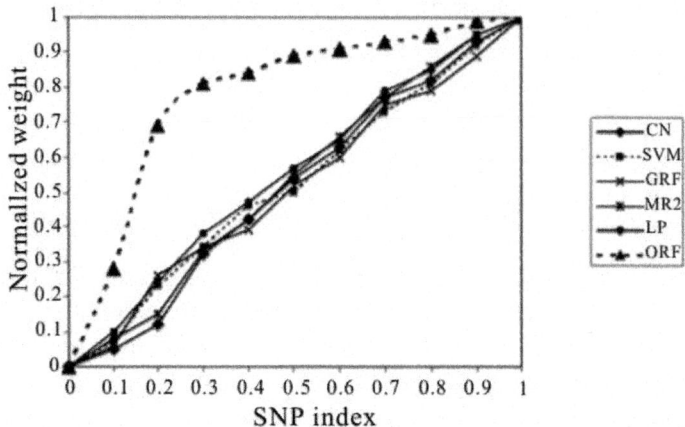

Figure 5: ROC curve for 6 prediction methods.

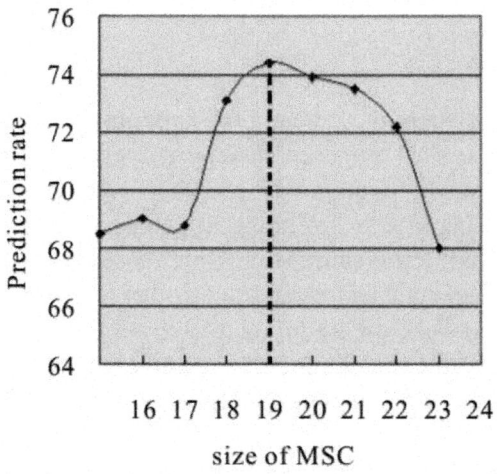

Figure 6: Best MSC size.

REFERENCE

1. National Digestive Diseases Information Clearinghouse (NDDIC), http:// digestive.niddk.nih.gov/ddiseases/pubs/crohns.

2. Cardon, L.R. & Bell, J.I. Association Study Designs for Complex Diseases. Nature Reviews: Genetics 2001, 2:91-98.

3. Hirschhorn, J.N. & Daly, M.J. Genome-wide Association Studies for Common Diseases and Complex Diseases. Nature Reviews: Genetics 2005, 6:95-108.

4. Merikangas, KR. & Risch, N. Will the Genomics Revolution Revolutionize Psychiatry. American Journal of Psychiatry, 2003, 160: 625-635.

5. Botstein, D. & Risch, N. Discovering Genotypes Underlying Human Phenotypes: Past Successes for Mendelian Disease, Future Approaches for Complex Disease. Nature Genetics 2003, 33: 228-237.

6. Clark, A.G., Boerwinkle E., Hixson J. & Sing C.F. Determinants of the success of whole-genome association testing. Genome Res. 2005, 15:1463-1467.

7. He, J. & Zelikovsky, A. Tag SNP Selection Based on Multivariate Linear Regression. Proc. of International Conference on Computational Science 2006, LNCS 3992:750-757.

8. Brinza, D., He, J. & Zelikovsky, A. Combinatorial Search Methods for Multi-SNP Disease Association. Proc. of International Conference of the IEEE Engineering in Medicine and Biology 2006, pages 5802-5805.

9. Cook N.R., Zee R.Y. & Ridker P.M. Tree and Spline Based Association Analysis of gene-gene interaction models for ischemic stroke. Stat Med 2004, 23(9):439-453.

10. York T.P. & Eaves L.J. Common Disease Analysis using Multivariate Adaptive Regression Splines (MARS): Genetic Analysis Workshop 12 simulated sequence data. Genetic Epidemiology 2001, 21 (S I):649-654.

11. Ritchie M.D., Hahn L.W., Roodi N., Bailey L.R., Dupont W.D., Parl F.F. & Moore J.H. Multifactor-dimensionality reduction reveals high-order interactions among estrogen-metabolism genes in sporadic breast cancer. Am J Hum Genet. 2001, 69: 138-147.

12. Hahn L.W., Ritchie M.D. & Moore J.H. Multifactor dimensionality reduction software for detecting gene-gene and gene-environment interactions. Bioinformatics 2003, 19:376-382.

13. Lunetta, K., Hayward, L., Segal, J. & Van Eerdewegh, P. Screening Large-scale Association Study Data: Exploiting Interactions Using Random Forests", BMC Genetics 2004, pages 5:32.

14. Daly, M., Rioux, J., Schaffner, S., Hudson, T. & Lander, E. High resolution haplotype structure in the human genome. Nature Genetics 2001, 29:229-232.

15. Mao, W., He, J., Brinza, D. & Zelikovsky, A. A Combinatorial Method for Predicting Genetic Susceptibility to Complex Diseases. Proc. International Conference of the IEEE Engineering In Medicine and Biology Society 2005, pages 224-227.

16. Mao, W., Brinza, D., Hundewale, N., Gremalschi, S. & Zelikovsky, A. Genotype Susceptibility and Integrated Risk Factors for Complex Diseases. Proc. IEEE International Conference on Granular Computing 2006, pages 754-757.

17. Kimmel, G. & Shamir R. A Block-Free Hidden Markov Model for Genotypes and Its Application to Disease Association. J. of Computational Biology 2005, 12(10): 1243-1260.

18. Listgarten, J., Damaraju, S., Poulin B., Cook, L., Dufour, J., Driga, A., Mackey, J., Wishart, D., Greiner,R. & Zanke, B. Predictive Models for Breast Cancer Susceptibility from Multiple Single Nucleotide Polymorphisms. Clinical Cancer Research 2004, 10:2725-2737.

19. Breiman, L. & Cutler, A. http://stat.berkeley.edu/breiman.

20. Waddell, M., Page,D., Zhan, F., Barlogie, B. & Shaughnessy, J., Predicting Cancer Susceptibility from SingleNucleotide Polymorphism Data: A Case Study in Multiple Myeloma. Proc. of the 5th international workshop on Bioinformatics 2005, pages 21-28.

21. Chang, C. and Lin, C. http://www.csie.ntu.edu.tw/libsvm.

22. Brinza, D. & Zelikovsky, A. 2SNP: Scalable Phasing Based on 2-SNP Haplotypes. Bioinformatics 2006, 22(3):371-373.

Chapter 2

IMMUNOHISTOCHEMICAL AND GENETIC ANALYSIS OF NON–SMALL CELL AND SMALL CELL GALLBLADDER CARCINOMA AND THEIR PRECURSOR LESIONS

Anil V Parwani M.D., Ph.D.[1], Joseph Geradts M.D.[4], Eric Caspers M.D.[5], G Johan Offerhaus M.D.[5], Charles J Yeo M.D.[2,3], John L Cameron M.D.[2], David S Klimstra M.D.[6], Anirban Maitra M.D.[1], Ralph H Hruban M.D.[1,3] and Pedram Argani M.D.[1]

[1]Department of Pathology, The Johns Hopkins Hospital, Baltimore, Maryland

[2]Department of Surgery, The Johns Hopkins Hospital, Baltimore, Maryland

[3]Department of Oncology, The Johns Hopkins Hospital, Baltimore, Maryland

[4]Department of Pathology and Laboratory Medicine, Roswell Park Cancer Institute, Buffalo, New York

[5]The Academic Medical Center, Amsterdam, The Netherlands

[6]Department of Pathology, Memorial Sloan-Kettering Cancer Center, New York

ABSTRACT

Gallbladder carcinomas can be highly lethal neoplasms. Relatively little is known about the genetic abnormalities that underlie these tumors, particularly with respect to their timing in neoplastic progression. The authors evaluated 5 noninvasive dysplasias and 33 invasive gallbladder carcinomas (6 small cell carcinomas, 27 non–small cell carcinomas, of which 16 were accompanied by an in situ carcinoma component) for expression of the protein products of the p16, p53, Dpc4, and pRB tumor suppressor genes by immunohistochemistry. Neoplasms were also evaluated for the presence of activating K-ras oncogene mutations. Seventy-five percent of non–small cell gallbladder carcinomas demonstrated loss of p16 expression, whereas 63% accumulated high levels of p53. Loss of Dpc4 and pRB expression was less frequent, seen in 19% and 4% of the neoplasms, respectively. Thirty percent of neoplasms harbored activating K-ras mutations. In contrast, 100% of the small cell carcinomas of the gallbladder demonstrated inactivation of the pRB/p16 pathway; 67% showed loss of pRB expression, and the other 33% lost p16 expression. Eighty-three percent of small cell carcinomas accumulated high levels of p53, whereas loss of Dpc4 expression and activating K-rasmutations were not found. Among 15 evaluable in situ components, 13 harbored the same alterations found in the invasive component. Inactivation of p16 and p53 occur in the majority of non–small cell gallbladder carcinomas. Dpc4 inactivation and K-ras mutations occur in a significant minority of cases. pRB loss is uncommon in non–small cell gallbladder carcinoma, but virtually all small cell carcinomas inactivate the p16/pRB pathway, usually by retinoblastoma protein loss. It is noteworthy that all of these alterations occur at the level of carcinoma in situ.

INTRODUCTION

Carcinoma of the gallbladder is a relatively uncommon, poorly understood, but highly lethal malignancy that tends to present at an advanced stage (1, 2, 3). There are approximately 5000 newly diagnosed cases per year in the United States, with a female predominance. The incidence of gallbladder carcinoma demonstrates marked geographic variation; for example, it is the single largest cause of cancer death for women in Chile but accounts for <0.5% of cancers in women in the United States (1). Much of the geographic variation correlates with the tendency to form gallstones, a recognized risk factor in gallbladder carcinogenesis. Other risk factors include an abnormal junction of the choledochopancreatic ducts (AJCPD); in which the bile duct and pancreatic duct unite well above the sphincter of Oddi, promoting reflux of pancreatic juice into the bile duct), familial adenomatous polyposis, and ulcerative colitis (1, 4). Overall 5-year survival is <5% (1).

The molecular pathogenesis of gallbladder carcinoma remains enigmatic. Loss of heterozygosity (LOH; 5, 6, 7, 8) and cytogenetic (9) studies of non–small cell gallbladder carcinoma have identified several loci of recurrent genetic loss. These loci likely harbor tumor suppressor genes that are inactivated and include chromosome arms 17p, 5q, 9p, 13q, and 18q. Alterations in several specific tumor suppressor genes that map to these loci have been identified (*i.e.*, *p53* at 17p13, *p16* at 9p21, *RB* at 13q), but the data remain cloudy to date as differing studies have yielded differing results (5, 7, 8, 10, 11, 12, 13, 14, 15, 16, 17, 18). Some of the variation in results of immunohistochemical studies may be due to use of different techniques and cutoff points for positivity in different studies. The loci on 18q involved in gallbladder carcinoma have not been identified, with the *DCC* and *DPC4* genes being prime candidates. Finally, mutational activation of the K-*ras* gene has been demonstrated in 5–59% of cases (7, 14, 15, 16, 17, 18), with suggestions that this frequency may be increased in Japanese patients, particularly those with AJCPD. Small cell carcinomas of the gallbladder have been studied even less frequently (19).

In the present study, we analyzed our large series of resected non–small cell and small cell gallbladder carcinomas for Dpc4, pRB, p16, and p53 protein expression and for K-*ras* gene mutation.

METHODS

Case Selection

This study was approved by the Johns Hopkins Institutional Review Board. The computerized files of the Surgical Pathology Division of the Department of Pathology of The Johns Hopkins Hospital were searched over the years 1985–2000 for cases coded as "gallbladder" and "carcinoma." Cases were selected for study on the basis of availability of a paraffin-embedded, formalin-fixed tissue block. An additional case (Case 31) was obtained from Memorial Sloan-Kettering Cancer Center. For each case, a representative formalin-fixed, paraffin-embedded tissue block containing carcinoma and normal tissue was chosen for labeling. For cases in which an *in situ* carcinoma component was noted, a block containing this component was specifically chosen for study.

Cases

We identified five cases of gallbladder dysplasia or carcinoma *in situ* unassociated with invasive carcinoma (Table 1) and identified 33 cases of invasive gallbladder carcinoma for which blocks were available. Patient ages ranged from 37 to 86 years (mean age = 62.8 y; median = 63.5 y), and the male-female ratio was 13:25. Twenty-one of 34 evaluable gallbladders harbored gallstones.

Table 1:- Clinical and Molecular Features of Five Patients with Pure Dysplasia or Carcinoma *In Situ*

Case Number	Age	Sex	Stone	p53	Dpc4	p16	pRB	K-*ras*
1 (LGD)	39	M	Yes	NL	NL	NL	NL	NL
2 (LGD)	37	F	Yes	NL	NL	NL	NL	NL
3 (CIS)	71	M	No	OE	NL	LOSS	NL	MUT D (Val)
4 (CIS)	63	F	Yes	OE	NL	LOSS	NL	NL
5 (CIS)	57	F	Yes	OE	NL	LOSS	NL	NL

NL. = normal pattern (intact labeling for p16, Dpc4, and pRB, less than 30% expression of p53, wild type K-*ras*); OE = overexpression; LOSS = loss of expression; LGD = low-grade dysplasia; CIS = carcinoma in situ; MUT = K-*ras* mutation; Stone = gallstone present in specimen.

Two of the pure dysplasia cases were low grade, whereas three were high grade or carcinoma *in situ*. Of the invasive carcinomas, 16 were associated with an *in situ* component; of these, the *in situ* component had a flat component in 10 and was purely papillary in 6.

Twenty-seven of the invasive carcinomas of the gallbladder were non–small cell carcinomas (Cases 1–27); these included 21 adenocarcinomas, 5 adenosquamous carcinomas (defined as showing variable mixture of malignant squamous and glandular components; 1), and 1 purely squamous carcinoma. Of these 27 neoplasms, the primary tumor was evaluated in 24, and a metastasis was the only tissue available in 3. Of these 24 primary tumors, 15 were moderately differentiated (defined as 40–94% of the tumor forming glands; 1), and 9 were poorly differentiated (defined as 5–39% of the tumor forming glands; 1). Six other neoplasms contained a small cell (high-grade neuroendocrine) carcinoma component (Cases 28–33). Two of these six were pure small cell carcinomas (Cases 29, 30), whereas one was predominantly small cell carcinoma with focal squamous differentiation (Case 31). One tumor was associated with flat carcinoma in situ (Case 28), whereas two other invasive carcinomas had both small cell carcinoma and adenocarcinoma components along with papillary adenocarcinoma in situ (Cases 32, 33).

Several patients› tumors were associated with unusual clinical presentations or pathologic findings. One patient›s gallbladder carcinoma arose within a septate gallbladder (Case 8). Another patient (Case 20) had concurrent colorectal adenocarcinoma and gallbladder carcinoma, whereas another patient›s gallbladder carcinoma extended into the extrahepatic bile ducts, which harbored a carcinoid tumor (Case 13).

Patients tended to present at advanced stage. Using AJCC criteria (20), 29 of 30 patients› primary tumors presented at Stage 3 or above. At least 23 of 30 primary tumor resections were associated with positive surgical margins microscopically, indicating incomplete excision of the neoplasm.

Immunohistochemistry for p53 and Dpc4

Unstained 4-μm sections were cut from the selected paraffin block and deparaffinized by routine techniques. The slides were steamed for 20 minutes in sodium citrate buffer (diluted to 1 × from 10 × heat-induced epitope retrieval buffer; Ventana-Bio Tek solutions, Tucson, AZ). After cooling for 5 minutes, the slides were labeled for 40 minutes at room temperature with either a 1:100 dilution of a monoclonal antibody to Dpc4 (clone B8, Santa Cruz Biotechnology, Santa Cruz, CA) or a 1:250 dilution of a monoclonal antibody to p53 (clone DO-7, DAKO, Carpinteria, CA) using the Bio Tek 1000 automated stainer (Ventana). Labeling was detected by adding biotinylated secondary antibodies, avidin-biotin complex, and 3, 3›-diaminobenzidine. Sections were then counterstained with hematoxylin. Dpc4 and p53 immunolabeling were evaluated jointly by two authors (AVP, PA) using a multiobserver microscope, with agreement on all cases. For p53 labeling, a percentage of positive nuclei was determined. Carcinomas were divided into two groups: normal (<30% nuclear labeling) and positive for high-level accumulation of p53 protein (>30% nuclear labeling). The labeling cutoff was chosen based on previous studies that have demonstrated that this cutoff point correlates best with the status of the *p53* gene in colorectal carcinomas (21). For Dpc4 labeling, any area of uniform cytoplasmic labeling and focal nuclear labeling was considered positive. In statistical analysis, any carcinoma showing even focal nuclear and cytoplasmic labeling was considered positive (expressor), whereas carcinomas demonstrating no expression in a background of intact expression by non-neoplastic cells (desmoplastic stroma, normal peribiliary glands, etc., which served as internal controls) were considered negative (nonexpressors). The rationale for considering carcinomas that labeled only focally as positives is based on the study of Wilentz *et al.* (22), which found that pancreatic tumors with this focal staining pattern proved to have an intact *DPC4* gene.

Immunohistochemistry for p16, pRB

Immunohistochemistry for p16 and pRB was performed in the laboratory of one of the authors (JG) and interpreted by three of the authors (AVP, PA, and JG). Mouse monoclonal anti-RB antibody 3C8 was purchased from QED (San Diego, CA), and mouse monoclonal anti-p16 antibody 16P07 was obtained from LabVision/NeoMarkers (Fremont, CA). Nonspecific mouse IgG was used as a negative antibody control. Standard ABC peroxidase assays were performed as described in detail elsewhere (22, 23, 24, 25, 26). For detection of pRB, deparaffinized sections were incubated with anti-RB antibody 3C8 at 2 μg/mL for 2 hours, after an antigen retrieval step in 0.01 m citrate buffer (95–100° C). For detection of p16, sections were incubated with the anti-p16 monoclonal antibody 16P07 at 1 μg/mL at 4° C overnight, after antigen retrieval in 0.1 m EDTA pH 8.0 (20 min at 95–100° C). The detection reactions for both markers used the Vectastain Elite ABC kit using conditions recommended by the manufacturer. Diaminobenzidine with hematoxylin counterstain was used for color development. Negative controls were labeled under identical conditions. The following external positive controls were used: normal colonic mucosa for pRB and p16 and a p16-positive lung cancer xenograft for p16. In addition, non-neoplastic stromal cells served as internal positive controls for pRB and p16 in every tumor section.

Each case was scored for pRB and p16 reactivity using previously published criteria (19, 23, 24, 25, 26). Briefly, sections were examined for evidence of nuclear labeling above any cytoplasmic background; cytoplasmic labeling itself was disregarded. If there was nuclear labeling in a diffuse or mosaic distribution throughout the lesion, it was considered "positive" (normal) for the respective protein. If all of the neoplastic nuclei failed to label whereas admixed non-neoplastic cells reacted positively, the lesion was scored as "negative" (abnormal). Cases were scored initially by one author (JO), and scoring was subsequently reviewed by two authors (AVP, PA) using a multiobserver microscope with agreement on all cases.

K-*ras* Gene Codon 12 Analysis

For the K-*ras* gene analysis, tumor was microdissected from unstained paraffin-embedded section slides to ensure ≥50% neoplastic cells in the sample. K-*ras*codon 12 analysis was performed with allele-specific oligonucleotide (ASO) hybridization method, as described elsewhere (27). DNA was isolated from microdissected tumor tissue by overnight incubation at 56° C in proteinase K solution, followed by inactivation of proteinase K at 95° C for 10 minutes. DNA was amplified by PCR using K-*ras* codon 12 specific primers (27). PCR products were then digested with *Mva*I, which only recognizes wild-type K-*ras* codon 12. Subsequently, a second-round PCR was performed on both the digested and undigested first-round PCR products. Cell suspensions with mutant–wild-type ratios of 1:100 and 1:1000 were used as positive controls in every PCR procedure. The cell suspensions were made of the human colon cancer cell line SW 480 with a homozygous GGT to GTT mutation at codon 12 of K-*ras* and the human colon cancer cell line HT 29 with wild type K-*ras*. Water was used as a negative control, and placental DNA was used as a control for nonspecific hybridization. After denaturation, the undigested and digested (mutant-enriched) PCR products were spotted onto a nylon membrane and hybridized to each of the K-*ras* codon 12 mutation-specific oligodeoxynucleotides. Final stringency washes were carried out at 63° C, followed by autoradiography. K-*ras* codon 12 mutational analysis was performed twice, in independent experiments.

In all cases, K-*ras* codon 12 mutations identified by ASO hybridization were confirmed by direct fluorescent sequencing using the dideoxy chain termination method (27). Mutant-enriched PCR products (second-round PCR products of ASO-test) were purified using the Qiaquick PCR purification kit (Qiagen Inc.). DNA was sequenced with the DNA sequencing kit, BigDye-Terminator Cycle sequencing Ready Reaction (Applied Biosystems, Warrington, United Kingdom). The reaction products were analyzed on an ABI Prism 3100 Genetic Analyser (Hitachi, Applied Biosystems). Selected wild-type cases were also sequenced.

RESULTS

Pure Dysplasias (*n* = 5)

Neither of the two low-grade dysplasias demonstrated abnormal expression of p53, Dpc4, and p16 or pRB protein, and neither contained a K-*ras* gene mutation. In contrast, all three flat carcinoma *in situ* lesions demonstrated accumulation of p53 protein at high levels and demonstrated loss of p16 expression. pRB expression was intact and diffuse in these three cases, and Dpc4 was detectable in all three (Fig. 1). One of the three carcinoma *in situ* lesions harbored a K-*ras* mutation, involving conversion to valine (Table 1).

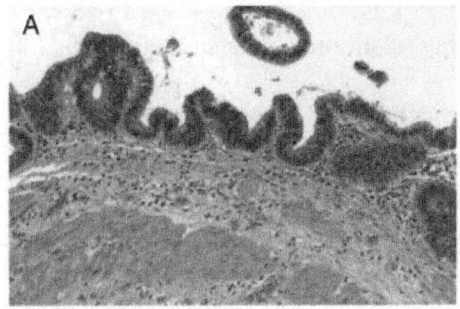

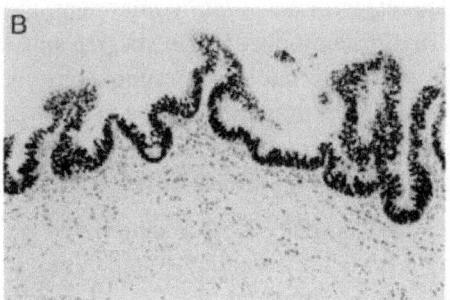

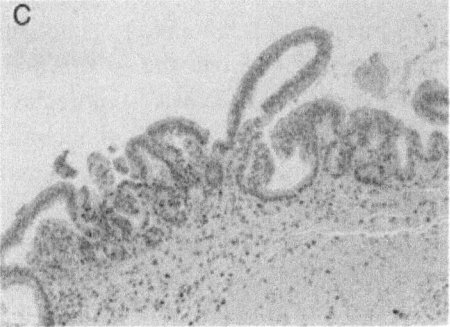

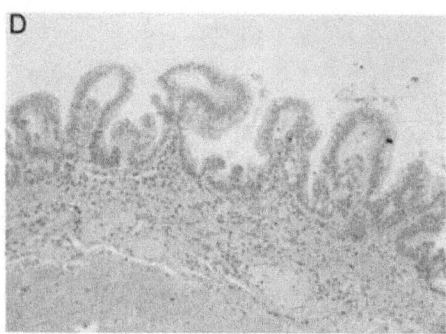

Figure 1: Flat carcinoma *in situ* (A) demonstrating p53 overexpression (B); intact, diffuse pRB expression (C); and loss of p16 expression (D). Note the intact patchy labeling of stromal and endothelial cells for p16.

Invasive Non–Small Cell Carcinomas (*n* = 27)

This group included 14 pure invasive non–small cell carcinomas and 13 invasive non–small cell carcinomas associated with an *in situ* carcinoma component. Loss of p16 expression and abnormal accumulation of p53 were the most common abnormalities noted in this group. Eighteen (75%) of 24 evaluable tumors demonstrated loss of p16 expression. This included 10 of 12 carcinomas unassociated with an *in situ* component and 8 of 12 carcinomas associated with an *in situ* component. The six p16-positive tumors generally displayed a weak to moderate nuclear immunoreactivity in <50% of cells in a mosaic pattern. Of note, in the case (Case 13) with concurrent gallbladder carcinoma and bile duct carcinoid tumor, p16 expression was intact in the carcinoid tumor but was lost in the poorly differentiated gallbladder carcinoma.

Seventeen of 27 carcinomas (63%) accumulated p53 at high levels; this included 8 of 14 pure invasive carcinomas and 9 of 13 invasive carcinomas associated with an *in situ* component. Five of 27 invasive carcinomas (19%) demonstrated loss of Dpc4 expression, including 3 of 14 pure invasive carcinomas and 2 of 13 invasive carcinomas associated with carcinoma *in situ*. Of note, Dpc4 expression was intact in the *in situ* and invasive carcinoma of the gallbladder but absent in the invasive colorectal carcinoma that were concurrently resected (Case 20), supporting the clinical and morphologic impression that these were independent primary neoplasms. Only 1 of 27 non–small cell carcinomas (4%) showed loss of pRB expression; in this tumor the *in situ* component also demonstrated loss of pRB expression. Most tumors with intact pRB expression demonstrated extensive labeling for pRB, particularly those demonstrating loss of p16 expression (Fig. 2).

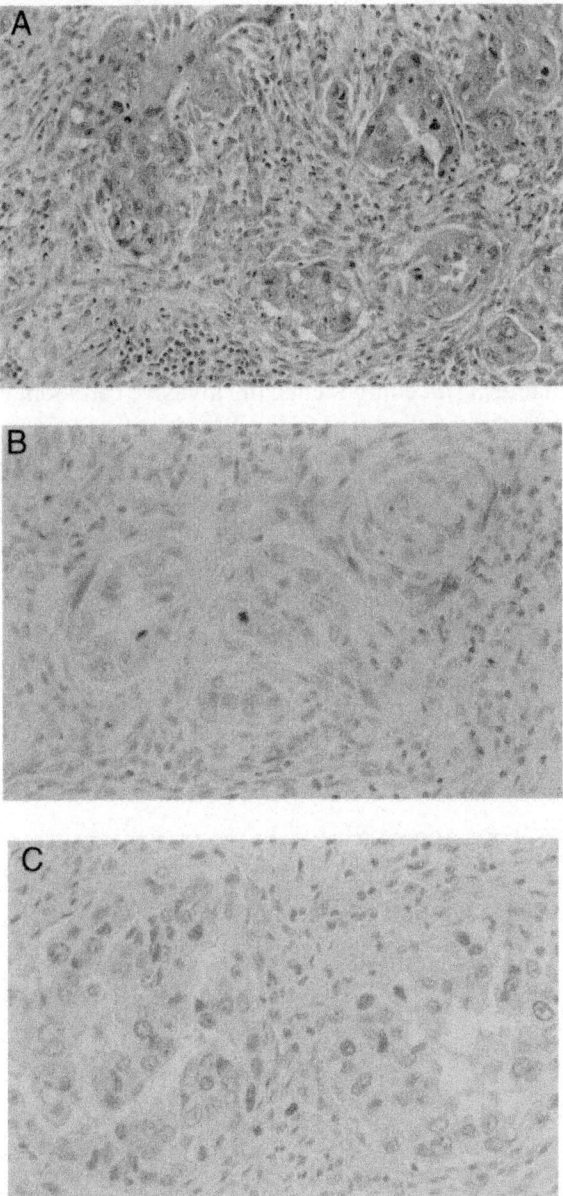

Figure 2: Infiltrating non–small cell carcinoma (A) demonstrating loss of p16 expression (B) and intact, diffuse pRB expression (C). Note the intact labeling of endothelial cells in Figure 2B.

K-*ras* gene mutations were identified in 8 of 27 invasive non–small cell carcinomas (30%). Of the eight invasive adenocarcinomas harboring a

K-*ras*gene mutation, six were primary tumors and two were metastases. Five mutations involved conversion to aspartate, and three, to serine.

There were no significant differences in results for any tested marker among cases classified as moderately *versus* poorly differentiated.

In Situ Carcinoma Components Associated with Invasive Non–Small Cell Carcinoma (*n* = 13)

There was a remarkable concordance between the immunohistochemical labeling of the *in situ* and the invasive non–small cell carcinoma components for p16, pRB, p53, and Dpc4 in the above cases. We identified p16 loss in 7 of 11 (64%) evaluable *in situ* components; the invasive carcinomas corresponding to all of these tumors had shown p16 loss. We identified accumulation of p53 at high levels in 10 of 13 *in situ* lesions (77%) associated with invasive carcinoma; this included all 9 cases in which the invasive component accumulated p53 at high levels. There was one case with discordant findings: in this case, the papillary *in situ* carcinoma component demonstrated high levels of p53, whereas the invasive component showed no p53 expression. We identified loss of Dpc4 expression in both of the *in situ* lesions associated with invasive carcinomas that showed loss of Dpc4 expression (2 of 13 *in situ* lesions overall, 15%; Fig. 3), whereas we found loss of pRB expression in the *in situ* component of the one invasive carcinoma that showed loss of pRB (1 of 13 cases overall, 8%).

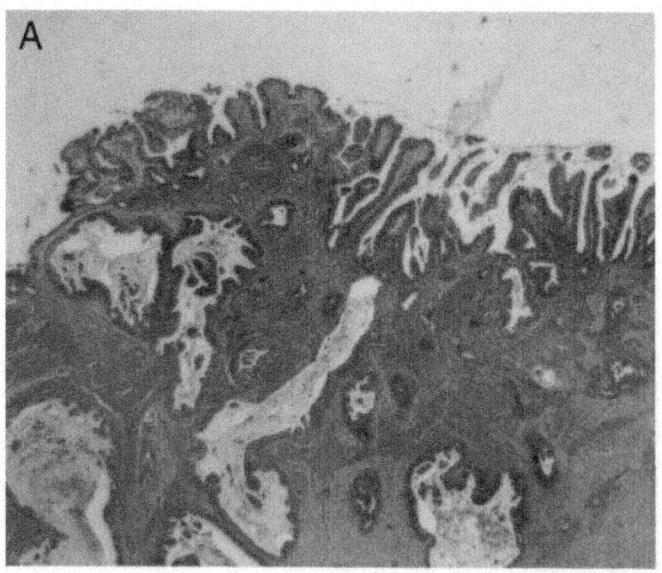

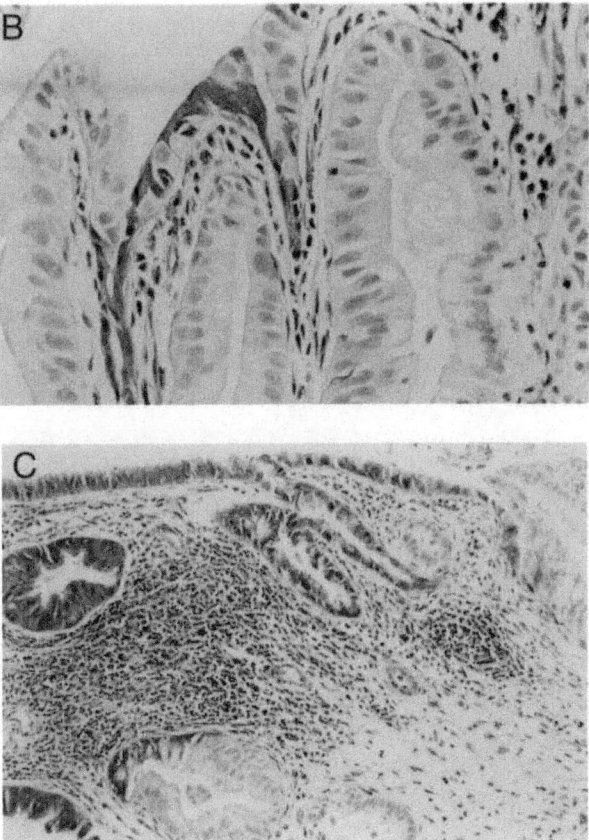

Figure 3: Papillary *in situ* and infiltrating carcinoma (A) demonstrating loss of Dpc4 expression in the *in situ* component (B) and in the invasive component (C). Note the intact labeling of normal stroma and benign gallbladder epithelium (Fig. 3C).

Invasive Small Cell Carcinomas (*n* = 6)

Accumulation of p53 at abnormally high levels was the most common abnormality noted, identified in five of six invasive small cell carcinomas. All six small cell carcinomas demonstrated evidence of inactivation of the pRB/ p16 pathway. In contrast to non–small cell carcinomas, loss of pRB expression more common in this tumor type, identified in four of six cases (67%), whereas the other two small cell carcinomas with intact pRB expression demonstrated loss of p16 expression (Fig. 4). Dpc4 protein expression was intact in all six tumors, whereas K-*ras* gene mutations were not identified in any of the three tumors evaluated.

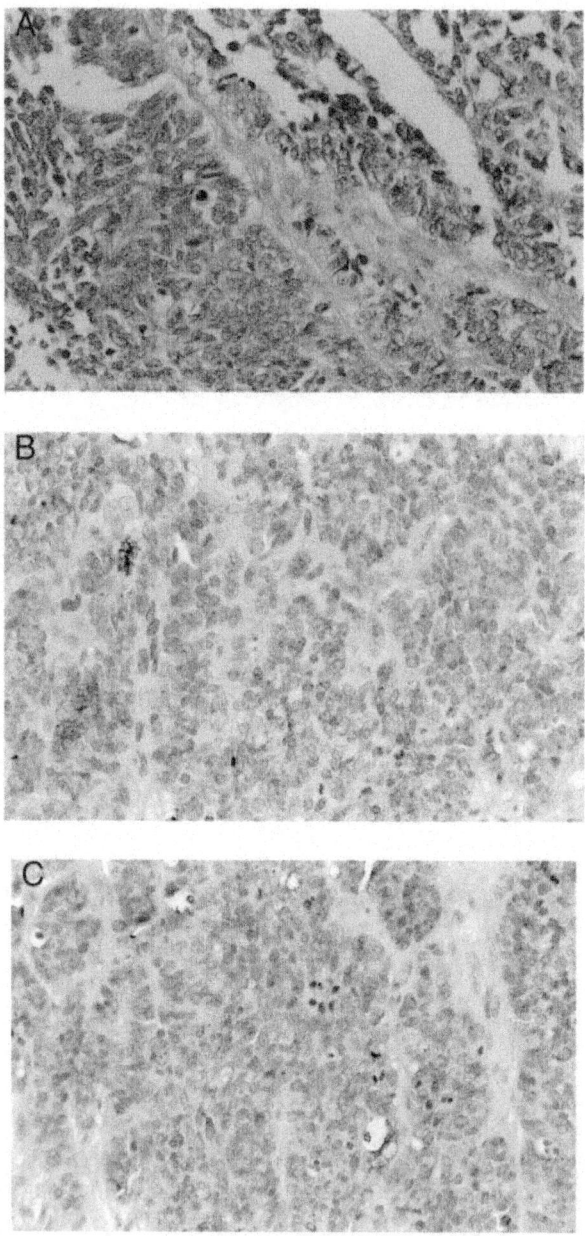

Figure 4: Small cell carcinoma (A) demonstrating loss of pRB expression (B) and intact, diffuse p16 expression (C). Note the intact labeling of endothelial cells for pRB in Figure 4B.

Non–Small Cell Components Associated with Small Cell Carcinomas (*n* = 3)

In two of three cases, the results were concordant. The flat carcinoma *in situ* associated with one small cell carcinoma and the papillary *in situ* and invasive adenocarcinoma associated with another each demonstrated p53 accumulation at high levels and p16 loss like their small cell components. In the other case, only the small cell carcinoma component demonstrated loss of p16 expression; the papillary *in situ* carcinoma and invasive carcinoma with which it was associated expressed p16.

The results of immunohistochemical assays are summarized in Table 2.

Table 2:- Immunolabeling and K-ras Mutation Analysis: Summary

| | Dysplasia Only Cases | | In Situ Component of Non-Small Cell Carcinomas | | Invasive Carcinoma Cases | |
| | | | | | Invasive Component | |
	Low Grade	CIS	Flat	Papillary	Non-Small Cell Carcinoma	Small Cell Carcinoma
p53	0/2	3/3	7/9	3/4	17/27 (63%)	5/6 (83%)
p16	0/2	3/3	6/8	1/3	18/24 (75%)	2/5 (40%)
pRB	0/2	0/3	1/9	0/4	1/27 (4%)	4/6 (67%)
Dpc4	0/2	0/3	1/9	1/4	5/27 (19%)	0/6 (0%)
K-ras	0/2	1/3	ND	ND	8/27 (30%)	0/3 (0%)

ND = not determined.
* The data are expressed as the number of abnormal results/number of normal and abnormal results. These data do not include the non-small cell carcinoma components associated with the small cell carcinomas.

DISCUSSION

We studied a large series of resected gallbladder carcinomas for expression of the protein products of the *p16, DPC4, RB*, and *p53* tumor suppressor genes and for mutation of the K-*ras* oncogene.

Our immunohistochemical results for p53 generally parallel those previously reported in the literature. A number of prior studies have shown that accumulation of p53 at high levels is frequent in invasive non–small cell carcinomas of the gallbladder and occurs at the level of carcinoma *in situ* (7, 10,11). Although most of these older studies have used primary antibodies other than the D0–7 clone (which is currently used by most laboratories), several have shown good correlation of p53 expression with *p53* gene mutation (10) and LOH at 17p (7). Our study is slightly different from those previously published in that we show a slightly higher frequency of high-level accumulation of p53 in *in situ* carcinoma than in invasive carcinomas. In contrast, most previous studies have shown progressive overexpression of p53 as dysplastic lesions progress to invasion (5, 7). Possible reasons for this difference could include variability in the antibody clone or immunohistochemical technique used, as well as criteria for distinction of low-grade dysplasia from carcinoma *in situ*. For example, it is possible that some of the lesions we designated low-grade dysplasia might be termed carcinoma *in situ* by others, which would reduce the frequency of p53 alterations somewhat. Nonetheless, because of the concordance between *in situ* and invasive carcinoma, our results suggest the possibility that gallbladder carcinomas may arise via two pathways, with those arising from flat carcinoma *in situ* overexpressing p53 and with those that do not being less likely to overexpress p53. The high frequency of p53 overexpression that we observed in small cell carcinomas of the gallbladder (83%) is similar to that reported elsewhere (19).

To our knowledge, this is the first study to assess Dpc4 protein expression in non–small cell gallbladder carcinomas. The 18% frequency of loss of protein expression that we observed in non–small cell carcinomas is concordant with results from our previous studies of biliary tract carcinomas. Previously, we showed that loss of Dpc4 expression is as frequent in distal (intrapancreatic) bile duct carcinomas (55%) as it is in pancreatic adenocarcinoma. In contrast, the frequency of Dpc4 loss was lower (overall, 15%) in more proximal (perihilar and intrahepatic) bile duct carcinomas (28). Interestingly, we observed loss of Dpc4 protein in a similar percentage of *in situ* gallbladder carcinomas associated with invasive carcinomas (2 of 13, or 15%), suggesting that *DPC4* inactivation occurs early in these lesions. The 18% frequency of Dpc4 protein loss in these tumors may account for a subset of the approximately 30% frequency of LOH (5, 7) that has been reported at chromosome 18q in prior allelotyping studies of gallbladder carcinomas.

The significant role of *p16* alterations in the pathogenesis of gallbladder carcinomas has emerged in the recent literature. Although studies have shown a rate of mutation that ranges from 30.7 (13) to 80% (12), LOH at the p16 locus on chromosome 9p has been reported to range between 0 and 68% (7, 8, 12). To our knowledge, only two other studies have evaluated p16 protein immunolabeling in gallbladder carcinomas. Shi *et al.* (29) have recently shown loss of p16 expression in 75.7% of gallbladder carcinomas, whereas Kim *et al.*(13) showed loss of labeling in only 23%. Our study is concordant with the results of Shi *et al.* (29) in that we saw frequent loss of p16 protein expression (69%). It is interesting that the two adenocarcinomas associated with small cell carcinomas in this study maintained p16 expression, suggesting that they are biologically different from pure adenocarcinomas.

We found pRB loss to be infrequent in invasive non–small cell gallbladder carcinomas; in fact, most tumors expressed pRB diffusely. Diffuse pRB expression in these tumors likely reflects the sequelae of p16 inactivation, because p16 negatively regulates cellular levels of RB (30). In fact, inverse relationship of p16 and pRB has been demonstrated in several malignancies including bladder carcinomas, lung carcinomas, pancreatic carcinomas, and gliomas (30, 31, 32), and, most recently, gallbladder carcinomas (29). In our series, the inverse relationship was also apparent; no invasive carcinoma demonstrated loss of both p16 and pRB, whereas an inverse relationship (loss of p16 or Rb, but not both) was evident in 25 of 30 tumors. One would not expect inactivation of both of these tumor suppressor genes within the same tumor, because they each function in the same growth-regulatory pathway, and therefore there would be no selective pressure to inactivate both.

In contrast, we found that pRB loss was common (4 of 6 cases) in the small cell carcinomas available for study. Although pRB expression has, to our knowledge, never been evaluated in small cell carcinoma of the gallbladder before, pRB loss is common in pulmonary small cell carcinomas (33) and has been reported in small cell carcinomas in extrapulmonary sites (34, 35). Hence, gallbladder carcinomas appear to fit the paradigm that small cell (high grade) neuroendocrine carcinomas of any site are consistently characterized by loss of pRB expression. It is noteworthy that the two cases that retained pRB showed loss of p16, indicating that all small cell carcinomas of the gallbladder target the p16/pRB pathway.

The role of K-*ras* gene mutation in gallbladder carcinomas is not clear from the literature. Several groups have considered K-*ras* mutations to be "rare and late events" (7) in gallbladder carcinomas. Wistuba *et al.* (36) found K-*ras* gene mutations in only 1 of 21 gallbladder carcinomas, and the mutation was confined to the poorly differentiated area of this tumor. This same group found a higher frequency of K-*ras* gene mutation (25%) in gallbladder adenomas and used this finding to support the idea that adenomas are not the precursor of most invasive carcinomas (36). In contrast, studies from Japan have shown a higher frequency of K-*ras* gene mutation in gallbladder carcinomas and occurrence in precursor lesions. For example, Ajiki *et al.* (15) found K-*ras* gene mutations in 73% of gallbladder dysplasias and 59% of invasive carcinomas. The presence of K-*ras*gene mutations does correlate with the anatomic finding of anomalous junction of the choledochopancreatic ducts (AJCPD) outside of the ampulla of Vater, which is more common in Japan. Hanada *et al.* (10) showed that the rate of K-*ras*mutation was higher in gallbladder carcinomas of patients with AJCPD than in gallbladder carcinomas of patients without AJCPD (50% *versus* 6%, $P < .05$). Our study shows an intermediate rate of mutation in our cohort of 30%. Two of our three metastatic carcinomas harbored K-*ras* gene mutations; however, we demonstrated a mutation in one patient with pure flat carcinoma *in situ*, indicating that this genetic alteration may be an early event in some cases.

In summary, our results with p53, p16, and K-*ras* are consistent with those of previous studies in the literature that have shown a significant rate of alterations in these genes in gallbladder carcinoma. Loss of p16 expression and accumulation of p53 at high levels are the most common abnormalities associated with these carcinomas. Our study is the first to show Dpc4 protein loss in gallbladder carcinoma. We also show that loss of pRB expression is rare in non–small cell gallbladder carcinoma but is common in small cell carcinoma of the gallbladder. Indeed, the p16/pRB pathway was inactivated in all six small cell carcinomas studied. Perhaps most important, using immunohistochemistry, we were able to localize all of these alterations to the stage of carcinoma *in situ*. Indeed, of 15 invasive carcinomas associated with an evaluable *in situ* component, there was complete concordance between the results in 13 cases. Along these lines, one case of pure *in situ* carcinoma in our study contained three distinct alterations: p53 accumulation, p16 loss, and K-*ras* gene mutation.

Hence, carcinoma *in situ* of the gallbladder, although clinically occult, is a genetically advanced lesion. These results suggest that patients with pure carcinoma *in situ* of the gallbladder should receive careful clinical follow-up, particularly if the status of the cystic duct margin is unclear. At the same time, these results highlight our ignorance of mechanisms and markers of progression in this lethal neoplasm.

ACKNOWLEDGEMENTS

Supported by the Margaret Lee Fund for Gallbladder and Bile Duct Cancer research at The Johns Hopkins Hospital.

REFERENCES

1.   Albores-Saavedra J, Henson DE, Klimstra DS. *Tumors of the gallbladder, extrahepatic bile ducts, and ampulla of Vater. Atlas of tumor pathology.* 3rd series. Fascicle 27. Washington, D.C.: Armed Forces Institute of Pathology; 2000.

2.   Levin B. Gallbladder carcinoma. *Ann Oncol* 1999; 10: S129–S130.

3.   Henson DE, Albores-Saavedra J, Corle D. Carcinoma of the gallbladder. Histologic types, stage of disease, grade, and survival rates. *Cancer* 1992;70: 1493–1497.

4.   Nugent KP, Spigelman AD, Talbot IC, Phillips RKS. Gallbladder dysplasia in patients with familial adenomatous polyposis. *Br J Surg* 1994; 81: 291–292.

5.   Chang HJ, Kim SW, Kim Y-T, Kim WH. Loss of heterozygosity in dysplasia and carcinoma of the gallbladder. *Mod Pathol* 1999; 12: 763–769. |

6.   Wistuba II, Albores-Saavedra J. Genetic abnormalities involved in the pathogenesis of gallbladder carcinoma. *J Hepatobiliary Pancreatic Surg*1999; 6: 237–244.

7.   Wistuba II, Sugio K, Hung J, Kishimto Y, *et al.* Allele-specific mutations involved in the pathogenesis of endemic gallbladder carcinoma in Chile. *Cancer Res* 1995; 55: 2511–2515.

8.   Wistuba II, Tang M, Maitra A, *et al.* Genome-wide allelotyping analysis reveals multiple sites of allelic loss in gallbladder carcinoma. *Cancer Res*2001; 61: 3795–3800.

9.   Gorunova L, Parada LA, Limon J, *et al.* Nonrandom chromosomal aberrations and cytogenetic heterogeneity in gallbladder carcinomas. *Genes Chromosomes Cancer* 1999; 26: 312–321.

10. Hanada K, Itoh M, Fujii K, Tsuchida A, Ooishi H, Kajiyama G. K-*ras* and p53 mutations in stage 1 gallbladder carcinoma with an anomalous junction of the pancreaticobiliary duct. *Cancer* 1996; 77: 452–458.

11. Wistuba II, Gadzar AF, Roa I, Albores-Saavedra J. p53 protein expression in gallbladder carcinoma and its precursor lesions. An immunohistochemical study. *Hum Pathol* 1996; 27: 360–365.

12. Yoshida S, Todoroki T, Ichikawa Y, *et al.* Mutations of $p16^{Ink4}$/CDKN2 and$p15^{Ink4B}$/MST2 genes in biliary tract cancers. *Cancer Res* 1995; 55: 2756–2760.

13. Kim Y-T, Kim J, Jang YH, *et al.* Genetic alterations in gallbladder adenoma, dysplasia, and carcinoma. *Cancer Lett* 2001; 169: 59–68.

14. Kim SW, Her K-H, Jang J-Y, Kim W-H, Kim Y-T, Park Y-H. K-*ras* oncogene mutation in cancer and precancerous lesions of the gallbladder. *J Surg Oncol* 2000; 75: 246–251.

15. Ajiki T, Fujimori T, Onoyama H, *et al.* K-*ras* gene mutation in gall bladder carcinomas and dysplasia. *Gut* 1996; 38: 426–429.

16. Watanabe M, Asaka M, Tanaka J, Kurosawa M, Kasai M, Miyazaki T. Point mutation of the K-*ras* gene codon 12 in biliary tract tumors. *Gastroenterology* 1994; 107: 1147–1153.

17. Imai M, Hoshi T, Ogawa K. K-*ras* codon 12 mutations in biliary tract tumors detected by polymerase chain reaction denaturing gel electrophoresis.*Cancer* 1994; 73: 2727–2733.

18. Itoi T, Watanabe H, Ajioka Y, *et al.* APC, K-*ras* codon 12 mutations and p53 gene expression in carcinoma and adenoma of the gall-bladder suggest two genetic pathways in gall-bladder carcinogenesis. *Pathol Int* 1996; 46: 333–340.

19. Maitra A, Tascilar M, Hruban RH, Offerhaus GJ, Albores-Saavedra J. Small cell carcinoma of the gallbladder. A clinicopathologic, immunohistochemical, and molecular pathology study of 12 cases. *Am J Surg Pathol* 2001; 25: 595–601.

20. Fleming ID, Cooper JS, Henson DE, *et al.*, editors. *AJCC cancer staging handbook*. Philadelphia, PA: Lippincott-Raven; 1997.

21. Baas IO, Mulder J-WR, Offerhaus GJA, Vogelstein B, Hamilton SR. An evaluation of six antibodies for immunohistochemistry of mutant p53 gene product in archival colorectal neoplasms. *J Pathol* 1994; 172: 5–12.

22. Wilentz RE, Su GH, Dai JL, *et al.* Immunohistochemical labeling for Dpc4 mirrors genetic status in pancreatic adenocarcinomas: a new marker of *DPC4* inactivation. *Am J Pathol* 2000; 156: 37–43.

23. Geradts J, Kratzke RA, Crush-Stanton S, Wen SF, Lincoln CE. Wild-type and mutant retinoblastoma protein in paraffin sections. *Mod Pathol* 1996; 9: 339–347.

24. Geradts J, Hruban RH, Schutte M, Kern SE, Maynard R. Immunohistochemical p16[INK4a] analysis of archival tumors with deletion, hypermethylation, or mutation of the *CDKN2/MTS1* gene: a comparison of four commercial antibodies. *Appl Immunohistochem Mol Morphol* 2000; 8: 71–79.

25. Maitra A, Roberts H, Weinberg AG, Geradts J. Loss of p16[INK4a] expression correlates with decreased survival in pediatric osteosarcomas. *Int J Cancer* 2001; 95: 34–38.

26. Wilentz RE, Geradts J, Maynard R, *et al.* Inactivation of p16 (INK4A) tumor-suppressor gene in pancreatic duct lesions: loss of intranuclear expression. *Cancer Res* 1998; 58: 4740–4744.

27. Hruban RH, Sturm PDJ, Slebos RJC, *et al.* Can K-*ras* mutations be used to distinguish benign bile duct proliferations from metastases to the liver? A molecular analysis of 101 liver lesions from 93 patients. *Am J Pathol* 1997; 151: 943–949.

28. Argani P, Shaukat A, Kaushal M, *et al.* Differing rates of Dpc4 expression and of p53 overexpression among carcinomas of the proximal and distal bile ducts. Evidence for a biologic distinction. *Cancer* 2001; 91: 1322–1341.

29. Shi Y-Z, Hui A-M, Li X, Tayayama T, Makuuchi M. Overexpression of retinoblastoma protein predicts decreased survival and correlates with loss of p16^{INK4} protein in gallbladder carcinomas. *Clin Cancer Res* 2000; 6: 4096–4100.

30. Benedict WF, Lerner SP, Zhou J, Shen X, Tokunaga H, Czerniak B. Level of retinoblastoma protein expression correlates with p16 (MTS-1/INK4A/CDKN2) status in bladder cancer. *Oncogene* 1999; 18: 1197–1203.

31. Sakaguchi M, Fujii Y, Hirabayashi H, *et al*. Inversely correlated expression of p16 and RB protein in non-small cell lung cancers: an immunohistochemical study. *Int J Cancer* 1996; 8: 442–445.

32. Ueki K, Ono Y, Henson JW, Efird JY, von Deimling A, Louis D. CDKN2/p16 or RB alterations occur in the majority of glioblastomas and are inversely correlated. *Cancer Res* 1996; 56: 150–153.

33. Shimizu E, Coxon A, Otterson GA, *et al*. RB protein status and clinical correlation from 171 cell lines representing lung cancer, extrapulmonary small cell carcinoma, and mesothelioma. *Oncogene* 1994; 9: 2441–2448.

34. Herrington CS, Graham D, Southern SA, Bramdev A, Chetty R. Loss of retinoblastoma protein expression is frequent in small cell carcinoma of the uterine cervix and is unrelated to HPV type. *Hum Pathol* 1999; 30: 906–910.

35. Takubu K, Nakamura K, Sawabe M, *et al*. Primary small cell undifferentiated carcinoma of the esophagus. *Hum Pathol* 1999; 30: 216–221.

36. Wistuba II, Miquel JF, Gadzar AF, Albores-Saavedra J. Gallbladder adenomas have molecular abnormalities different from those present in gallbladder carcinomas. *Hum Pathol* 1999; 30: 21–25.

Chapter 3

ORIGINS OF ALBINO AND HOODED RATS: IMPLICATIONS FROM MOLECULAR GENETIC ANALYSIS ACROSS MODERN LABORATORY RAT STRAINS

Takashi Kuramoto[1], Satoshi Nakanishi[1], Masako Ochiai[2], Hitoshi Nakagama[2], Birger Voigt[1], Tadao Serikawa[1]

[1]Institute of Laboratory Animals, Graduate School of Medicine, Kyoto University, Sakyo-ku, Kyoto, Japan

[2]National Cancer Center Research Institute, Chuo-ku, Tokyo, Japan

ABSTRACT

Albino and hooded (or piebald) rats are one of the most frequently used laboratory animals for the past 150 years. Despite this fact, the origin of the albino mutation as well as the genetic basis of the hooded phenotype remained unclear. Recently, the albino mutation has been identified as the Arg299His missense mutation in the Tyrosinase gene and the hooded (*H*) locus has been mapped to the ~460-kb region in which only the *Kit* gene exists. Here, we surveyed 172 laboratory rat strains for the albino mutation and the hooded (*h*) mutation that we identified by positional cloning approach to investigate possible genetic roots and relationships of albino and hooded rats. All of 117 existing laboratory albino rats shared the same albino missense mutation, indicating they had only one single ancestor. Genetic fine mapping followed by *de novo* sequencing of BAC inserts covering the *H* locus revealed that an endogenous retrovirus (ERV) element was inserted into the first intron of the *Kit* gene where the hooded allele maps. A solitary long terminal repeat (LTR) was found at the same position to the ERV insertion in another allele of the *H* locus, which causes the so called Irish (*h^i*) phenotype. The ERV and the solitary LTR insertions were completely associated with the hooded and Irish coat patterns, respectively, across all colored rat strains examined. Interestingly, all 117 albino rat strains shared the ERV insertion without any exception, which strongly suggests that the albino mutation had originally occurred in hooded rats.

INTRODUCTION

The rat (*Rattus norvegicus*) was the first mammalian species domesticated for scientific research with work dating back to before 1850. Since that time, the rat has a leading role in various research fields, such as physiology, pharmacology, neurosciences, genetics, and medical sciences. Among the rat strains available now, the albino and piebald (hooded) are most common, which is the same situation as in the earliest days when rats were firstly used for scientific research.

Albino rats have played a pioneering role in animal experimentation since its inception. Gregor Mendel reported his famous laws on 'Mendelian inheritance' in 1866. Hugo Crampe was the first scientist to confirm the validity of these laws in animals using some 15,000 white, grey, black, and piebald rats between 1877 and 1885 [1]. These albino rats were in fact the first animals to be domesticated for the purpose of scientific research.

The "hooded" phenotype is also one of the oldest mutations in the rat. Hooded rats have a pattern in which the entire ventral surface is white. Dorsally pigmentation is limited to the head and shoulders (the "hood") and a mid-dorsal stripe extending back to the tip of the tail. In addition to this hooded (*h*), other modifiers such as Irish and notch alleles are known in the so named *H* locus [2]. The Irish (*h^i*) causes a white spot on the belly between and behind the front legs. Notch (*h^n*) rats have a white body with pigmented fur on the sides of the head. The wild-type allele against the *h*, *h^i*, or *h^n* is called "self" (*H*). Self animals are pigmented all over.

Despite their undisputed importance to scientific progress, the origin of albino and hooded rats has never been clearly determined. The question remains unanswered, is there only one single spontaneous mutation that has been transferred and bred for more than 100 years into hundreds existing albino rat strains? Or, did several different mutations such as those in mice[3] contribute to the albino phenotype that is now seen in millions of laboratory rodents around the world?

Henry H Donaldson at the Wistar Institute in Philadelphia wrote in 1915: "The Norway rat, *Mus norvegicus* (now *Rattus norvegicus*), is the one mammal easily obtainable both wild and as a domesticated form. This latter is represented by either the albino or the pied rats so common in our laboratories. We do not know whether the common albino variety had a single or multiple origins, or whether the colonies found in Europe are directly related to those now existing here"[4].

A deeper look into the historic records on rodents and a detailed molecular genetic analysis of the genetic patterns of all available coat color variations in rats are the measures of choice for solving these questions. The molecular nature of the albino rat has been identified as Arg299His missense mutation in the rat tyrosinase (*Tyr*) gene [5]. The Arg299His mutation has already been described in humans affected by ocular cutaneous albinism 1A (OCA1A) without any tyrosinase activity and consequential complete lack of pigmentation [6]. The hooded (*H*) locus was mapped to the rat Chr 14 [7]. Recently, Torigoe et al. mapped the *H* locus to the ~460-kb region in which only the *Kit* gene existed but did not found any mutations in the *Kit*coding region of the hooded LEA rats [8].

The aim of this study is to clarify and characterize the origin and nature of albino and hooded mutations using molecular genetic approaches and combine the findings with historical records. The albino related question was addressed by genotyping the *Tyr* gene mutation Arg299His for 172 laboratory strains from different regions of the world. In parallel, we confirmed the mapping of the hooded locus to the *Kit* gene on the rat Chr 14 by genetic studies and subsequently identified genetic variations in the *Kit* gene that are unique for hooded rats.

RESULTS

Genotyping of the Albino Allele

Preliminary sequencing of several albino strains confirmed the Arg299His mutation. Subsequent screening across all 172 strains examined was performed by direct digestion of the PCR product with the specific *Sna*BI restriction endonuclease to genotype the Arg299His mutation in the rat *Tyr* gene. All 55 colored rat strains had the wild-type allele (299Arg) and all of the 117 albino rat strains had the missense mutation (299His) (Table S1).

Fine Mapping of the Rat Hooded (H) Locus

Torigoe et al. mapped the *H* locus between *D14Rat84* and *D14Got40* [8]. Genotyping of the backcross progeny for both SSLP markers revealed six animals carrying the recombinant chromosome between *D14Rat84* and *D14Got40*. By using five SSLP markers, we narrowed the *H* locus down to between *D14Rat13* (position 34,941 kb) and *D14Got40* (position 35,259 kb). Haplotype analysis revealed that the *H* locus spanned from *ENSRNOSNP2799339* (position 34,910 kb) to *ENSRNOSNP2799341* (position 35,033 kb) (Fig. 1). These findings clearly indicated that the *H* locus spanned about 92 kb defined by *D14Rat13* and*ENSRNOSNP2799341* (Fig. 2). This hooded critical region contained the five exons (1 to 5) of the *Kit* gene and the 48.8 kb genomic region upstream of the *Kit* gene.

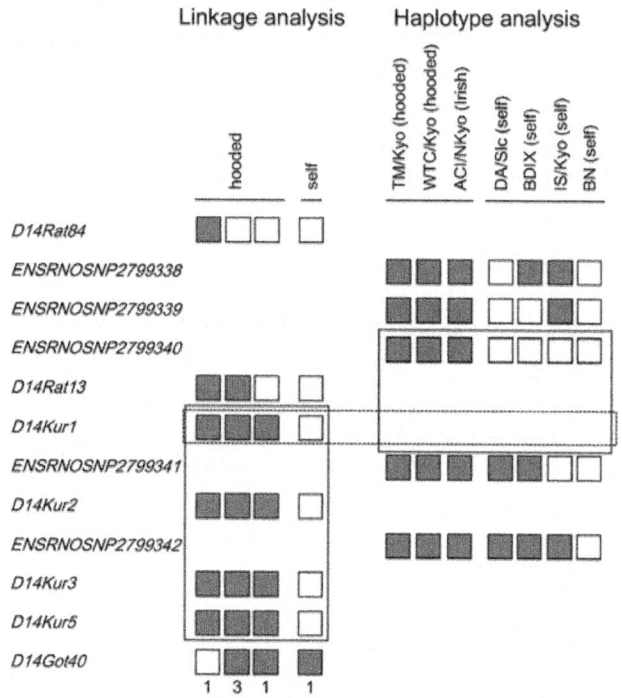

Figure 1: Linkage and haplotype mapping of the hooded (H) locus.

Left: Haplotypes of backcross progeny carrying the recombinant chromosome between D14Rat84 and D14Got40. The grey boxes represent rats homozygous for genetic markers, while the white boxes represent rats heterozygous for genetic markers. The coat pattern phenotypes of the progeny and the number of the progeny for each haplotype are described above and below the haplotypes, respectively. The H locus was narrowed down from D14Rat13 to D14Got40 (boxed with a solid line). Right: SNP haplotypes around the H locus in representative inbred rat strains. Two hooded rat strains and one Irish rat strain shared the identical haplotype of the genomic region between ENSRNOSNP2799338 and ENSRNOSNP2799342. The genomic region that was shared among the four self strains was narrowed down from ENSRNOSNP2799339 to ENSRNOSNP2799341 (boxed with a solid line). The linkage and haplotype analyses demonstrated that the hooded locus was mapped to the region defined by D14Rat13 and ENSRNOSNP2799341 (boxed with a dashed line).

doi:10.1371/journal.pone.0043059.g001

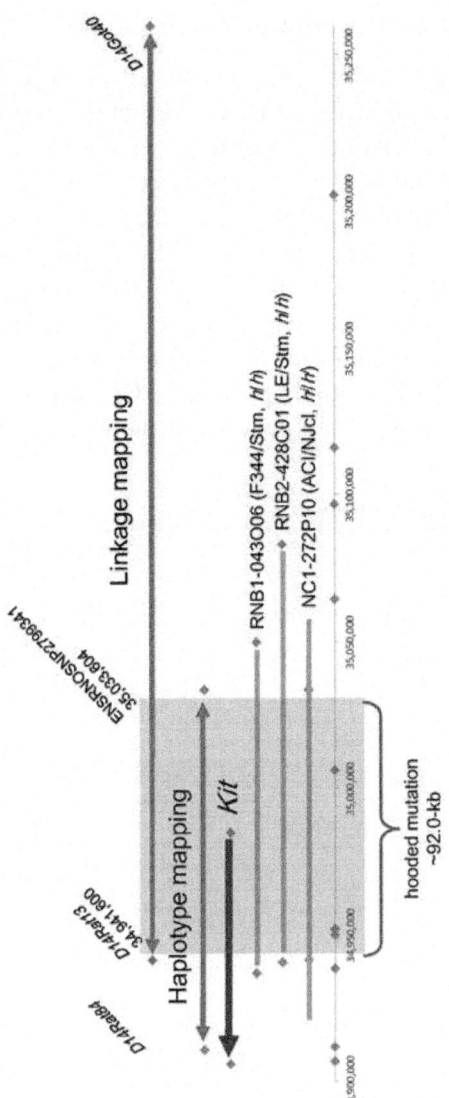

Figure 2: Physical position of the hooded mutation.

The hooded mutation spanned to the 92 kb region defined by D14Rat13 and ENSRNOSNP2799341 (boxed in red). This region contains the upstream region and five exons of the rat Kit gene (black arrow). The hooded critical region was covered by BACs that contain F344/Stm, LE/Stm and ACI/NJcl genomic DNAs. Grey bars represent the regions covered by BACs. Grey squares represent positions of the SSLP or SNP makers examined.

doi:10.1371/journal.pone.0043059.g002

Endogenous Retrovirus Sequences in the Rat Kit Gene

De novo sequencings of the entire inserts of the BAC clones derived from F344/Stm (*c/c, h/h*), and LE/Stm (*C/C, h/h*) rats covering the hooded critical region demonstrated that the 7,098 bp fragment was inserted into intron 1 of the rat *Kit* gene. The position of the insertion was 24,508-bp downstream of exon 1 and 8,054 bp upstream of exon 2 of the *Kit* gene. The 585-bp sequences on the both ends of the insertion showed significant similarities to the LTR of rat ERV (RLTR01_Rn) [9]. The remaining 5,930-bp sequence showed significant similarity to the rat ERV element (RAL_Rn_I) [10]. The ERV insertion was oriented on the minus strand for the*Kit* gene (Fig. 3A). Moreover, the BAC insert containing the Irish allele from the ACI/NJcl rat included the 584 bp LTR element only at the same position of the insertion found in the hooded allele (Fig. 3A).

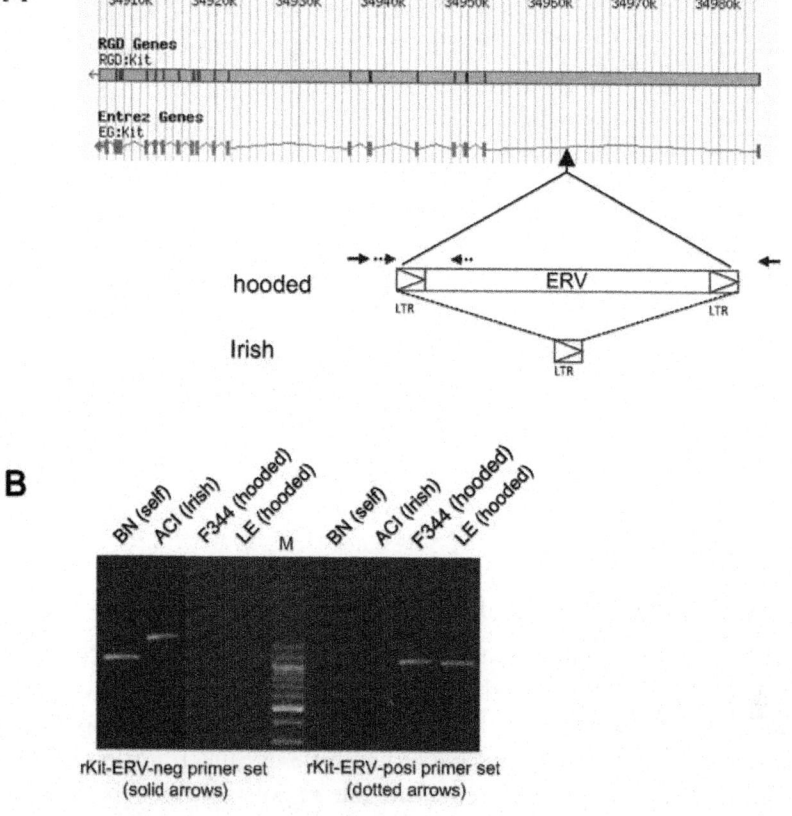

Figure 3: Insertion of the ERV into intron 1 of the rat Kit gene.

A: Position and organization of the ERV insertion found in intron 1 of the rat Kit gene. The insertion was located 8,054 bp upstream of exon 2. The ERV sequence with 2 LTRs on both ends was found in the "hooded" strains. The solitary LTR sequence was found in the "Irish" strains. A set of primers (indicated by solid arrows) was designed to detect the absence of the ERV insertion and the presence of the solitary LTR sequence. A set of primers (indicated by dotted arrows) was designed to detect the presence of the ERV insertion. B (Left): PCR product obtained with the rKit-ERV-neg primer set (solid arrows). A 1,006 bp product was obtained from the BN (self) genome and a 1,590 pb product was obtained from the ACI (Irish) genome. Meanwhile, no products were obtained from the F344 (albino, hooded) genome and the LE (hooded) genome. (Right): PCR products obtained with the rKit-ERV-posi primer set (dotted arrows). Products 997 bp in size were obtained from the F344 genome and the LE genome, while no products were obtained from the BN genome and the ACI genome. M; 100-bp ladder molecular marker.

doi:10.1371/journal.pone.0043059.g003

Genotyping of the 7,098 bp Insertion and the 584 bp Insertion

To find the correlation of the insertions with the hooded or Irish phenotype and survey the prevalence of the insertions among rat strains, we determined the presence of the insertions for 172 rat strains (Fig. 3B). Among 55 colored rat strains, the 18 strains with the self-coat color pattern had no insertion. The 29 strains with the hooded pattern had the ERV (7,098 bp) sequence. Finally, the eight strains with the Irish pattern had the LTR (584 bp) element only. As for the albino rat strains, all of the 117 strains had the ERV (7,098 bp) sequence, indicating the shared uniform haplotype for the albino and hooded mutations (Table S1).

DISCUSSION

The *Kit* gene encodes a cell surface receptor, c-Kit, (molecular weight 145–160 kd) that belongs to the immunoglobulin gene family and carries an intrinsic tyrosine kinase activity in its cytoplasmic portion. The interaction of c-Kit with its ligand, steel factor, leads to receptor dimerization, kinase activation, and tyrosine phosphorylation of cytoplasmic proteins. The c-Kit gene is expressed on melanocytes, gametocytes, mast cells, hematopoietic stem cells, and interstitial cells of Cajal. Thus, the Kit tyrosine kinase membrane receptor is essential for melanogenesis, gametogenesis and hematopoiesis during embryonic development and postnatal life.

c-Kit is expressed in melanoblasts during the melanogenesis starting from the time they leave the neural crest. Expression continues during embryonic development [11]. It is also expressed in the melanocytes of postnatal animals. Thus, mutations of the murine *Kit* gene manifest as dominant white spotting (*W*). In addition to the *Kit* gene, several other genes, such as *Pax3*, *Mitf*, and *Sox10*, could also be involved in the white-spotting phenotype. Mutations in these genes are related to defects of melanocyte development [12]. *W* mutations either alter the coding sequence of the Kit receptor tyrosine kinase, resulting in a receptor with impaired kinase activity, or affect *Kit* expression. *W* mutations that affect *Kit* expression are often located in the regulatory sequence. For example, the Kit^{W-57J} allele affects the temporal and the spatial pattern of the *Kit* expression, so that Kit^{W-57J}/Kit^{W-57J} mice have an irregular band of spotting, lack pigmentation in their feet and tails, and have a head blaze. The Kit^{W-57J} allele comprises an 80 kb deletion located at the 5′ end of the *Kit*-coding sequence [13]. The Kit^{W-bd} and Kit^{W-sh} alleles also affect the *Kit* expression pattern during the developmental stage and the both $Kit^{W-bd}/+$ and $Kit^{W-sh}/+$ mice show a white band in the trunk region [13], [14]. Both alleles are associated with a genomic inversions located at the 5′ region of the *Kit* gene [13]. These findings indicate that the dysregulation of the *Kit* expression affects the development of melanoblasts and thereby the white spotting appears in mutant mice.

We found the insertion of the ERV sequence in the intron 1 of the rat *Kit* gene. ERV sequences have resulted from both ancient and modern infections of exogenous retroviruses, which have successfully colonized the germ line of their host [15]. Insertion of the ERV disrupts host protein-coding genes or alters gene expression by affecting splicing or by providing novel signals for initiation, regulation or termination of transcription [15]. The ERV insertions of the first intron with antisense orientation lead the small and lower level of transcripts [16], or aberrant splicing [17], [18] in mouse mutants. Thus, we consider that the ERV insertion we found in the hooded allele may provoke the dysregulation of the *Kit*expression and thereby causes the specific hooded pattern.

We also found the solitary LTR sequence in the Irish allele. In mice, it has been known that the internal ERV sequence is deleted via homologous recombination between the 5′ and the 3′ LTRs so that a solitary LTR is left behind [19]. Such deletion reverses the mutant phenotype to the wild-type or, occasionally, attenuates the mutant phenotype. The latter case is found in the reversion of the mouse nonagouti (*a*) to the black-and-tan (*a^t*) or the white-bellied agouti (*A^w*). The *a* insertion consists of a 5.5-kb VL30 element that has incorporated 5.5 kb of additional sequence internally; this internal sequence is flanked by 526-bp direct repeats. Homologous recombination utilizing the

526-bp direct repeats generates the a^t allele, containing only the VL30 element with a single internal 526-bp repeat. Homologous recombination utilizing the VL30 LTRs generates the A^w allele, containing only a solitary LV30 LTR [16]. The rat Irish (h^i) causes a white spot on the belly between and behind the front legs. Therefore, we conclude that the solitary LTR in the Irish allele also may have been generated by homologous recombination between the 5' and 3' LTRs of the rat ERV that is inserted in the hooded allele and that the Irish allele is a partial revertant of the hooded allele due to the residual solitary LTR.

Coat color variations have historically played important roles in genetic studies to understand the basics of inheritance. Thus, the identification of the causative mutation of the color variations followed by the survey of them for existing strains provides insights into the origin of coat color mutations. Since the coat color variations were noted in the early days of the domestication of the rat, such molecular genetic approach could give new insights on the establishment of laboratory rat strains. In the present study, we focused on the oldest color variations in the rat, albino and hooded.

We collected tissues or DNAs of rat strains from across the world in order to cover all possible albino or hooded mutations that are still available in laboratory rat strains. Thus, it appears very likely that all albino laboratory rat strains developed so far share only one common mutation: 299His in the *Tyr* gene. Most existing albino rat strains are derived from albino strains or stocks of the Wistar Institute or from crosses between Wistar albino rats and other rats including wild rats. The remaining albino rats were not directly derived from the Wistar Institute (Table S1). DON strains, Ihara's rat strains, and TO strains were established in Japan, while the F344 strains and HTX strain were established in the USA, and Yagil's rat strains were established in Israel. They also carry the same *Tyr* mutation. These findings suggest that the heredity of all albino rats can be traced back to one rat with the albino mutation.

Moreover, we found that all albino strains we examined share the 7,098-bp ERV insertion in the*Kit* gene without any exception. According to Donaldson [4], there must have been both albino and hooded stocks before the establishment of the Wistar albino stock. Given the uniform genotypes for the albino and hooded mutations in the albino rat strains, we suggest two possible scenarios for the establishments of the albino and hooded strains. The most likely one is that the albino mutation occurred in a rat of a hooded stock. From that colony, the first albino rats were discovered and were used as founders of albino rats. Some albino rats were introduced into the Wistar Institute and some were used for developments of Wistar-independent albino stocks (Fig. 4). Another scenario is that albino rat strains had been developed independently of hooded stock and subsequent crossing occurred between the albino and the hooded stocks. In the resultant stocks of this cross, albino rats that carried either the hooded (*h*) or self (*H*) allele must be present. Given the evidence provided in this study, albino rats with the particular genotype (*h/h*, *c/c*) were "by chance" selected and used to establish the albino rat stock. Such newly established albino stock was introduced into the Wistar Institute and some rats were used for the development of the Wistar-independent albino stocks (Fig. 4). Since our survey could not detect any albino rat without the hooded ERV insertion, the second scenario appears very unlikely and is just of hypothetical nature. Therefore, it is very likely that the hooded stock had been developed earlier than the albino strains and the rat albino missense mutation occurred in one hooded rat.

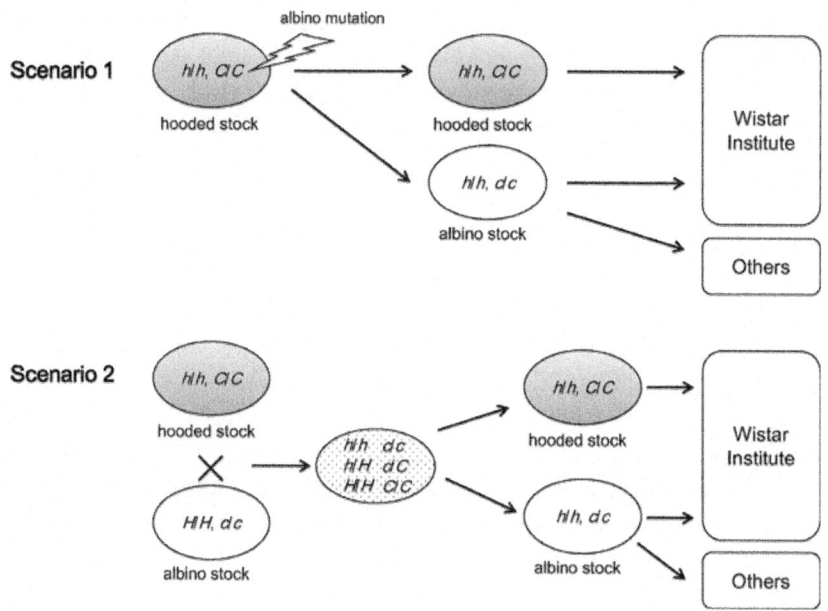

Figure 4: Possible scenarios for the establishment of the albino and hooded rat strains.

Scenario 1: The rat albino (299His) mutation occurred in a rat of the hooded (piebald) rat stock. From the stock, the albino stock was established. Then they were introduced into the Wistar Institute and some albino rats were used for the establishment of the Wistar-independent stocks. Scenario 2: The albino stock that can be traced back to an albino rat that carried the 299His mutation and the hooded (piebald) stock were independently established. Crossing occurred between the albino and the hooded stocks. In the resultant stocks of this cross, albino rats that carried either the hooded (h) or self (H) allele must be present. The albino rats with the particular genotype (h/h, c/c) were "by chance" selected and used to establish the albino rat stock. Such new-established albino stock was introduced into the Wistar Institute and some rats from it were used for the development of the Wistar-independent albino stocks.

doi:10.1371/journal.pone.0043059.g004

MATERIALS AND METHODS

Ethics Statement

Animal experimentation in this study was carried out in strict accordance with the Regulation on Animal Experimentations at Kyoto University. The protocol was approved by the Animal Research Committee of Kyoto University.

Genomic DNAs

Genomic DNAs from 172 rat strains were used. They consist of 55 colored rat strains and 117 albino rat strains (Table S1).

Linkage Analysis

A total of 897 backcross progeny were genotyped for the hooded (*H*) locus by identifying the appearance of the characteristic hooded mark. They consisted of 300 rats from the (ACI × WTC)F1 × WTC cross [20], 207 rats from the (BN × TM)F1 × TM cross, and 390 rats from the (BN × GRY)F1 × GRY cross [21]. All of the rats were genotyped for *D14Rat84* and *D14Got40* and rats that carried the recombinant chromosome between them were used for fine mapping of the *H* locus. Five simple sequence length polymorphism (SSLP) markers (*D14Rat13,D14Kur1, D14Kur2, D14Kur3*, and *D14Kur5*) were used for the fine mapping.

Haplotype Analysis

Two hooded strains (TM/Kyo and WTC/Kyo) and four "self" strains (BDIX, BN/NSlc, DA/Slc, and IS/Kyo) were genotyped for five single nucleotide polymorphism (SNP) markers that were mapped between *D14Rat84* and *D14Got40* [22]. Their names were as follows: ENSRNOSNP2799338 (position: 34,823,877), ENSRNOSNP2799339 (34,910,749), ENSRNOSNP2799340 (34,938,580), ENSRNOSNP2799341 (35,033,604), and ENSRNOSNP2799342 (35,096,682).

BAC Sequencing

RNB1-043O06 BAC containing the hooded allele of the rat strain F344/Stm, RNB2-428C01 BAC containing the hooded allele of the rat strain LE/Stm and NC1-272P10 BAC containing the Irish allele of the rat strain ACI/NJcl, were used for BAC sequencing. High molecular weight BAC DNAs were prepared by using a Large-Construct Kit (QIAGEN) and physically fractionated into sections several hundred base pairs in length by using a Covaris S-series (Covaris, MT, USA). The BAC inserts were sequenced by using a Genome Sequencer FLX system (Roche Diagnostics) and the sequences were assembled by using a GS *De Novo* assembler (Roche Diagnostics).

Genotyping

The genotyping of the Arg299His missense mutation in the *Tyr* gene was performed by the PCR-RFLP analysis [5]. The PCR products amplified with a set of primers (rTyr12&13 TTTCATTCATATGTAAGTCCCTTG and GCTTAGCATTGCAAAACTCACA) were digested with *Sna*BI restriction endonuclease (New England Biolabs). In this assay, the PCR products from the wild-type (299Arg) are digested, while those from the albino-type (299His) are not digested.

To determine the presence of the ERV insertion or the LTR insertion in intron 1 of the *Kit* gene, primer sets were designed. The rKit-ERV-posi (GGCCTGTGAGTGTGAATTTG and GGACGAGCCCCCATAAATA) were used to detect the ERV insertion and the rKit-ERV-neg (ACTTAAAGACCACTGAGGACA and AATGCGGAACATCTTTCAA) were used to detect the LTR insertion.

Supporting Information

Table_S1.xlsx

Table S1: Coat color phenotypes and the genotypes for Tyr and Kit genes in laboratory rat strains.

doi:10.1371/journal.pone.0043059.s001

(XLSX)

	1	2	3	4	5	6	7	8	9
1									
2	Table S1. Coat color phenotypes and the geno-types for Tyr and Kit genes in laboratory rat strains								
3									
4	Strain family	Sub strain name	Source1	NBRP No.2	Coat color pheno-type	Geno-type of the hood-ed locus	Tyr gene	Kit gene	Origin3
5							missence muta-tion Arg 299 His	ERV in-sertion type	

ACI and its derived strains								
6								Established from a cross beteen August and COP strains by MR Curtiss and WF Dunning at Columbia University, Institute for Cancer Research, USA in 1926
7	ACI/NHok	NBRP-Rat	33	agouti hooded Irish	hood-ed Irish	Arg	LTR	From the NIH, USA, to Tokyo Biochemical Research Institute, Japan, and to Hokkaido University, Japan in 1967.
8	ACI/NJcl	CLEA	NA	agouti hooded Irish	hood-ed Irish	Arg	LTR	From the NIH, USA, to National Institute of Genetics, Japan, to Central Institute of Experimental Animals, Japan in 1980, then to CLEA Japan Inc in 1986.
9	ACI/NKyo	NBRP-Rat	1	agouti hooded Irish	hood-ed Irish	Arg	LTR	From the NIH, USA to Kyoto University, Japan in 1988.
10	ACI-Lystbg/Kyo	NBRP-Rat	2	beige hooded Irish	hood-ed Irish	Arg	LTR	Derived from spontaneous beige coat colour mutation found in ACI/NKyo at Institute of Laboratory Animals, Kyoto University in 1999, and maintained as a congenic strain.
11	ACI/NMna	NBRP-Rat	203	agouti hooded Irish	hood-ed Irish	Arg	LTR	Derived from ACI/NMs by M Matsushima at Fujita Health University School of Medicine, Japan
12	ACI/NSlc	NBRP-Rat	158	agouti hooded Irish	hood-ed Irish	Arg	LTR	From National Institute of Infectious Disease, Tokyo, Japan to Japan Slc Inc in 1992
13	ACI-Chib/Kyo	NBRP-Rat	465	agouti hooded Irish	hood-ed Irish	Arg	LTR	Derived from small body size mutation found in ENU-mutagenesis using ACI/NsIc, and maintained as congenic line.
14	ACIS/Hok	NBRP-Rat	34	agouti with white face	hood-ed Irish	Arg	LTR	Derived from a spontaneous mutation showing wider white pigmentation found in ACI/Hok in 1981, and maintained by crossing its heterozygous rats with wild type rats.

No.	Strain	Substrain	Source	Count	Phenotype			LTR	Description
15	Albany rat strain								Albany rat colony derived from an unknown source, supplied by CE Bills at the Mead Johnson Company of Evansville, Indiana, USA, breeding started by A Knudson at the Department of Biochemistry, the Albany Medical College, New York, USA in 1930. Representatives of the Albany stock were transferred to the NIH by JM Wolfe and AM Wright at Albany Medical CollegeUSA, where inbreeding started.
16		ALB/Hok	NBRP-Rat	35	non-agouti dilute brown	self	Arg	-	To Hokkaido University in 1956 (F8), to National Institute of Genetics, Mishima, Japan in 1958, then to the Center for Experimental Plants & Animals Chromosome Research Unit, Faculty of Science, Hokkaido University, Japan
17	Bio Breeding Laboratory derived rat strains								Established from spontaneous mutants, which develop autoimmune Type 1 diabetes, found in a Wistar rat stock at Bio-Breeding Laboratories, Ontario, Canada
18		BB/OK	NBRP-Rat	260	albino	N.D.	His	LTR-ERV1-LTR	Introduced from the Bio Breeding Laboratories at University of Greifswald, Karlsburg, Germany in 1983, and established as BB subline by I Klötting.
19		BB/WorTky	NBRP-Rat	30	albino	N.D.	His	LTR-ERV1-LTR	Introduced from the Bio Breeding Laboratories to the massachusetts medical school, Worcester in 1977, where inbreeding began. Imported to Tokyo Medical College, Japan, by K Komeda in 1983.
20	Berlin-Druckrey strains								BDI was derived from a pair of pink-eyed yellow rats obtained from Kröning, Göttingen, and the inbred strain was served as progenitor strain for further Berlin-Druckrey strains.
21		BDII	HAN-NOVER	NA	albino	N.D.	His	LTR-ERV1-LTR	Derived from an outbred colony of Wistar rats obatained from the Unilever Research Institute, Rotterdam, The Netherlands.

#		Strain	Source	No.	Color				Description
22		BDIX/NemOda	NBRP-Rat	304	agouti	self	Arg	-	Derived from a cross between BDI and BDVIII. BDVIII was derived from F2 cross between BDII and BDIII with the same origin as BDI.
23		KB/Oda	NBRP-Rat	306	albino	N.D.	His	LTR-ERV1-LTR	A segregating inbred strain for albino locus, derived from a cross between Wistar rat of a breeder Nihon rat Ltd. in Japan and BDI.
24	Brown Norway rat strains								Established from a brown coat color rat found in a pen-bred colony of Wistar Insitute maintained by HD King and P Aptekman. The origin are derived from wild rats captured in the vicinity of Philadelphia. F35 by 1934.
25		BN/1Hok	NBRP-Rat	37	brown	self	Arg	-	BN substrain maintained at the Chromosome Research Unit, Hokkaido University in 1987 (F15).
26		BN/2Hok	NBRP-Rat	38	brown	self	Arg	-	BN substrain maintained at the Chromosome Research Unit, Hokkaido University in 1987 (F16).
27		BN/fMailHok	NBRP-Rat	36	brown	self	Arg	-	Introduced to Hokkaido University through Aichi Colony institute from Kyoto University in 1976, and then transferred from the Laboratory of Experimental Animal Science to the Center for Experimental Plants & Animals at Hokkaido University, Japan in 1982.
28		BN/KtsSlc	NBRP-Rat	165	brown	self	Arg	-	To Japan Slc Inc from Baba at Kitasato University School of Medicine, Japan in 2002.
29		BN/KunKtsSlc	NBRP-Rat	164	brown	self	Arg	-	To Japan Slc Inc from Baba at Kitasato University School of Medicine, Japan in 2001.
30		BN/Seac	KYUDO	NA	brown	self	Arg	-	BN substrain maintained at Seac Yoshitomi Co Ltd.

No.	Group	Strain	Supplier		Coat				Notes
31		BN/SsNHsd	HAR-LAN	NA	brown	self	Arg	-	To the NIH, USA in 1972 from WK Silvers at F34. Silvers began brother-sister matings with selection for histocompatibility in 1958 from a brown mutation in a stock of wild rats maintained by HD King in a penbred colony.
32		BN/SsNSlc	NBRP-Rat	149	brown	self	Arg	-	To Japan SLC Inc from the NIH in 1994
33	Buffalo rat strains								Derived from Buffalo stock of H Morris, presumably from Wistar commercial stock in 1946.
34		BUF/Mna	NBRP-Rat	200	albino	N.D.	His	LTR-ERV1-LTR	Established from sib mating of Buffalo rat colony transferred from Biological Associates in America, USA. To Department of Pathology, Nagoya City University in 1964. At F18 generation, the rats showed 100% thymoma and were named as BUF/Mna by M Matsuyama at Fujita Health University, Japan in 1977.
35		BUF/NacJcl	CLEA	NA	albino	N.D.	His	LTR-ERV1-LTR	To Japan Slc inc from National Cancer Center Research Institute, Tokyo, Japan.
36	Cohen diabetes rat strains								Derived from the Hebrew University albino rats, and genetically selected for blood glucose levels by AM Cohen in Israel at about 1965.
37		CDR/Ygl	YAGIL	NA	albino	N.D.	His	LTR-ERV1-LTR	Cohen diabetic-resistant strain developed by selective inbreeding for a lower blood glucose level.
38		CDS/Ygl	YAGIL	NA	albino	N.D.	His	LTR-ERV1-LTR	Cohen diabetic-sensitive strains developed by selective inbreeding for a higher blood glucose level.

No.	Strain group	Substrain	Supplier	No.	Coat color	Marking	Allele	LTR	Notes
39	Copenhagen rat strains								Original rats were obtained from a breeder in Copenhagen by J Rosenstirn and introduced to MR Curtis at Columbia University Institute for Cancer Research in 1920.Inbreeding began in 1922 by WF Dunning at Crocker Reserach Institute of Columbia University, USA.
40		Iar:Copenhagen Rat	IAR	NA	brown hooded	hood-ed	Arg	LTR-ERV1-LTR	To Institue of Animal Reproduction from the NIH, USA, in 1986.
41	DA strain								Derived from heterogenous stocks of unknown origin, and developed by TT Odell at OakRidge National Laboratory. It may be related to COP. It was named DA because it express the Ag-D blood group allele.
42		DA/Slc	NBRP-Rat	157	agouti	self	Arg	-	To Japan SLC Inc from Kumamoto University School of Medicine, Japan in 1983.
43	Dahl rat strains								Derived from Sprague-Dawley outbred rats, and selected for salt sensitive hypertension by LK Dahl at Brookhaven National Laboratories, New York, USA.
44		DahlS/Seac	KYUDO	NA	albino	N.D.	His	LTR-ERV1-LTR	Salt sensivive hypertension substrain
45		SR/JrNgs	NBRP-Rat	485	albino	N.D.	His	LTR-ERV1-LTR	Control strain of SS/JrNgs
46		SS/JrNgs	NBRP-Rat	484	albino	N.D.	His	LTR-ERV1-LTR	Salt sensivive hypertension substrain
47	DOB strain								Derived from wild rats trapped at goat shed at the Shitara field, Graduate School of Bioagricultural Sciences and School of Agricultural Sciences Nagoya University, Japan in 2000.

48		DOB/Oda	NBRP-Rat	307	agouti	self	Arg	-	From S Oda at Nagoya University to Institute of Laboratory Animals, Kyoto University, Japan, in 2007 (F14), where this inbred strain was established in 2009.
49	Donryu rat strains								Derived from albino rats purchased from a local breeder in Ohota city, Gumma, Japan, by R Sato in 1950.
50		ALD/Hyo	NBRP-Rat	576	albino	N.D.	His	LTR-ERV1-LTR	Derived from a male rat showing hepatomegaly and splenomegaly, acid lipase deficiency, in non-inbred Donryu rat found by H Yoshida at Ehime University, Japan in 198, and maintained in heterozygous condition.

Coat color phenotypes and the genotypes for Tyr and Kit genes in laboratory rat strains.

ACKNOWLEDGMENTS

We thank the National Bio Resource Project – Rat (http://www.anim.med. kyoto-u.ac.jp/nbr/) for providing the rat genome DNA. We are grateful to Dr. Dirk Wedekind for providing DNA from rat strains PAR, R33, and BDII, Prof. Yoram Yagil and Dr. Chana Yagil for providing DNA for rat strains CDR/Ygl, CDR/Ygl, SBH/Ygl, and SBN/Ygl, Dr. Pernilla Strindh-Forsgren for providing tissue samples of the M520 rat strain. The RNB1 and RNB2 BAC clones are available from the RIKEN Bio Resource Center (http://dna.brc.riken.jp/index. html). The genome sequences of BAC inserts are available from the DDBJ with accession numbers (AP012486 for NC1-272P10, AP012487 for RNB1-043O06, and AP012488 for RNB2-482C01).

AUTHOR CONTRIBUTIONS

Conceived and designed the experiments: TK TS. Performed the experiments: TK SN. Analyzed the data: TK BV TS. Contributed reagents/materials/analysis tools: MO HN. Wrote the paper: TK BV TS.

REFERENCES

1.   Crampe H (1885) Landwirthschaftliche Jahrbücher.

2.   Castle WE (1951) Variation in the hooded pattern of rats, and a new allele of hooded. Genetics 36: 254–266.

3.   Beermann F, Orlow SJ, Lamoreux ML (2004) The *Tyr* (albino) locus of the laboratory mouse. Mamm Genome 15: 749–758. doi: 10.1007/s00335-004-4002-8

4.   Donaldson HH (1915) The Rat data and reference tables.

5.   Blaszczyk WM, Arning L, Hoffmann KP, Epplen JT (2005) A Tyrosinase missense mutation causes albinism in the Wistar rat. Pigment Cell Res 18: 144–145. doi: 10.1111/j.1600-0749.2005.00227.x

6.   Gershoni-Baruch R, Rosenmann A, Droetto S, Holmes S, Tripathi RK, et al. (1994) Mutations of the tyrosinase gene in patients with oculocutaneous albinism from various ethnic groups in Israel. Am J Hum Genet 54: 586–594.

7.   Serikawa T, Kuramoto T, Hilbert P, Mori M, Yamada J, et al. (1992) Rat gene mapping using PCR-analyzed microsatellites. Genetics 131: 701–721.

8.   Torigoe D, Ichii O, Dang R, Ohnaka T, Okano S, et al. (2011) High-resolution linkage mapping of the rat hooded locus. J Vet Med Sci 73: 707–710. doi: 10.1292/jvms.10-0529

9. Jurka J (2009) Long terminal repeat of ERV1 enodogenous retrovirus from rat: consensus. Repbase Reports 9: 1214.

10. Nakamuta M, Furuich M, Takahashi K, Suzuki N, Endo H, et al. (1989) Isolation and characterization of a family of rat endogenous retroviral sequences. Virus Genes 3: 69–83. doi: 10.1007/bf00301988

11. Keshet E, Lyman SD, Williams DE, Anderson DM, Jenkins NA, et al. (1991) Embryonic RNA expresson patterns of the c-Kit receptor and its cgnate ligand suggest multiple functional roles in mouse development. EMBO J 10: 2425–2435.

12. Baxter LL, Hou L, Loftus SK, Pavan WJ (2004) Spotlight on spotted mice: a review of white spotting mouse mutants and associated human pigmentation disorders. Pigment Cell Res 17: 215–224. doi: 10.1111/j.1600-0749.2004.00147.x

13. Kluppel M, Nagle DL, Bucan M, Bernstein A (1997) Long-range genomic rearrangements upstream of Kit dysregulate the developmental pattern of Kit expression in W^{57} and W^{banded} mice and interfere with distinct steps in melanocyte development. Development 124: 65–77.

14. Duttlinger R, Manova K, Chu TY, Gyssler C, Zelenetz AD, et al. (1993) W-sash affects positive and negative elements controlling c-kit expression: ectopic c-kit expression at sites of kit-ligand expression affects melanogenesis. Development 118: 705–717.

15. Stocking C, Kozak CA (2008) Murine endogenous retroviruses. Cell Mol Life Sci 65: 3383–3398. doi: 10.1007/s00018-008-8497-0

16. Bultman SJ, Klebig ML, Michaud EJ, Sweet HO, Davisson MT, et al. (1994) Molecular analysis of reverse mutations from nonagouti (a) to black-and-tan (a^t) and white-bellied agouti (A^w) reveals alternative forms of agouti transcripts. Genes Dev 8: 481–490. doi: 10.1101/gad.8.4.481

17. Bowes C, Li T, Frankel WN, Danciger M, Coffin JM, et al. (1993) Localization of a retroviral element within the rd gene coding for the beta subunit of cGMP phosphodiesterase. Proc Natl Acad Sci U S A 90: 2955–2959. doi: 10.1073/pnas.90.7.2955

18. Hofmann M, Harris M, Juriloff D, Boehm T (1998) Spontaneous mutations in SELH/Bc mice due to insertions of early transposons: molecular characterization of null alleles at the nude and albino loci. Genomics 52: 107–109. doi: 10.1006/geno.1998.5409

19. Maksakova IA, Romanish MT, Gagnier L, Dunn CA, van de Lagemaat LN, et al. (2006) Retroviral elements and their hosts: insertional mutagenesis in the mouse germ line. PLoS Genet 2: e2. doi: 10.1371/journal.pgen.0020002

20. Kuramoto T, Yokoe M, Hashimoto R, Hiai H, Serikawa T (2011) A rat model of hypohidrotic ectodermal dysplasia carries a missense mutation in the *Edaradd* gene. BMC Genet 12: 91. doi: 10.1186/1471-2156-12-91

21. Tokuda S, Kuramoto T, Tanaka K, Kaneko S, Takeuchi IK, et al. (2007) The ataxic groggy rat has a missense mutation in the P/Q-type voltage-gated Ca^{2+} channel α1A subunit gene and exhibits absence seizures. Brain Res 1133: 168–177. doi: 10.1016/j.brainres.2006.10.086

22. Saar K, Beck A, Bihoreau MT, Birney E, Brocklebank D, et al. (2008) SNP and haplotype mapping for genetic analysis in the rat. Nat Genet 40: 560–566. doi: 10.1038/ng.124

Chapter 4

MOLECULAR GENETIC ANALYSIS OF CENTRAL NERVOUS SYSTEM GERM CELL TUMORS WITH COMPARATIVE GENOMIC HYBRIDIZATION

Dominik T Schneider[1], Susanne Zahn[1], Sonja Sievers[1,2], Katayoun Alemazkour[1], Guido Reifenberger[3], Otmar D Wiestler[4], Gabriele Calaminus[1], Ulrich Göbel[1] and Elizabeth J Perlman[5]

[1]Clinic of Paediatric Oncology, Haematology and Immunology, Heinrich-Heine-University, Düsseldorf, Germany

[2]Max Planck Institute of Molecular Physiology, Dortmund, Germany

[3]Department of Neuropathology, Heinrich-Heine-University, Düsseldorf, Germany

[4]German Brain Tumor Reference Center Institute of Neuropathology, University of Bonn, Bonn, Germany

[5]Department of Pathology, Children's Memorial Hospital, Chicago, IL, USA

ABSTRACT

The limited information available to date regarding the genetic alterations in germ cell tumors of the central nervous system has raised concerns about their biologic relationship to other germ cell tumor entities. We investigated fresh-frozen or archival tumor samples from 19 patients with central nervous system germ cell tumors (CNS-GCTs), including seven germinomas, eight malignant nongerminomatous germ cell tumors and four teratomas, using chromosomal comparative genomic hybridization to determine recurrent chromosomal imbalances. All 15 malignant CNS-GCTs and two of four teratomas showed multiple chromosomal imbalances. Chromosomal gains (median: 4 gains/tumor, range: 0–9 gains/tumor) were observed more frequently than losses (median: 1.6 losses/tumor, range: 0–6 losses/tumor). Gain of 12p, which is considered characteristic for germ cell tumors of the adult testis, was detected in 11 of 19 tumors and 10 of 15 malignant CNS-GCTs. In one tumor, gain of 12p was confined to an amplicon at 12p12, corresponding to the commonly amplified region on 12p. Other common gains were found on chromosome arms 1q and 8q (n=9, each). Among the chromosomal losses, parts of chromosome 11 (n=5), 18 (n=4), and 13 (n=3) were deleted most frequently. Notably, we observed no difference in the genetic profiles of germinomatous and nongerminomatous CNS-GCTs; however, the average number of imbalances was higher in the latter group. A meta-analysis comparing 116 malignant gonadal and extragonadal germ cell tumors revealed that the genomic alterations in CNS-GCTs are virtually indistinguishable from those found in their gonadal or other extragonadal counterparts of the corresponding age group. These data strongly argue in favor of common pathogenetic mechanisms in gonadal and extragonadal germ cell tumors.

During childhood and adolescence, the majority of germ cell tumors arise outside of the gonads, and beyond early childhood, the central nervous system and the mediastinum constitute the most frequent sites of extragonadal germ cell tumors.[1] The vast majority of central nervous system germ cell tumors (CNS-GCTs) develop in the pineal gland or the suprasellar region, and approximately 10% of all CNS-GCTs present as bifocal tumors.[2,3] The extragonadal appearance of germ cell tumors is likely related to errors in germ cell migration during early embryonal development. Accordingly, imprinting studies of gonadal and extragonadal germ cell tumors show loss of the methylation imprint at imprinting control regions, correlating with the methylation status within early stages of primordial germ cell development.[4,5,6] However, the molecular mechanisms that interfere with normal homing of germ cells to the gonadal ridge and that allow for a sustained survival of germ cells in an extragonadal environment remain unclear. Furthermore, the molecular mechanisms of the apparent tropism to the pineal gland and suprasellar region in particular have to be elucidated.

Morphologically, gonadal and extragonadal germ cell tumors are virtually indistinguishable. Apart from spermatocytic seminomas, which may develop in the testis of elderly patients, all histologic entities of germ cell tumors may also arise in the central nervous system. Among malignant germ cell tumors, germinomas are distinguished from nongerminomatous germ cell tumors, which include embryonal carcinoma, yolk sac tumor and choriocarcinoma. In addition, teratomas can be distinguished as a distinct group with a histologically benign appearance. These may include completely differentiated organoid (mature teratoma) or immature structures, the latter resembling immature steps of embryonal and fetal organ development. Characteristically, malignant nongerminomatous germ cell tumors present as tumors with composite histology that include different histologic subentities. It is generally assumed that this phenomenon is related to the cell of origin, the totipotent primordial germ cell, which may differentiate along many different pathways.

Among all germ cell tumors, malignant testicular germ cell tumors of young adults constitute the most intensively studied entity. These are characterized by a specific cytogenetic aberration, the isochromosome 12p (i (12p)), which is considered pathognomonic and can be detected in about 80% of malignant testicular germ cell tumors.[7] In the remaining tumors, gain of chromosomal material of 12p is related to either marker chromosomes or homogeneously staining regions that contain chromosomal material derived from chromosomal band 12p11.2–12.1.[8, 9] Additional chromosomal imbalances have also been reported in testicular germ cell tumors. However, these are less frequent and less consistent as compared to gain of 12p.[10, 11, 12] Notably, testicular germ cell tumors that arise during infancy and childhood show a divergent genetic profile that is characterized by loss of 1p and 6q, and gain of 1q and 20, while gain of 12p11–12 is not detectable.[13, 14, 15, 16] In extragonadal germ cell tumors, similar age-dependent cytogenetic and molecular-genetic patterns have been identified.[13, 17]

There are only limited data regarding cytogenetic and molecular genetic patterns of intracranial germ cell tumors. The largest genetic study published to date includes 15 pineal germ cell tumors analyzed with comparative genomic hybridization (CGH). In this study, only a minority of germinomas showed gain of 12p.[18] Others detected gain of 12p at varying frequencies.[19, 20, 21, 22, 23] Therefore, several authors concluded that gain of 12p, which constitutes the biologic hallmark of germ cell tumors in adolescents and adults, does not play a major role in the biology of CNS-GCTs, thus distinguishing these from their counterparts at gonadal or other extragonadal sites, such as the mediastinum.[18,23] Of note, the genetic profiles of pineal germ cell tumors were different from pineal parenchymal tumors, which frequently showed gain of 12q and 22.[24]

Considering these conflicting data, we have performed chromosomal CGH analysis of 19 tumors from representative histologic groups in order to define characteristic patterns of chromosomal imbalances in CNS-GCTs. In addition, we compared the CGH profiles of CNS-GCTs to those of 101 malignant germ cell tumors from gonadal and different extragonadal sites, including tumors of different age groups and all histologic entities.

MATERIALS AND METHODS

Patients

Fresh-frozen tumor tissue (n=14) or formalin-fixed and paraffin-embedded archival tumor tissue (n=6) was obtained from the germ cell tumor bank of the Pediatric Oncology Group, the German Brain Tumor Reference Center in Bonn, Germany, the Department of Neuropathology in Düsseldorf, Germany, and the German Children's Tumor Registry in Kiel. In one case, both fresh frozen and paraffin embedded tissue was available for analysis, and both experiments yielded identical profiles. All tumors were classified according to the WHO classification of brain tumors and germ cell tumors.[25, 26] Hematoxylin–Eosin stained sections were evaluated prior to DNA extraction. Only those tumor samples that showed more than 70% viable tumor cells were analyzed. Germinomas with pronounced lymphocytic infiltration comprising greater than 30% of the specimen cellularity were excluded. A brain metastasis of a testicular nonseminomatous germ cell tumor was included for comparison to primary CNS-GCTs. None of the patients had a history of a previous, synchronous or metachronous testicular, ovarian or extracranial extragonadal germ cell tumor. The study was reviewed and approved by the Local Ethics Committee of the Heinrich-Heine University, Düsseldorf, Germany, and the Joint Committee on Clinical Investigations of the Johns-Hopkins Medical Institutions, Baltimore, MD, USA.

DNA Extraction from Fresh-Frozen Tissue

DNA was salt-extracted from twenty 5 μm sections after sodium dodecyl sulfate (SDS)/proteinase K (GibcoBRL, Rockville, MD, USA) digestion for 24 h, and purified with phenol–chloroform–isoamyl alcohol extraction followed by ethanol precipitation using standard protocols.[27] Alternatively, DNA was extracted with the DNeasy Tissue Kit (Qiagen, Hilden, Germany), according to the manufacturer's instructions. Reference DNA was extracted from normal male lymphocytes.

DNA Extraction from Archival Tissue

Five 10 μm sections were deparaffinized with xylene and rehydrated through an alcohol series. For tumor dissection, five to 10 10-μm sections on glass slides were manually dissected after deparaffinization with a razor blade using a stained section as a guide. The tissue was incubated at 55°C for 72 h with SDS/proteinase K (1.5 mg/ml, new enzyme added after 24 and 48 h).[28] DNA was extracted according to the phenol–choloform–isoamyl alcohol method[27] and ethanol-precipitated.

Labeling of Tumor and Reference DNA

Prior to labeling, the presence of high-molecular weight DNA and absence of overwhelming DNA degradation (in archival tissues) were confirmed by 1% agarose gel electrophoresis, and the amount of DNA was determined with UV spectroscopy. Tumor and reference DNA (2 μg) were directly labeled with Spectrum-Green- and Spectrum-Red-dUTP (Vysis, Downers Grove, IL, USA) using nick translation, as previously described.[14, 17, 29] The amount of DNase and

DNA-Polymerase (GibcoBRL) and the reaction time were carefully adjusted in order to obtain labeled DNA fragment lengths of 500–2000 base pairs on a 1% agarose gel.

Comparative Genomic Hybridization

In total, 500 ng of labeled tumor and reference DNA were ethanol coprecipitated with 20 μg of Cot-1 DNA (GibcoBRL). The precipitate was resuspended in hybridization buffer containing 50% formamide, 2 × SSC, and 10% dextran sulfate. The probe was denatured at 75°C for 10 min and partially reannealed at 37°C for 30 min. The probe was hybridized at 37°C for 3 days to methotrexate-synchronized male metaphase cells that previously had been aged for 5 days, stored at -20°C, washed in 2 × SSC, and denatured at 70°C for 2 min in 70% formamide, 2 × SSC, pH 7.0. Posthybridization washes were performed for 5 min each in 2 × SSC at 70°C, 37°C, and in water at room temperature. The chromosomes were counterstained with 0.1 μg/ml 4'6/diamidino-2-phenylindole-2HCl (DAPI) in Antifade (Oncor, Gaithersburg, MD, USA). The experimental protocol has been validated using control samples that have been characterized by conventional cytogenetic analysis.

Microscopic Analysis

Gray-level images were aquired for each fluorescent dye with a charge-coupled device camera on a Zeiss Axioscope epifluorescence microscope, using the Applied Imaging Corporations's dedicated Cytovision soft- and hardware for 'high-resolution CGH' (according to the protected designation by Cytovision Inc., Santa Clara, CA, USA).[30, 31] The chromosomes of at least 15 metaphase cells were identified using reverse DAPI banding. The background fluorescence was subtracted, and the green:red ratio of each entire metaphase was normalized to 1.0. According to the designated high-resolution modus for CGH analysis, average fluorescence ratios and 95% confidence intervals were calculated from at least 12 representative chromosomes and compared to dynamic standard reference intervals,[32] thus increasing the sensitivity and spatial resolution of chromosomal CGH to approximately 5 MB.[33]

RESULTS

Fourteen tumors were from male and five tumors from female patients. Age at diagnosis ranged from a neonate to 25 years, with a median of 11.5 years. This distribution is in accordance with the general epidemiologic pattern of CNS-GCTs, indicating that this panel of CNS-GCTs constitutes a representative group.[1] Among 18 patients with complete clinical data, pineal tumors were most frequent (*n*=11), followed by the suprasellar region (*n*=5) and the spine (*n*=2). In a 3-year-old child (Table 1, patient no. 12), a ventricular recurrence of a primarily suprasellar tumor was analyzed. Nine patients were prospectively enrolled onto the German MAKEI (Maligne Keimzelltumoren) 89 or SIOP CNS GCT 96 protocol, allowing for analysis of clinical follow-up data. The patient with tumor recurrence died as a result of tumor progression, whereas the others are in continuous remission. Table 1 summarizes the documented clinical and pathologic data in relation to the chromosomal gains and losses detected by CGH analysis.

Table 1:- Summary of the clinical, pathologic and genetic data of the 19 patients with CNS-GCTs, sorted by histology and age

#	Tissue	Age	Sex	Site	Histology	Status	Gains	Losses
1	FFT	Neonate	F	Spinal cord	MT	NED	None	None
2	FFT	6 y	M	Pineal gland	IT	NED	1q, 8, 17q, X	20
3	FFT+FFPET	8 y	M	Pineal gland	IT	NED	1q, 8, 12p, X	9q, 11, 18
4	FFT	20 y	M	Pineal gland	IT	ND	None	None
5	FFPET	10 y	F	Pineal gland	GER, TER	NED	1q, 8, 21	—
6	FFPET	11 y	M	Pineal gland	GER, TER	NED	12p, 17q	—
7	FFPET	12 y	M	Pineal gland	GER	ND	1, 2, 8, 10, 12	4, 13q, 22q
8	FFPET	14 y	M	Pineal gland	GER	NED	1q21–q24, 7q22–qter, amp12p12, 17q24–qter, 21, 22	Y
9	FFT	15 y	F	Suprasellar region	GER	ND	8q11–q21	—
10	FFPET	25 y	M	Suprasellar region	GER	NED	7, 12p	13
11	FFPET	Adult	F	Pineal gland	GER	ND	3p, 12p	1p, 6, 11
12	FFT	3 y	M	Relapse (3rd ventr.)	EC, YST	DOD	1p21–qter, 2pter–2q22, 2q34–qter, 3	11q24–qter, 20q
13	FFPET	9 y	M	Pineal gland	IT, YST	NED	1q, 5q, 8, 10q22–q23, 12p, 16q23–qter, 22	11q14–q22, 18, Y
14	FFPET	10 y	M	Pineal gland	IT, YST	NED	3,12p,14,18,21	—
15	FFPET	11 y	M	Pineal gland	YST, EC, CHC	NED	2q32–qter, 8, 12, amp12p, 14, 21, 22, X, Y	4, 9p, 10, 11
16	FFPET	13 y	F	Suprasellar region	YST	ND	1q21–q32, 2p16–p22, 3p, 8, 10, 12p, 17q, 20q, 21	—
17	FFPET	17 y	M	Suprasellar region	EC, TER	ND	X, Y	3p13–p14, 5q32–qter, 8q24–qter. 10q25–qter, 13, 18q
18	FFPET	20 y	M	Spinal cord	EC, YST	ND	1q, 5	—
19	FFPET	20 y	M	Pineal gland	YST	NED	2, 7, 8q, 12p, 14, 18p, 21, X, Y	8p21–pter, 15, 18q

Four patients presented with pure mature (*n*=1) or immature (*n*=3) teratomas, whereas all other patients suffered from malignant germ cell tumors. Seven patients had germinomas, which showed additional teratomatous elements in two patients. Eight patients were diagnosed with malignant nongerminomatous germ cell tumors. In these tumors, yolk sac tumor was the most common histologic type, followed by embryonal carcinoma and teratoma, while choriocarcinoma was detected in one patient only.

Of 19 analyzed tumors, 17 showed chromosomal imbalances (Figure 1). In general, chromosomal gains (mean: 4 gains/tumor, range: 0–9 gains/tumor) were more frequent than losses (mean: 1.6 losses/tumor, range: 0–6 losses/tumor). The most frequent chromosomal imbalance, gain of 12p, was detected in 11 of 19 tumors and 10 of 15 malignant CNS-GCTs. In two of these tumors, the whole chromosome 12 was gained, with one of these two tumors showing a high level gain of 12p. Notably, in one germinoma, gain of 12p was limited to the chromosomal band 12p12 (Figure 2). Gain of 12p was absent in tumors of the three patients who were younger than 8 years of age.

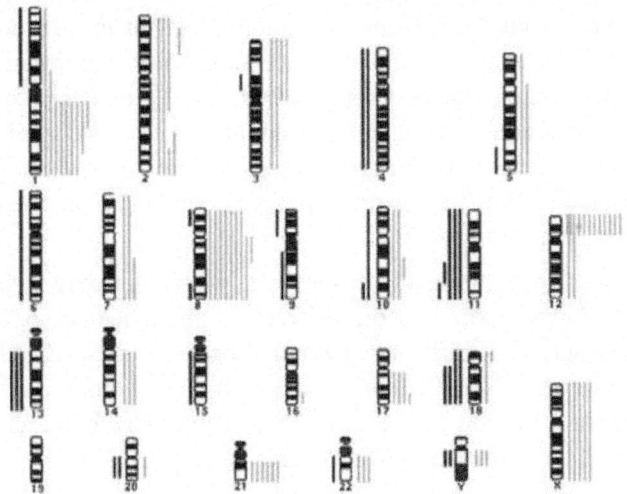

Figure 1: High-resolution CGH analysis of 19 CNS-GCTs: Each line to the right or left side of a chromosome represents a region gained or lost, respectively. Chromosomal amplifications (>1.5:1 ratio) are indicated as thick bars.

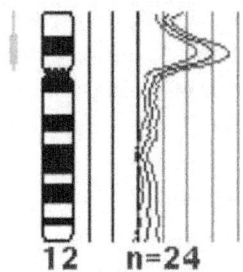

12   n=24

Figure 2: CGH profile of a pineal germinoma showing a circumscribed high-level amplification restricted to chromosomal band 12p12.

The second most common chromosomal gains were found at 1q, with a region of overlap mapping to 1q21–q24, and chromosome 8, with a commonly gained region at 8q11–q21. In six patients, chromosome 21 was gained, and five tumors showed gain of the X chromosome. Due to restricted tumor material and lack of normal reference tissue, no analysis defining the constitutional sex chromosomal status could be performed. Chromosome arms

11q and 18q as well as chromosome 13 were most frequently affected by losses of genetic material. The only tumors with balanced CGH profiles were two pure teratomas, while all malignant germ cell tumors showed chromosomal imbalances. However, the CGH profile of the two imbalanced teratomas resembled those found in malignant CNS-GCTs. Among malignant germ cell tumors, chromosomal imbalances were more frequent in nongerminomatous germ cell tumors (mean: 8.1 imbalances/tumor) than in germinomas (mean: 4.1 imbalances/tumor, Student's t-test, $P=0.03$). Particularly, chromosomal gains were more frequently detected in nongerminomatous germ cell tumors (mean: 5.9 gains/tumor) than in germinomas (mean: 3 gains/tumor, $P=0.045$), whereas chromosomal losses were found at comparable frequencies (mean: 2.3 vs 1.1 losses/tumor, $P=0.27$).

Otherwise, no significant differences were detected when specific imbalances were compared to the different malignant histologic types. We also did not observe a significant correlation between a specific chromosomal imbalance and patient gender, tumor site or clinical outcome. Last, there was no significant correlation between the tissue type (fresh frozen or paraffin embedded) and chromosomal profiles detected with CGH.

Meta-Analysis of Gonadal and Extragonadal Germ Cell Tumors

We compared the chromosomal patterns observed in the 15 malignant CNS-GCTs of this series to those published in other major CGH studies of malignant gonadal and extragonadal germ cell tumors, in which the chromosomal profiles of individual tumors were provided, sorted by age and tumor site ([14, 17, 18, 29, 34]plus 16 unpublished tumors). Other studies, in which the CGH profile cannot be attributed to individual tumors, were excluded (eg Korn et al[11] and Mostert et al[34]). In total, 116 malignant germ cell tumors were included in this meta-analysis.

Similar to the study of Chibon et al,[35] we divided the genome into 100 datapoints resulting in chromosomal segments of approximately 30–40 Megabases. Since most of the studies included did not specifically indicate chromosomal amplification, we only designated chromosomal gain as +1 and loss as -1, while balanced regions were counted as 0. In Figure 3, all chromosomal gains are illustrated as green bars, and losses as red bars, respectively.

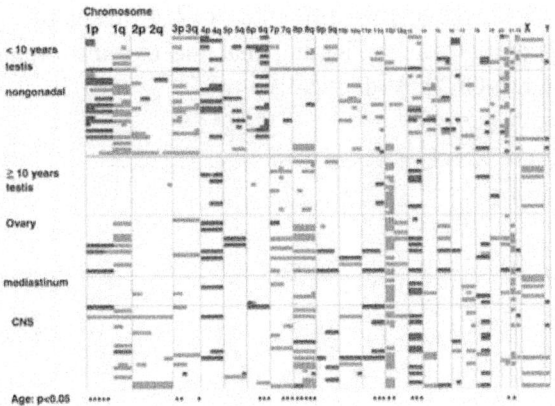

Figure 3: Meta-analysis of 116 malignant gonadal and extragonadal GCTs, separated by age and sorted by site[14, 17, 18, 29, 34] (including this series and 16 unpublished tumors). Each line represents 100 chromosomal data points of a single tumor, with chromosomal gains illustrated as green bars and losses as red bars, respectively. At the bottom, chromosomal regions showing statistically significant differences that correlated with age are indicated by an asterix.

This comparison revealed no detectable difference between seminomas or malignant nonseminomatous germ cell tumors. Instead, significant differences according to site and age are obvious. Loss of 1p is most frequent in sacrococcygeal and mediastinal germ cell tumors of infancy and childhood ($P<0.01$), while gain of 1q can be found in both age groups. Loss of 6q is virtually exclusive to malignant childhood germ cell tumors ($P<0.001$), irrespective of site. Moreover, gain of 20q represents a distinctive imbalance of childhood germ cell tumors ($P<0.001$). In contrast, gain of 12p represents the most important marker of both gonadal and extragonadal malignant germ cell tumors of adolescent and adult patients ($P<0.001$), but is only exceptionally found in children. In addition, gain of 8/8q is also common in this age group ($P<0.01$). Lastly, tumors arising in adolescents and adults are characterized by more frequent gain of 7q, 21q and loss of 13 ($P<0.001$), while gain of 3p is more commonly observed in pediatric tumors/($P<0.05$). Taken together, this meta-analysis of 116 germ cell tumors clearly demonstrates the reproducibility of patterns detected by CGH across different laboratories.

Last, a specific comparison of the 34 CNS-GCTs analyzed by Rickert *et al* and our group with 36 testicular germ cell tumors was performed (Figure 4). This analysis included 18 children younger than 10 years. This analysis indicates that with regard to chromosomal imbalances detectable with CGH, there are no apparent differences according to site. The only detectable difference appears to be a higher percentage of gain of chromosome 8 in CNS-

GCTs ($P<0.05$); however, this difference is not detectable, if age is use as a stratum for comparison, indicating that the chromosomal constitution of testicular and CNS-GCTs is identical.

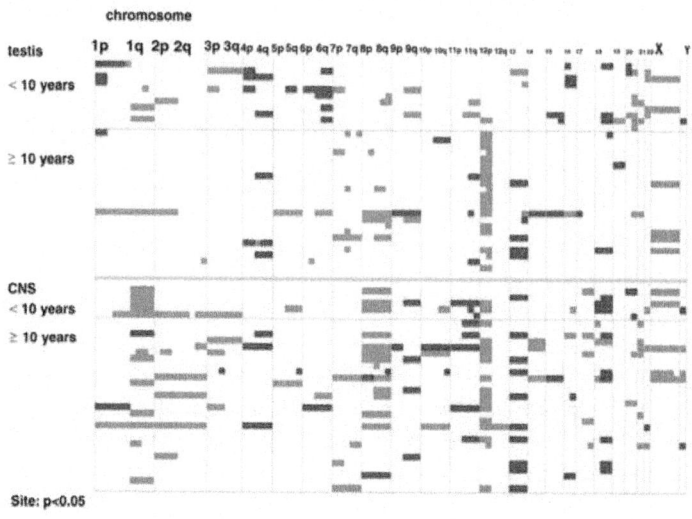

Figure 4: Meta-analysis of 36 testicular and 34 CNS-GCTs, separated by site and sorted by age[15, 18, 34] (including this series and eight unpublished tumors). Each line represents 100 chromosomal data points of a single tumor, with chromosomal gains illustrated as green bars and losses as red bars, respectively. As indicated at the bottom, no chromosomal regions showed statistically significant differences correlated with site.

DISCUSSION

The last 25 years have been characterized by a significant breakthrough in the treatment of malignant germ cell tumors, which can mainly be attributed to the introduction of cisplatin-based chemotherapy and the optimization of multidisciplinary therapeutic strategies.[36] Recently, this success has been translated to the treatment of CNS-germinomas and, to a lesser extent, to nongerminomatous germ cell tumors.[2, 37, 38, 39, 40, 41] To a certain degree, the less favorable prognosis of CNS-GCTs that secrete alpha fetoprotein or human chorionic gonadotrophin (embryonal carcinoma, yolk sac tumor, choriocarcinoma) compared to secreting germ cell tumor at other sites might be related to anatomic problems that interfere with surgical resection. Nevertheless, it remains possible that specific biologic aspects of CNS-GCTs may additionally contribute to the poorer prognosis compared to malignant germ cell tumors at other sites.

In general, it can be assumed that lessons learned in testicular germ cell tumors might also be transferred to CNS-GCTs. This assumption is supported by imprinting studies implicating a common precursor cell for both gonadal and extragonadal germ cell tumors.[5, 6] However, there are several lines of evidence that suggest certain biologic differences between central nervous system and non-CNS-GCTs. Testicular germ cell tumors show a characteristic association with a preinvasive carcinoma *in situ*, the testicular intratubular neoplasia.[7, 31, 42]No such *in situ* lesion has been seen in the brain of patients with CNS-GCT. Furthermore, it has been demonstrated that certain histologic types express markers associated with distinct maturational stages of germ cell development, which presumably are dependent on a specific gonadal environment.[7,43,44,45]Lastly, there are some data suggesting that the chromosomal marker characteristic of testicular germ cell tumors, the isochromosome 12p, plays only a minor role in CNS-GCTs.[18, 22] As a consequence, it must be debated whether the information generated in the more frequent (and more easily assessable) testicular germ cell tumors can indeed be transferred to CNS-GCTs without scrutiny. This concern relates not only to our general understanding of CNS-GCT biology but also to the development of future therapeutic strategies.

In this context, the meta-analysis of germ cell tumors of the different clinical and histologic groups (Figures 3 and 4) provides important information. While age-dependent differences become clearly apparent, there is no obvious correlation with either site or histologic subtype of malignant germ cell tumor.[13, 14, 46] In particular, our meta-analysis unequivocally indicates that there is no apparent difference in the chromosomal constitution of testicular and CNS-GCTs (Figure 4). In accordance with the observation that the vast majority of CNS-germ cell tumors develop in adolescents and adults,[1] CNS-GCTs also display CGH profiles characteristic of malignant germ cell tumors of this age group. In fact, gain of 12p, which in most cases is related to an isochromosome 12p, constitutes the most frequent chromosomal imbalance in our series of CNS-GCT. Furthermore, the additional chromosomal imbalances such as gain of 1q or 8 also occur at comparable frequencies as in other gonadal and nongonadal germ cell tumors of adults (Figures 3 and 4). Although CNS-GCTs show no bimodal age distribution as testicular or mediastinal germ cell tumors,[1] our analysis suggests that also in the CNS, germ cell tumors of infancy and early childhood are genetically distinct from those of adolescence and adulthood, as indicated by the analysis of patients 1 and 12, which show profiles characteristic of teratoma and malignant germ cell tumors of childhood.[47] However, in the patients 2, 3 and 13, who were 6, 8 and 9 years old, profiles characteristic of adolescent patients have been found. Therefore, in contrast to other sites such as the mediastinum, an age cutoff somewhat younger than 10 years might be more appropriate for CNS-GCTs, indicating for modulating local environmental factors. Nevertheless, our observation and our meta-analysis substantiate the notion that age and not site constitutes the primary distinguishing parameter.

In other extragonadal germ cell tumors (in particular mediastinal germ cell tumors), there is a significant association with constitutional Klinefelter syndrome that distinguishes these tumors from testicular germ cell tumors, for which this association is not seen.[48, 49] Furthermore, there have been single reports of CNS-GCTs in the presence of Klinefelter syndrome.[50, 51, 52] In our series of patients, we have no clinical data indicating constitutional Klinefelter syndrome. A previous FISH analysis of six suprasellar teratomas (age 2–25 years) revealed a high frequency of sex chromosomal abnormalities, with a normal constitutional karyotype in one analyzed patient.[53] In accordance to this study, both teratomas with chromosomal imbalances and three malignant nongerminomatous germ cell tumors showed gains of the X chromosome. Unfortunately, we did not have access to extratumoral tissue allowing for an analysis of the constitutional karyotype.

The Role of Gain of 12p in CNS-GCTs

The question has been raised whether those CNS-GCTs that lack gain of 12p constitute a different biologic subtype.[18] In contrast to our series, in which five of six germinomas displayed gain of 12p, Rickert et al found gain of 12p in only one out of eight central nervous system germinomas. This statistically significant difference (x^2-test, $P=0.02$) could probably be related to methodical differences such as sample selection. Germinomas often show a pronounced lymphocytic infiltration that could significantly dilute the tumor DNA content, and therefore reduce the ability to detect chromosomal imbalances. In accordance with this hypothesis, Rickert et al detected no chromosomal imbalances in two of eight germinomas, whereas in our series, all malignant tumors showed genomic alterations. Furthermore, it might be argued whether the high-resolution CGH algorithm might additionally increase the heterogeneity tolerance as it has been postulated for array CGH.[54] In this context, it has to be noted that although the spatial resolution of chromosomal CGH is limited, even with array CGH approaches no discrepant or additional frequently recurrent imbalances have been detected in extracranial germ cell tumors[16] or CNS-GCTs (Chin C Lau, oral presentation on the Second International Symposium in CNS-GCTs in Los Angeles, November 2005[55]).

Furthermore, other issues such as a patient selection bias must also be considered. First, the three youngest patients included in this series (and the youngest patient of the series published by Rickert et al) do not show gain of 12p but a CGH profile comparable to that of testicular, sacrococcygeal or mediastinal CGH at this age with imbalances of chromosome 1 and gain of 1q or 20q in malignant germ cell tumors (Figures 1 and 3). Second, three of the tumors that lacked gain of 12p are teratomas. Characteristically, the vast majority of teratomas indeed do not show chromosomal aberrations on cytogenetic, CGH or array CGH analyses.[13, 16, 17, 47]

Nevertheless, it is very important to note that two of the teratomas included in our study (and both immature central nervous system teratomas analyzed by Rickert *et al*) show chromosomal imbalances resembling those of malignant germ cell tumors. This might indicate that in contrast to teratomas in infants, teratomas of adolescents and adults might be related to a maturational differentiation of malignant germ cell tumors with a preservation of genetic aberrations that may result in a potentially malignant clinical presentation (accordingly, patient no. 3 showed tumor cell dissemination in the cerebrospinal fluid). In previous classification systems, this phenomenon had been attributed the term 'malignant teratoma, differentiated'. In accordance, we and other have found significant cytogenetic and CGH aberrations in residual or growing teratomas of patients with malignant nonseminomas (unpublished observation and van Echten *et al*[56]). Therefore, the histologic diagnosis of teratoma within the brain of a postpubertal patient should not result in confidence of benignity.

Considering these exceptions, 10 out of 14 malignant germ cell tumors of patients older than 5 years showed gain of 12p. The Figures 3 and 4 illustrate that this frequency is almost identical to that found in testicular[11, 34] and mediastinal germ cell tumors of adolescents and adults.[17] In addition, our meta-analysis underlines that CNS-GCTs without 12p gain do not show chromosomal patterns distinct from other germ cell tumors.

Restricted Amplification of 12p12

We observed one patient whose tumor demonstrated a circumscribed amplification of 12p12. To date, comparable profiles have only been detected in testicular and ovarian (Riopel *et al*) germ cell tumors but not in extragonadal germ cell tumors. Albrecht *et al*[57] published a cytogenetic analysis of a central nervous system germinoma that showed a homogeneously staining region (HSR), the chromosomal origin of which could not be identified by FISH with painting probes to the chromosomes 6 or 12.

In testicular tumors, gain of chromosomal material at 12p in the absence of an isochromosome 12p is also related to the presence of either an HSR or double minutes. Microarray and FISH analyses helped to delineate the commonly amplified region at 12p12. It is generally assumed that the amplification of tumor genes at this region triggers malignant transformation at the transition from TIN to invasive malignant germ cell tumors.[8, 9, 31] Since in our patient the amplicon is located at the identical chromosomal region as in testicular germ cell tumors, this observation provides additional evidence in favor of a common pathogenesis of testicular and CNS-GCTs.

In conclusion, this CGH data obtained in our present series of 19 CNS-GCTs and their comparison to previous findings in gonadal and extragonadal germ cell tumors clearly indicate that at the resolution level of CGH, CNS-GCTs are virtually indistinguishable from malignant germ cell tumors at other sites presenting in the same age group. This observation strongly argues for a similar genetic constitution and biology of germ cell tumors at different sites.

ACKNOWLEDGEMENTS

This study has been supported by a Max-Eder Grant of the German Cancer Aid (Deutsche Krebshilfe), the Children Cancer Foundation of Maryland and the American Cancer Society, Grant Number RPG-97-113-01-CCE. We thank Dieter Harms for providing tissue samples, Amy E Schuster for technical assistance and Renate Groth and Susanne Koch for data management.

REFERENCES

1. Schneider DT, Calaminus G, Koch S, *et al.* Epidemiological analysis of 1442 children and adolescents registered in the german germ cell tumor protocols. *Pediatr Blood Cancer* 2004;42:169–175.

2. Calaminus G, Bamberg M, Baranzelli MC, *et al.* Intracranial germ cell tumors: a comprehensive update of the European data. *Neuropediatrics* 1994;25:26–32.

3. Calaminus G, Bamberg M, Harms D, *et al.* AFP/beta-HCG secreting CNS germ cell tumors: long-term outcome with respect to initial symptoms and primary tumor resection. Results of the cooperative trial MAKEI 89.*Neuropediatrics* 2005;36:71–77.

4. Bussey KJ, Lawce HJ, Himoe E, *et al.* SNRPN methylation patterns in germ cell tumors as a reflection of primordial germ cell development. *Genes Chromosomes Cancer* 2001;32:342–352.

5. Schneider DT, Schuster AE, Fritsch MK, *et al.* Multipoint imprinting analysis indicates a common precursor cell for gonadal and nongonadal pediatric germ cell tumors. *Cancer Res* 2001;61:7268–7276

6. Sievers S, Alemazkour K, Zahn S, *et al.* IGF2/H19 imprinting analysis with the methylation-sensitive single nucleotide primer extension method in human germ cell tumors reflects their origin from different stages of primordial germ cell development. *Genes Chromosomes Cancer* 2005;44:256–264.

7. Oosterhuis JW, Looijenga LH. Testicular germ-cell tumours in a broader perspective. *Nat Rev Cancer* 2005;5:210–222.

8. Zafarana G, Gillis AJ, van Gurp RJ, *et al.* Coamplification of DAD-R, SOX5, and EKI1 in human testicular seminomas, with specific overexpression of DAD-R, correlates with reduced levels of apoptosis and earlier clinical manifestation. *Cancer Res* 2002;62:1822–1831.

9. Zafarana G, Grygalewicz B, Gillis AJ, *et al* 12p-Amplicon structure analysis in testicular germ cell tumors of adolescents and adults by array CGH.*Oncogene* 2003;22:7695–7701.

10. van Echten J, Oosterhuis JW, Looijenga LH, *et al.* No recurrent structural abnormalities apart from i(12p) in primary germ cell tumors of the adult testis. *Genes Chromosomes Cancer* 1995;14:133–144.

11. Korn WM, Oide Weghuis DE, Suijkerbuijk RF, *et al.* Detection of chromosomal DNA gains and losses in testicular germ cell tumors by comparative genomic hybridization. *Genes Chromosomes Cancer*1996;17:78–87.

12. Oosterhuis JW, Looijenga LH, van Echten J, *et al.* Chromosomal constitution and developmental potential of human germ cell tumors and teratomas.*Cancer Genet Cytogenet* 1997;95:96–102.

13. Bussey KJ, Lawce HJ, Olson SB, *et al.* Chromosome abnormalities of eighty-one pediatric germ cell tumors: sex- , age-, site-, and histopathology-related differences – a Children's Cancer Group study. *Genes Chromosomes Cancer* 1999;25:134–146.

14. Perlman EJ, Hu J, Ho D, *et al.* Genetic analysis of childhood endodermal sinus tumors by comparative genomic hybridization. *J Pediatr Hematol Oncol* 2000;22:100–105.

15. Mostert M, Rosenberg C, Stoop H, *et al.* Comparative genomic and *in situ*hybridization of germ cell tumors of the infantile testis. *Lab Invest*2000;80:1055–1064.

16. Veltman I, Veltman J, Janssen I, *et al.* Identification of recurrent chromosomal aberrations in germ cell tumors of neonates and infants using genomewide array-based comparative genomic hybridization. *Genes Chromosomes Cancer* 2005;43:367–376.

17. Schneider DT, Schuster AE, Fritsch MK, *et al.* Genetic analysis of mediastinal nonseminomatous germ cell tumors in children and adolescents. *Genes Chromosomes Cancer* 2002;34:115–125.

18. Rickert CH, Simon R, Bergmann M, *et al.* Comparative genomic hybridization in pineal germ cell tumors. *J Neuropathol Exp Neurol*2000;59:815–821.

19. de Bruin TW, Slater RM, Defferrari R, *et al.* Isochromosome 12p-positive pineal germ cell tumor. *Cancer Res* 1994;54:1542–1544.

20. Losi L, Polito P, Hagemeijer A, *et al.* Intracranial germ cell tumour (embryonal carcinoma with teratoma) with complex karyotype including isochromosome 12p. *Virchows Arch* 1998;433:571–574.

21. Dal Cin P, Dei Tos AP, Qi H, *et al.* Immature teratoma of the pineal gland with isochromosome 12p. *Acta Neuropathol (Berl)* 1998;95:107–110.

22. Lemos JA, Barbieri-Neto J, Casartelli C. Primary intracranial germ cell tumors without an isochromosome 12p. *Cancer Genet Cytogenet* 1998;100:124–128.

23. Okada Y, Nishikawa R, Matsutani M, *et al.* Hypomethylated X chromosome gain and rare isochromosome 12p in diverse intracranial germ cell tumors. *J Neuropathol Exp Neurol* 2002;61:531–538.

24. Rickert CH, Simon R, Bergmann M, *et al.* Comparative genomic hybridization in pineal parenchymal tumors. *Genes Chromosomes Cancer* 2001;30:99–104.

25. Kleihues P, Burger PC, Scheithauer BW. The new WHO classification of brain tumours. *Brain Pathol* 1993;3:255–268.

26. Mostofi FK, Sobin LH. *Histopathological Typing of Testis Tumors.* World Health Organization: Geneva, 1993.

27. Ausubel FM, Brent R, Kingston RE, *et al.* Preparation and analysis of DNA. In: Ausubel FM, Brent R, Kingston RE, Moore DD, Seidman JG, Smith JA, Struhl K (eds). *Short Protocols in Molecular Biology.* 3 edn. John Wiley & Sons, Inc.: New York, 1997, pp 2-1–2-43.

28. Isola J, DeVries S, Chu L, *et al.* Analysis of changes in DNA sequence copy number by comparative genomic hybridization in archival paraffin-embedded tumor samples [see comments]. *Am J Pathol* 1994;145:1301–1308.

29. Riopel MA, Spellerberg A, Griffin CA, *et al.* Genetic analysis of ovarian germ cell tumors by comparative genomic hybridization. *Cancer Res* 1998;58:3105–3110.

30. Ness GO, Lybaek H, Houge G. Usefulness of high-resolution comparative genomic hybridization (CGH) for detecting and characterizing constitutional chromosome abnormalities. *Am J Med Genet* 2002;113:125–136.

31. Ottesen AM, Skakkebaek NE, Lundsteen C, *et al.* High-resolution comparative genomic hybridization detects extra chromosome arm 12p material in most cases of carcinoma *in situ* adjacent to overt germ

cell tumors, but not before the invasive tumor development. *Genes Chromosomes Cancer* 2003;38:117–125.

32. du Manoir S, Schrock E, Bentz M, *et al.* Quantitative analysis of comparative genomic hybridization. *Cytometry* 1995;19:27–41.

33. Kirchhoff M, Rose H, Lundsteen C. High resolution comparative genomic hybridisation in clinical cytogenetics. *J Med Genet* 2001;38:740–744. |

34. Mostert MM, van de PM, Olde WD, *et al.* Comparative genomic hybridization of germ cell tumors of the adult testis: confirmation of karyotypic findings and identification of a 12p-amplicon. *Cancer Genet Cytogenet* 1996;89:146–152.

35. Chibon F, Mariani O, Mairal A, *et al.* The use of clustering software for the classification of comparative genomic hybridization data. an analysis of 109 malignant fibrous histiocytomas. *Cancer Genet Cytogenet* 2003;141:75–78.

36. Einhorn LH, Donohue JP. Chemotherapy for disseminated testicular cancer.*Urol Clin North Am* 1977;4:407–426.

37. Bamberg M, Kortmann RD, Calaminus G, *et al.* Radiation therapy for intracranial germinoma: results of the German cooperative prospective trials MAKEI 83/86/89. *J Clin Oncol* 1999;17:2585–2592.

38. Bouffet E, Baranzelli MC, Patte C, *et al.* Combined treatment modality for intracranial germinomas: results of a multicentre SFOP experience. Societe Francaise d'Oncologie Pediatrique. *Br J Cancer* 1999;79:1199–1204.

39. Balmaceda C, Finlay J. Current advances in the diagnosis and management of intracranial germ cell tumors. *Curr Neurol Neurosci Rep* 2004;4:253–262.

40. Kellie SJ, Boyce H, Dunkel IJ, *et al.* Primary chemotherapy for intracranial nongerminomatous germ cell tumors: results of the second international CNS germ cell study group protocol. *J Clin Oncol* 2004;22:846–853.

41. Calaminus G, Bamberg M, Jürgens H, *et al.* Impact of surgery, chemotherapy and irradiation on long term outcome of intracranial malignant non-germinomatous germ cell tumors: results of the German Cooperative Trial MAKEI 89. *Klin Pädiatr* 2004;216:141–149.

42. Jong JD, Stoop H, Dohle GR, *et al.* Diagnostic value of OCT3/4 for pre-invasive and invasive testicular germ cell tumours.*J Pathol* 2005;206:242–249.

43. Stoop H, van Gurp R, de Krijger R, *et al*. Reactivity of germ cell maturation stage-specific markers in spermatocytic seminoma: diagnostic and etiological implications. *Lab Invest* 2001;81:919–928.

44. Looijenga LH, Stoop H, De Leeuw HP, *et al*. POU5F1 (OCT3/4) identifies cells with pluripotent potential in human germ cell tumors. *Cancer Res*2003;63:2244–2250.

45. Honecker F, Stoop H, de Krijger RR, *et al*. Pathobiological implications of the expression of markers of testicular carcinoma *in situ* by fetal germ cells. *J Pathol* 2004;203:849–857.

46. Schneider DT, Perlman EJ, Harms D, *et al*. Mediastinal germ cell tumors (MGCT) in children and adolescents: age correlates with histological differentiation, genetic profiles and clinical outcome. In: Harnden P, Joffe JK, Jones WG (eds). *Germ Cell Tumours V*. Springer: London, 2002, pp 127–128.

47. Schneider DT, Schuster AE, Fritsch MK, *et al*. Genetic analysis of childhood germ cell tumors with comparative genomic hybridization. *Klin Pädiatr*2001;213:204–211.

48. Hasle H, Mellemgaard A, Nielsen J, *et al*. Cancer incidence in men with Klinefelter syndrome. *Br J Cancer* 1995;71:416–420.

49. Nichols CR, Heerema NA, Palmer C, *et al*. Klinefelter's syndrome associated with mediastinal germ cell neoplasms. *J Clin Oncol*1987;5:1290–1294.

50. Hashimoto M, Hatasa M, Shinoda S, *et al*. Medulla oblongata germinoma in association with Klinefelter syndrome. *Surg Neurol* 1992;37:384–387.

51. Casalone R, Righi R, Granata P, *et al*. Cerebral germ cell tumor and XXY karyotype. *Cancer Genet Cytogenet* 1994;74:25–29.

52. Prall JA, McGavran L, Greffe BS, *et al*. Intracranial malignant germ cell tumor and the Klinefelter syndrome. Case report and review of the literature. *Pediatr Neurosurg* 1995;23:219–224.

53. Yu IT, Griffin CA, Phillips PC, *et al*. Numerical sex chromosomal abnormalities in pineal teratomas by cytogenetic analysis and fluorescence*in situ* hybridization. *Lab Invest* 1995;72:419–423.

54. Garnis C, Coe BP, Lam SL, *et al*. High-resolution array CGH increases heterogeneity tolerance in the analysis of clinical samples. *Genomics*2005;85:790–793.

55. Lau CC. Genomic profiling of intracranial germ cell tumors [abstract]. *Neurooncol* 2005;7:514–515.

56. van Echten J, Sleijfer DT, Wiersema J, *et al.* Cytogenetics of primary testicular nonseminoma, residual mature teratoma, and growing teratoma lesion in individual patients. *Cancer Genet Cytogenet* 1997;96:1–6.

57. Albrecht S, Armstrong DL, Mahoney DH, *et al.* Cytogenetic demonstration of gene amplification in a primary intracranial germ cell tumor. *Genes Chromosomes Cancer* 1993;6:61–63.

Chapter 5

GENETIC ANALYSIS OF SINONASAL ADENO-CARCINOMA PHENOTYPES: DISTINCT ALTERATIONS OF HISTOGENETIC SIGNIFICANCE

Sue S Yom[1], Asif Rashid[2], David I Rosenthal[1], Danielle D Elliott[2], Ehab Y Hanna[3], Randal S Weber[3] and Adel K El-Naggar[2,3]

[1]Department of Radiation Oncology, The University of Texas MD Anderson Cancer Center, Houston, TX, USA

[2]Department of Pathology, The University of Texas MD Anderson Cancer Center, Houston, TX, USA

[3]Department of Head and Neck Surgery, The University of Texas MD Anderson Cancer Center, Houston, TX, USA

ABSTRACT

Sinonasal adenocarcinomas, a relatively rare entity, are composed of distinctly different morphologic subtypes with variable biological behavior. To investigate the genetic events associated with their development and clinicopathologic features, we analyzed the alterations in K-ras, APC, β-catenin, hMLH1 and hMSH2 and p53 genes expression in a cohort of 15 primary tumors comprising the two main sinonasal adenocarcinoma subtypes (enteric and seromucinous). The patients consisted of 13 men and two women, who ranged in age from 50 to 87 years. Tumors were predominantly located in the ethmoid sinus. Eight tumors were Enteric-type, and seven were seromucinous type. Nine patients were smokers and four were nonsmokers; and no information was available on two patients. Two of the eight enteric-type, had K-ras mutation at codons 12A and 12B, and one showed microsatellite instability at BAT-25. Two patients with enteric-type tumors had a history of wood-dust exposure, and one had a K-ras mutation at 12A codon as well as p53 overexpression. No patients with the seromucinous type had any genetic abnormalities, except for overexpression of p53 in two tumors. Our results show that (1) a subset of enteric-type sinonasal adenocarcinoma shares certain genetic alterations with colonic adenocarcinomas, (2) the seromucinous-type sinonasal adenocarcinoma lacks alterations and may develop through a different pathway, (3) high p53 expression is associated with aggressive tumor features in both subtypes and (4) the enteric-type runs a more malignant course than the seromucinous counterpart.

Primary sinonasal adenocarcinomas (SNACs), excluding those of salivary origin, are uncommon and represent approximately 10–20% of malignant neoplasms at these locations. Despite their common origin from an ectodermally derived respiratory mucosa, they manifest two distinct phenotypic categories, including the enteric and the seromucinous adenocarcinomas.[1, 2] The underlying mechanism for their histopathologic diversity is unknown. Our group has recently shown that the respiratory epithelium undergoes intestinal epithelial metaplasia (Figure 2a) prior to the development of the enteric-form of sinonasal adenocarcinoma, supporting histogenetic resemblance to primary colonic carcinoma.[3] We, therefore, hypothesize that the enteric-type shares common genetic alterations with primary colonic adenocarcinoma and differs from the seromucinous-type.

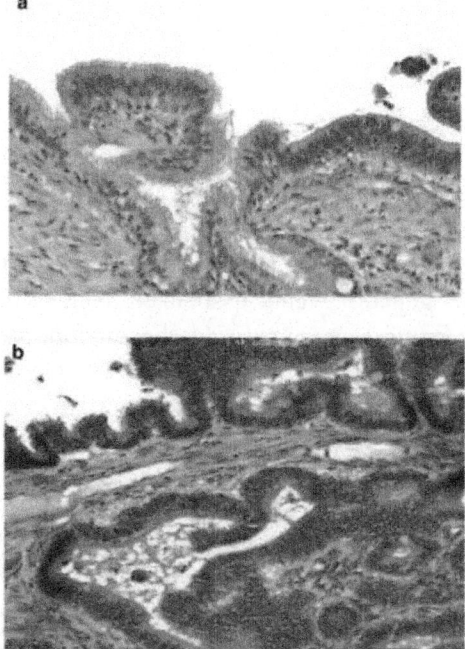

Figure 2: Intestinal metaplasia of the respiratory epithelium (a) and underlying intestinal-type sinonasal adenocarcinoma (b).

Previous investigations have shown that certain genetic alterations at the *adenomatous polyposis coli* (*APC*), ß-catenin and *K-ras* genes, and mutations and/or deletions of the *p53* suppressor gene characterize colorectal tumorigenesis.[4, 5, 6, 7, 8, 9, 10, 11, 12, 13, 14] Also, microsatellite instability (MSI) caused by alterations in nucleotide mismatch repair genes, including *hMSH2,hMLH1, PMS1, PMS2* and *GTBP*, has been reported to be associated with a subset of these tumors.[10, 11, 12] Molecular studies of SNAC are rare and limited to the intestinal type and have reported contradictory results. Wu *et al*[13] in a study of the enteric form, reported a lack of point mutations in the first or second exon of the K-*ras* oncogene, and only two p53 mutations in 11 specimens. However, a recent study by Saber *et al*[14] found four K-*ras* mutations at codons 12 or 13 in 28 patients with sinonasal adenocarcinoma, while only one of 13 enteric-type tumors had K-*ras* mutation.

To examine the molecular events associated with the phenotypic diversity of SNAC, we performed molecular and immunohistochemical analysis on 15 primary tumors of both types.

MATERIALS AND METHODS

Specimens

The surgical pathology database of MD Anderson Cancer Center from 1995 to 2003 was searched and 16 patients were identified who had sinonasal adenocarcinomas with available tissue blocks from curative resection or tumor biopsy. All slides were reviewed to confirm the original diagnosis and to select blocks for this study. Histologic classification was based on criteria of Kleinsasser[15] and Batsakis *et al.*[16] Histological slides were reviewed by two pathologists operating independently of clinical information. Patient information was extracted from medical records. Records were reviewed for age, sex, race, history of exposure to dust or smoking, site and stage of tumor, type of treatment, and follow-up status.

DNA Extraction

DNA was prepared from tissue specimens stored in paraffin-embedded, 4% formalin-fixed blocks. In a few cases for which this isolation technique was insufficient, genomic DNA was extracted by microdissection of hematoxylin- and eosin-stained slides without coverslip. A 27 1/2-gauge needle was used under low magnification ($\times$ 4) in selected areas where the neoplastic cellularity was >50%. Genomic DNA was extracted from the microdissected tissue, as described previously.[17] Briefly, three 5-μm-thick sections were extracted twice in 1000 μl of xylene for 30 min and twice in 1000 μl of 100% ethanol for 3–5 min. The solid residue was dried by evaporation, after which 75 μl of lysis buffer was added (0.25% Tween-20, TE9, 2 mg/ml Proteinase K). The mixture was incubated overnight at 56°C. The Proteinase K was deactivated for 10 min at 100°C. The condensate was separated from the insoluble material by centrifugation.

Molecules Analyses

DNA could not be isolated in one patient with enteric-type. Therefore, seven enteric-type and seven nonseromucinous tumors were analyzed.

Analysis of K-*ras*, APC, and β-catenin gene mutation was carried out as previously described.[5] Immunohistochemistry for microsatellite analysis of markers hMLH1 and hMSH2 was carried out according to the method of Alexander *et al.*[18]

Immunohistochemistry for p53 protein was also performed according to previously published method.[19] Overexpression of p53 protein was considered to be present when more than 50% of the nuclei of tumor cells were strongly stained.

RESULTS

Clinical Data

The clinicopathologic findings of all sinonasal adenocarcinoma cases are summarized in Table 1. Histopathologically, seven tumors were seromucinous-type (Figure 1a) and eight were enteric-type (Figures 1b and 2b). All of the seromucinous adenocarcinomas were of the low-grade category. Six of the eight enteric-type carcinomas were moderately differentiated villo-tubular adenocarcinomas and two were well-differentiated adenocarcinoma with mucinous component. Of the 15 primary tumors, 13 occurred in men and two in women. The ages ranged from 50 to 87 years, with a mean of 67 years. Ten primary tumors originated in the ethmoid sinus, one in the maxillary sinus and two in the nasal cavity; and the primary site was unknown for two cases. Six of eight patients with enteric-type carcinomas and three of seven with seromucinous form had a history of smoking. Two of the eight patients with the enteric-type carcinoma had a history of wood-dust exposure and were also smokers. Initial surgical treatment included five craniofacial resections, four medial maxillectomies, two endoscopic resections, one partial maxillectomy, one transpalatal resection, and two unknown.

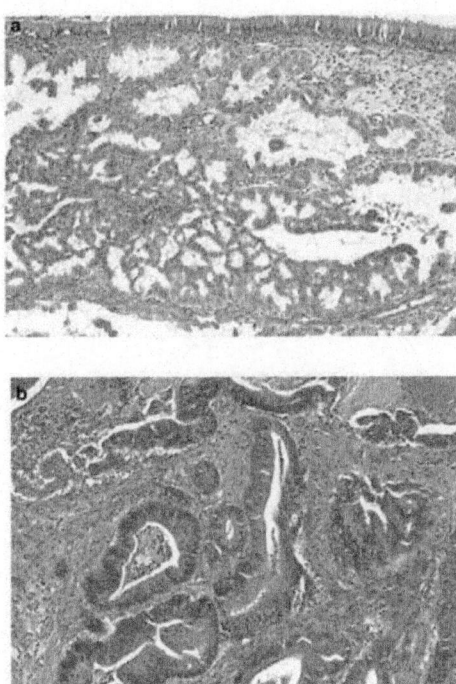

Figure 1: Photomicrographs of seromucinous (a) and intestinal-type (b).

Table 1:- Demographic, clinicopathologic and follow-up data of patients with sinonasal adenocarcinomas

Type	Age	Sex	Race	Risk factors	Site	P. Surgery	Stage	Margins	XRT	CHX	Follow-up
Enteric-type											
1	73	M	White	Smoking (20 py)	ES	CFR	4	Negative	N/A	N/A	N/A
2	67	M	White	None	ES	ER	2	Positive	Yes	Yes	LR
3	68	M	White	Smoking (20 py), chemicals, asbestos	ES	MM	2	Negative	Yes	No	NED
4	53	M	Black	Wood dust; smoking (20 py)	ES	CFR	4	Negative	No	No	NED
5	76	M	White	Smoking (20 py)	ES	ER	2	Positive	Yes	No	LR
6	71	F	N/A	N/A	N/A	N/A	N/A	N/A	N/A	N/A	N/A
7	67	M	White	Wood dust; smoking (175 py)	ES	PM	2	Positive	Yes	No	LR
8	67	M	White	Smoking (42 py)	ES	CFR	4	Negative	Yes	No	NED
Nonenteric											
9	71	M	White	None	NC	ER	2	Negative	Yes	No	NED
10	68	F	N/A	N/A	N/A	N/A	N/A	N/A	N/A	N/A	N/A
11	69	M	White	Smoking (90 py)	ES	MM	2	Negative	Yes	No	NED
12	67	M	Asian	Smoking (47 py)	NC	MM	2	Positive	Yes	No	NED
13	50	M	White	None	ES	CFR	2	Negative	Yes	No	NED
14	51	M	White	None	MS	CFR	4	Negative	Yes	Yes	DOD
15	87	M	Hispanic	Smoking (10 py)	ES	MM	2	Negative	N/A	N/A	NED

Genetic Alterations

Table 2 presents the molecular findings of sinonasal adenocarcinoma. Two K-*ras* mutations at codon 12 were identified in two of seven enteric-type carcinomas (29%) but in none of the seromucinous carcinomas (Figure 3). One of these tumors manifested associated p53 overexpression and MSI (14%). Two of seven (29%) seromucinous-type tumors had p53 overexpression, and none had any genetic alterations at the genes tested. No mutation at the APC or β-catenin genes was present in any tumor of either type. Alteration in hMLH1 expression was identified in two moderately differentiated enteric-type carcinomas, and one also had alteration of hMSH2. None of the seromucinous cancers showed MSI.

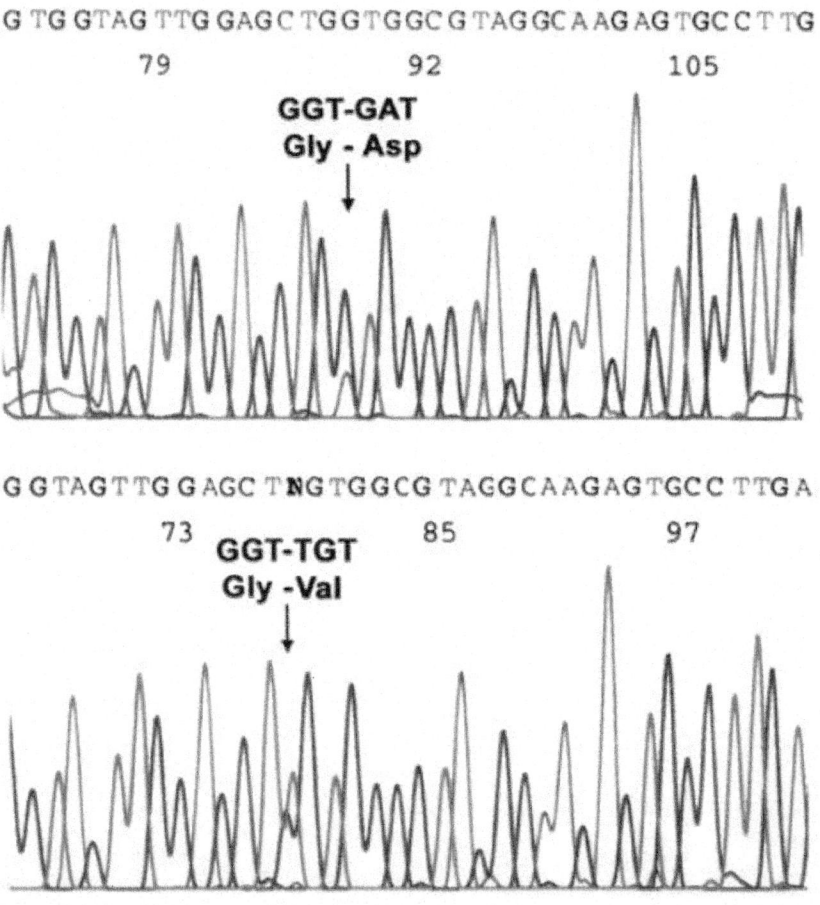

Figure 3: G → A transition mutation of the Ras gene in intestinal-type adenocarcinoma.

Table 2:- Molecular analysis of sinonasal adenocarcinomas

Type #	K-ras	APC	β-Catenin	p53	hMLH1	hMSH2	Bat 25	Bat 26
Enteric								
1	●	O	O	O	O	O	O	O
2	O	O	O	O	O	O	O	O
3	O	O	O	O	O	O	O	O
4	O	O	O	O	O	O	O	O
5	O	O	O	O	O	O	O	O
6	O	O	O	O	O	O	O	O
7	●	O	O	●	O	O	●	O
8	N/A	N/A	N/A	N/A	N/A	N/A	N/A	N/A
Non-enteric								
9	O	O	O	●	O	O	O	O
10	O	O	O	O	O	O	O	O
11	O	N/A	O	O	O	●	O	O
12	O	O	O	O	●	O	O	O
13	O	O	O	O	●	●	O	O
14	O	O	O	●	O	O	N/A	O
15	O	O	O	O	O	O	O	O

●, alteration; O, no alteration; N/A, not assessed.

DISCUSSION

Our results show K-*ras* mutations in 29% of the enteric-type carcinoma tumors and in none of the seromucinous-subtype. A similar incidence of mutations at this gene has previously been reported in studies of primary colorectal carcinomas.[14] This is further underscored by the detection of an MSI, a feature commonly present in a subset of primary enteric adenocarcinomas, in one of the enteric-type with K-*ras* mutation. Because of the phenotypic similarities between primary colonic and enteric-type sinonasal adenocarcinomas the presence of K-*ras* mutations in both indicates an early association with their development, regardless of the site of origin. In addition evidence that intestinal metaplasia of the sinonasal mucosa precede to the development of the enteric-type lends further credence to this hypothesis.[3]

The lack of other genetic features typically reported in primary colonic adenocarcinoma in our enteric-type tumors, however, suggest that specific events related to their site of origin underlie these differences. We contend that subsequent genetic events, especially in the former, are most likely acquired during the progression as a result of loco-regional and other epidemiological factors.[20, 21, 22, 23, 24, 25, 26, 27] These results, along with the lack of any alteration in the seromucinous subtype, support a distinct pathway for the evolution and progression of the intestinal and primary colonic adenocarcinomas.

In our cohort, only three high-grade tumors overexpressed p53, one of which was enteric in type and also had a K-*ras* mutation. Two patients with these tumors developed local recurrence or metastasis. Studies of colonic adenocarcinoma and a recent study of sinonasal adenocarcinoma have reported a high incidence of p53 abnormalities.[28, 29] Recent studies of p53 in enteric-type adenocarcinoma have associated this finding with sawdust exposure.[13, 28, 29]Similarly an association between dust exposure and epigenetic alterations at the CpG island of certain genes in these tumors has been reported.[28] The underlying factors for the differences between our findings and those of other investigators can be attributed to epidemiologic and/or patient populations variables. Only two patients in our cohort had a history of wood-dust exposure, and it is conceivable that sawdust contributes only to the development of a subset of the enteric-form tumors and that an alternative pathway may play a role in tumors' developing in nondust-exposed patients.

Our findings also suggest that patients with dust exposure develop via a different pathway than those without such history, and that the enteric-form are different molecularly from seromucinous adenocarcinoma. In our cohort, patients with the enteric-form had a more aggressive clinical course, than those with the seromucinous-type. However, because of the small number of tumors examined further studies are needed to determine the effect of tumor differentiation on the clinical behavior of these tumors. Sinonasal adenocarcinomas are generally unresponsive to medical management modalities and definitive surgical excision which achieves local control rates of 50% is the primary treatment of choice.[30]Recently, however, the addition of postoperative radiation following surgical resection have increased the 5-year local control rate to 59%.[31] Since our patient outcomes are consistent with these findings, we, recommend that patients with the enteric-type be treated with surgery and adjuvant radiation therapy.

ACKNOWLEDGEMENTS

This work was supported in part by the Kenneth D Müller Professorship and a National Cancer Institute Specialized Program of Research Excellence grant in head and neck cancer.

REFERENCES

1.   Batsakis JG, Rice DH, Solomon AR. The pathology of head and neck tumors: squamous and mucous-gland carcinomas of the nasal cavity, paranasal sinuses, and larynx, part 6. *Head Neck Surg* 1980;2:497–508.

2.   Franquemont DW, Fechner RE, Mills SE. Histologic classification of sinonasal–intestinal type adenocarcinoma. *Am J Surg Pathol* 1991;15:368–375.

3.   Choi H-R, Sturgis EM, Rashid A, *et al.* Sinonasal adenocarcinoma: evidence for histogenetic divergence of the Enteric and nonenteric phenotypes. *Hum Path* 2003;34:1101–1107.

4. Vogelstein B, Fearon ER, Hamilton SR, *et al.* Genetic alterations during colorectal-tumor development. *N Engl J Med* 1988;319:525–532.

5. Lee JH, Abraham SC, Kim HS, *et al.* Inverse relationship between APC gene mutation in gastric adenomas and development of adenocarcinoma. *Am J Pathol* 2002;161:611–618.

6. Kim IJ, Kang HC, Park JH, *et al.* Determination of tumor aggressiveness in colorectal cancer by K-ras-2 analysis. *Arch Surg* 1993;128:526–531.

7. Finkelstein SD, Sayegh R, Bakker A, *et al.* Prediction of biologic aggressiveness in colorectal cancer by p53/K-ras-2 topographic genotyping.*Mol Diagn* 1996;1:5–28.

8. Cunningham J, Lust JA, Schaid DJ, *et al.* Expression of p53 and 17p allelic loss in colorectal carcinoma. *Cancer Res* 1992;52:1974–1980.

9. Ohue M, Tomita N, Monden T, *et al.* A frequent alteration of p53 gene in carcinoma in adenoma of colon. *Cancer Res* 1994;54:4798–4804.

10. Kim IJ, Kang HC, Park JH, *et al.* Development and applications of a beta-catenin oligonucleotide microarray: beta-catenin mutations are dominantly found in the proximal colon cancers with microsatellite instability. *Clin Cancer Res* 2003;9:2920–2925.

11. Fukushima H, Yamamoto H, Itoh F, *et al.* Frequent alterations of the beta-catenin and TCF-4 genes, but not of the APC gene, in colon cancers with high-frequency microsatellite instability. *J Exp Clin Cancer Res*2001;20:553–559.

12. Fujiwara T, Stolker JM, Watanable T, *et al.* Accumulated clonal genetic alterations in familial and sporadic colorectal carcinomas with widespread instability in microsatellite sequences. *Am J Pathol* 1998;153:1063–1078.

13. Wu TT, Barnes L, Bakker A, *et al.* K-ras-2 and p53 genotyping of intestinal-type adenocarcinoma of the nasal cavity and paranasal sinuses. *Mod Pathol*1996;9:199–204.

14. Saber AT, Nielsen LR, Dictor M, *et al.* K-ras mutations in sinonasal adenocarcinomas in patients occupationally exposed to wood or leather dust. *Cancer Lett* 1998;126:59–65.

15. Kleinsasser O. Terminal tubulus adenocarcinoma of the nasal seromucous glands. A specific entity. *Arch Otorhinolaryngol* 1985;241:183–193.

16. Batsakis JG, Mackay B, Ordonez NG. Enteric-type adenocarcinoma of the nasal cavity. An electron microscopic and immunocytochemical study.*Cancer* 1984;54:855–860.

17. Goelz SE, Hamilton SR, Vogelstein B. Purification of DNA from formaldehyde fixed and paraffin embedded human tissue. *Biochem Biophys Res Commun* 1985;130:118–126.

18. Alexander J, Watanable T, Wu TT, *et al.* Histopathological identification of colon cancer with microsatellite instability. *Am J Pathol* 2001;158:527–535.

19. Baas IO, Mulder J-WR, Offerhaus GJA, *et al.* An evaluation of six antibodies for immunohistochemistry of mutant p53 gene product in archival colorectal neoplasms. *J Pathol* 1994;172:5–12.

20. Cecchi F, Buiatti E, Kriebel D, *et al.* Adenocarcinoma of the nose and paranasal sinuses in shoemakers and woodworkers in the province of Florence, Italy (1963–77). *Br J Ind Med* 1980;37:222–225.

21. Kleinsasser O, Schroeder HG. Adenocarcinomas of the inner nose after exposure to wood dust. Morphological findings and relationships between histopathology and clinical behavior in 79 cases. *Arch Otorhinolaryngol* 1988;245:1–5.

22. Wills JH. Nasal carcinoma in woodworkers: a review. *J Occup Med* 1982;24:526–530.

23. Klintenberg C, Olofsson J, Hellquist H, *et al.* Adenocarcinoma of the ethmoid sinuses. A review of 28 cases with special reference to wood dust exposure. *Cancer* 1984;54:482–488.

24. Hayes RB, Gerin M, Raatgever JW, *et al.* Wood-related occupations, wood dust exposure, and sinonasal cancer. *Am J Epidemiol* 1986;124:569–577.

25. Leclerc A, Martinez Cortes M, *et al.* Sinonasal cancer and wood dust exposure: results from a case–control study. *Am J Epidemiol* 1994;140:340–349.

26. Van den Oever R. Occupational exposure to dust and sinonasal cancer. An analysis of 386 cases reported to the N.C.C.S.F. Cancer Registry. *Acta Otorhinolaryngol Belg* 1996;50:19–24.

27. Lopez JI, Nevado M, Eizaguirre B, *et al.* Intestinal-type adenocarcinoma of the nasal cavity and paranasal sinuses. A clinicopathologic study of 6 cases. *Tumori* 1990;76:250–254.

28. Perrone F, Oggionni M, Birindelli S, *et al.* TP53, p14ARF, p16INK4a and H-ras gene molecular analysis in intestinal-type adenocarcinoma of the nasal cavity and paranasal sinuses. *Int J Cancer* 2003;105:196–203.

29. Bashir AA, Robinson RA, Benda JA, *et al.* Sinonasal adenocarcinoma: immunohistochemical marking and expression of oncoproteins. *Head Neck* 2003;25:763–771.

30. Perez P, Dominguez O, Gonzalez S, *et al.* Ras gene mutations in ethmoid sinus adenocarcinoma: prognostic implications. *Cancer* 1999;86:255–264.

31. Claus F, Boterberg T, Ost P, *et al.* Postoperative radiotherapy for adenocarcinoma of the ethmoid sinuses: treatment results for 47 patients. *Int J Radiat Oncol Biol Phys* 2002;54:1089–1094.

Chapter 6

AN ANALYSIS OF GENETIC CHANGES DURING THE DIVERGENCE OF DROSOPHILA SPECIES

Rui Sousa-Neves[1], Alexandre Rosas[2]

[1]Department of Biology, Case Western Reserve University, Cleveland, Ohio, United States of America,

[2]Departamento de Fı́sica, CCEN, Universidade Federal da Paraı́ba, Joã̃o Pessoa, Paraı́ba, Brazil

ABSTRACT

Background

It has been long appreciated that speciation involves changes in body plans and establishes genetic, reproductive, developmental and behavioral incompatibilities between populations. However, little is still known about the genetic components involved in these changes or the sequence and scale of events that lead to the differentiation of species.

Principal Findings

In this paper, we investigated the genetic changes in three closely related species of *Drosophila* by making pair-wise comparisons of their genomes. We focused our analysis on the modern relatives of the alleles likely to be segregating in pre-historic populations at the time or after the ancestor of *D. simulans* became separated from the ancestor of *D. melanogaster*. Some of these genes were previously implicated in the genetics of reproduction and behavior while the biological functions of others are not yet clear.

Conclusions

Together these results identify different classes of genes that might have participated in the beginning of segregation of these species millions of years ago in Africa.

INTRODUCTION

One of the greatest challenges in modern Biology is the identification of genes that operate during evolution and diversification of species. However, since species are already separated and preserved samples of the prehistoric species are scarce, the investigation of the scale and types of genetic changes occurred during and after speciation has been limited.

The problem of speciation has puzzled generations of biologists and is of great significance to a wide variety of fields in biology. For instance, during evolution, mutations modify the architecture of brains and external appearances, novel behaviors appear and reciprocal lethal/sterile genetic systems emerge to block gene flow. The recurrent appearance of these themes across different phyla suggests a conservation of genetic processes. However, little is still known about the scale of the genetic changes that occur during speciation.

To investigate this issue, we chose to use the genetic workhorse *Drosophila melanogaster* and two sequenced sibling species, *D. simulans* and *D. sechellia*. The fact that *D. melanogaster* is a close relative of *D. simulans* and *D. sechellia* and has a myriad of genetic and genomic resources, makes it an ideal model to study evolutionary processes. The latest estimates suggest that *D. simulans* diverged from *D. melanogaster* approximately 5.4 million years ago in Africa, while *D. sechellia* diverged from *D. simulans* 0.5 million years ago in the Indian island of Seychelles [1]. Similar estimates suggest that a fourth species, the incompletely sequenced*D. mauritiana*, diverged from *D. simulans* 0.1–0.3 million years ago in the island of Mauritius.

Anatomically, these sibling species are almost identical, except for the different appearance of the male genitalia in all four species and minor ambiguous morphological features [2]. All four species have a set of 4 chromosomes mostly homosequential, but *D. simulans*, *D. sechellia*and *D. mauritiana* have nearly identical rearrangements that distinguish them from *D. melanogaster* [3]–[6]. These rearrangements suggest that the common ancestor of these species had already diverged chromosomally from *D. melanogaster* between 0.5 and 5.4 Million years ago [3]–[6].

Despite their similarities, the three sibling species (*D. melanogaster*, *D. simulans* and *D. sechellia*) have an intriguing different biology. For instance, the mating between *D. simulans*males and *D. melanogaster* females results in a dramatic larval death of the male offspring that is accompanied by a reduction of the brain and lack of imaginal discs [7], [8], while the surviving adult females are sterile [3], [9]. The reciprocal mating between *D. melanogaster*males and *D. simulans* or *D. sechellia* females is rarely successful and results in embryonic lethality of female and sterility of male offspring [3], [10], [11]. Similar results are obtained with hybrids between *D. sechellia* and *D. melanogaster* and somewhat less extreme phenotypes with *D. sechellia* and *D. simulans* hybrids. In the latter case, both sexes survive, but the male progeny is sterile [3]. Behaviorally, these species exhibit a mating asymmetry and it has been proposed that females of the newest species (i.e. *D. simulans* and *D. sechellia*) reject males of the oldest species archetype (i.e. *D. melanogaster*). In contrast, females of the oldest species accept males of the newest species [3], [12], [13]. Similar mating asymmetries appear in a significant number of closely related species [14].

Together, the facts highlighted above suggest that *D. simulans* and *D. sechellia* are more closely related to each other than to *D. melanogaster*, and quite conceivably share incompatible genes that affect mitotic, embryonic, reproductive, sensory perception and behavioral systems. However, with few notable exceptions, the genes and alleles involved in these processes still remain elusive [15]–[19].

Here we screened the genome for genetic variants that might reveal the changes occurred during or after the divergence of the ancestor of *D. simulans* from the ancestor of *D. melanogaster*. For convenience, *D. melanogaster* is taken as the archetypical or ancestral form as previously suggested [5], [12]. In particular, we searched for alleles with little or no divergence between *D. simulans* and *D. sechellia* that greatly diverged from *D. melanogaster*. These alleles are expected to have appeared at the time or right after the divergence of the ancestor of *D. simulans* from the ancestor of *D. melanogaster*, but before the separation of *D. simulans* from *D. sechellia*. For this reason, we refer to them as *ancestral alleles*. It is noteworthy that ancestral alleles of *D. simulans* and *D. sechellia*, also happen to be fast evolving alleles, when the Melanogaster subgroup is used as a reference. The analysis of the predicted gene products of ancestral alleles reveal which classes of genes might have been involved in the segregation of these species.

RESULTS

Number of Coding Sequences Identified in *D. Simulans* and *D. Sechellia*

The major objective of this search was to quantify and identify alleles that might have been segregating in pre-historic populations of the ancestor of *D. simulans* that were inherited by the descendants *D. simulans* and *D. sechellia*. We expected that ancestral alleles should be informative of the developmental, reproductive and behavioral novelties that distinguish *D. simulans* and *D. sechellia* from *D. melanogaster*.

To begin addressing this issue, we extracted and compared the annotated coding sequences of *D. melanogaster* to the sequence of computationally predicted coding sequences of *D. simulans* and *D. sechellia* (Fig. 1A). A total of 13,740 predicted coding sequences were assembled from *D. simulans* and *D. sechellia* genomes: 2,226 on the X chromosome, 5,355 on the second chromosome, 6,074 on the third chromosome and 85 on the fourth chromosome.

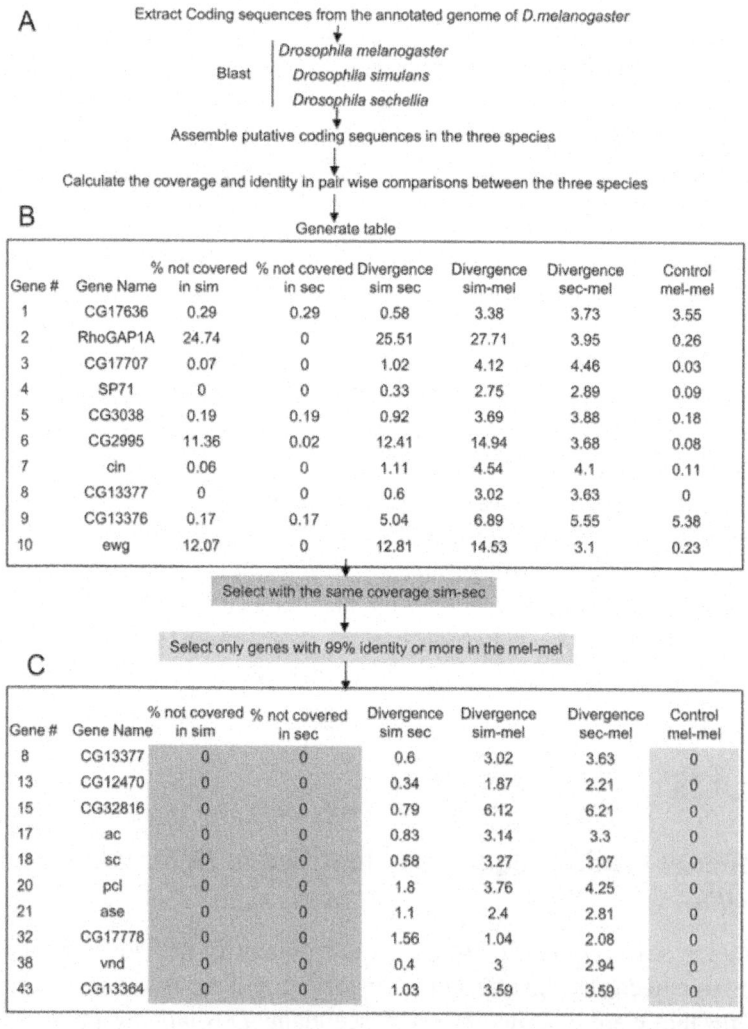

Figure 1: Overview of the data collection and sorting.

A) Exons of coding sequence were extracted from the annotated genome of *D. melanogaster* using Extractor and electronically joined using Analyst to obtain complete coding sequences. These coding sequences were then automatically blasted against the genome of *D. melanogaster*, *D. simulans* and *D. sechellia* with Megablast. Analyst scanned the resulting alignments for the best hits and assembled the coding sequences in the three species from them. B) Analyst also calculated the coverage, the percentage not covered, the divergence in sim-sec, sim-mel, sec-mel as well as the control mel-mel and organized this data in a table. C) To minimize artifacts due to incomplete clone representation in the genomic libraries, the coding sequences were filtered and only genes with the same coverage in *D. simulans* and *D. sechellia* retrieved. To avoid genes truncated by Megablast (i.e. usually genes with small exons), only genes with a mismatch up to 1% in the control mel-mel were retrieved. After these two filters were applied, a new table like the one exemplified in C) was generated for each chromosome.

doi:10.1371/journal.pone.0010485.g001

Identification of High Confidence Genes by Sorting and Filtering Data

The data collected from each chromosome arm was organized in a table, which consists of eight columns with the following information: (1) coding sequence number in *D. melanogaster*; (2) gene name; (3) percentage of bases not covered in *D. simulans* and (4) in *D. sechellia*; (5) divergence between *D. simulans* and *D. sechellia*; (6) divergence between *D. melanogaster* and *D. simulans*; (7) divergence between *D. melanogaster* and *D. sechellia*; and finally, (8) assembly control (i.e. percentage of mismatches between the actual coding sequences of *D. melanogaster* vs. the coding sequences assembled from Blast results in *D. melanogaster*).

To avoid false positives due to truncated fragments in the WGS libraries, we first applied a filter that discards all genes with different coverage in *D. simulans* and *D. sechellia* (Fig. 1C, Columns 3 and 4). The remaining genes were sorted using values of the control Blast mel vs. mel (Fig. 1C, Column 8) in ascending order. In addition, only genes with a mismatch of up to 1% were retrieved. Our control of automatic assembly of coding sequences assured that only high quality coding sequences assembled from Blast alignments (i.e. 99% match or greater) were analyzed. After filtering the data using the criteria above, 8,416 reliable coding sequences corresponding to 61% of the total number of coding sequences extracted from *D. melanogaster* were obtained: 1,039 on the X; 3,407 on the second; 3,951 on the third and 19 on the fourth chromosome.

Identification of Genes that vary the least between *D. Simulans* and *D. Sechellia* and the most in *D. Melanogaster*

To identify genes that diverged the least in the pair sim-sec and the most in the pairs sim-mel and sec-mel (ancestral alleles), we employed two strategies. The first strategy selects genes in the pair sim-sec that diverged less than the average plus the standard deviation of all genes in the same chromosome, and in addition that also diverged more than the average plus the standard deviation in sim-mel and sec-mel pairs. The second method, which will be explained in more detail in the next section, is based on the observation that most of the genes in *D. simulans*/*D. sechellia* diverge linearly from *D. melanogaster*, while few very similar genes diverge non-linearly.

Out of the total 8,416 reliable coding sequences selected previously, the first method led to the identification of 517 genes: 67 genes on the X; 112 on the left arm of the second; 106 on the right arm of the second; 88 on the left arm of the third; and 144 on the right arm of the third chromosome (Table 1, Fig. 2). No ancestral alleles were identified on the fourth chromosome due to the fact this chromosome has highly homogeneous divergences (data not shown).

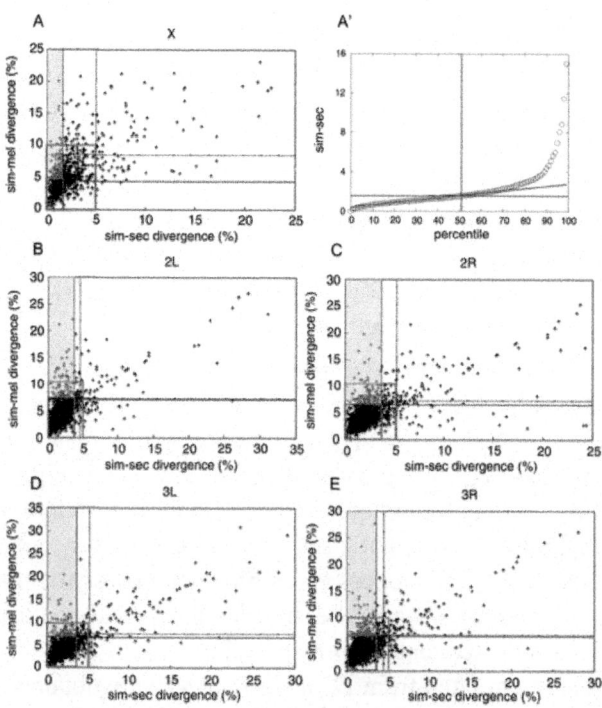

Figure 2: Patterns of divergence along chromosomes and two screening methods.

A, B, C, D and E) Each graph corresponds to a chromosome or chromosome arm (X, 2L, 2R, 3L and 3R), where the genes are ordered from the least divergent to the most divergent in sim-sec (abscissa) and sim-mel (ordinate). The horizontal and vertical dashed lines delimit the averages plus one standard deviation of the divergence between sim-sec (horizontal) and sim-mel (vertical). The upper left quadrants delimit genes found by the method of averages and standard deviations. Note that the divergences of most genes in all 5 graphs are clustered in a quadrant that can be roughly delimited between the abscissa values of 0%–5% and ordinate values 0%–10% (red rectangles). In this quadrant, the genes have a good fit with a linear distribution (P<0.0001). To better delimit the quadrant in which the divergence is linear in each chromosome, the data was divided in percentiles of divergences of sim-sec, sim-mel and sec-mel. (A›) exemplifies the percentiles of the X chromosome. Since each point in these curves represents one percentile, the percentage of genes that diverge linearly is equal to the number of points that can be transected by a straight line. Once this linear interval is defined, the values on the x and y axes become known and can be used to redefine the quadrant of linear divergences (inferior left quadrant in blue). The region where the genes in sim-sec vary the least and the genes in sim-mel vary the most is the adjacent upper quadrant to the left of the point where the horizontal and vertical lines cross (gray quadrant).

doi:10.1371/journal.pone.0010485.g002

	X	2L	2R	3L	3R	Totals
Screening 1	67	112	106	88	144	517
Screening 2	101	74	89	77	98	439
Common	14	73	76	61	96	320

The genes identified by method 1 have divergences inferior to the average plus standard deviation for the chromosomes in which they are located in sim-sec and higher divergences than the average plus standard deviation in sim-mel and sec-mel. Screening 2 selects genes that retained ancestral features and diverge significantly form D. melanogaster. Note that screening 2 appears more stringent than screening 1. Note also that a large number of common genes that can be found by both screenings in the autosomes, but considerably less on the X. No ancestral alleles could be found on the fourth chromosome (see text).
doi:10.1371/journal.pone.0010485.t001

Table 1: Number of genes identified by the screening methods 1 and 2.

doi:10.1371/journal.pone.0010485.t001

Genes that Diverge Linearly and Non-Linearly in *D. Simulans* and *D. Sechellia*

In the second strategy to identify ancestral alleles, we searched for patterns of divergence among the 8,416 coding sequences. In this case, the data was sorted in ascending order of identity, and the values of the sim-sec pair were plotted against the sim-mel pair for the X chromosome, 2R and 2L, 3R and 3L chromosome arms (Fig. 2). Since the divergences of sim-mel and sec-mel are approximately the same (data not shown), the graphs of the sec-mel pair were not included in the figure.

The graphs in Figure 2 show that the genes that diverge the least in sim-sec pair, but also diverge the most in sim-mel pair fall between values 0 to 5% of the abscissa. Moreover, it is clear that the majority of genes also fall roughly within the ordinate range of 0 to 10%, and that within this interval there is a linear fit (Fig. 2, P<0.0001). We refer this interval to as *initial linear interval* (Fig. 2 A–E, outlined in red). Conversely, for values above 10% in the ordinate there is no significant agreement with a linear fit (P>0.05). We refer this interval to as *initial non-linear interval* (Fig. 2 A–E, outside the blue region). These results suggest that within the initial linear interval, the sim-sec genes diverge from sim-mel fairly linearly, while within the initial non-linear interval, this linearity breaks down. Thus, the non-linear interval contains the genes that varied the least in sim-sec and the most in sim-mel.

The results above suggest the existence of at least two gene populations; one large group that changes at a similar pace over generations in *D. simulans* and *D. sechellia* and a smaller group with a high degree of divergences.

Delimiting a Quadrant with Genes that Diverged from *D. Melanogaster* and were Inherited in *D. Simulans* and *D. Sechellia*

In the graphs shown in Fig. 2, we noticed the presence of a linear and a non-linear interval, but it is difficult to determine the boundary between the two intervals. In order to define more precisely this limit, we divided the data of sim-mel, sim-sec and sec-mel in percentiles, as shown in Fig. 2A' for the X chromosome. Since each point in the graph represents one percentile, the line that transects the largest number of linear points along this curve reveals the percentage of genes that diverge linearly. On the X chromosome, this range corresponds to the 51st percentile. Thus, the linearly distributed genes fall between the abscissa values 0 to 1.56 (sim-sec) and ordinate values 0 to 4.35 (sim-mel) (Fig. 2A'). We also applied the same methodology for the autosomes (data not shown).

We identified a more approximate interval where the genes diverge linearly in the abscissa and ordinate (Fig. 2, the inferior left quadrants in blue), and selected the genes that are in the quadrant above in the ordinate (Fig. 2, highlighted in gray). Using the same methodology for the pairs sim-sec and sec-mel, we identified the common set of genes in both searches (i.e. common genes to the percentiles of sim-sec vs. sim-mel and sim-sec vs. sec-mel).

A total of 439 genes common to *D. simulans* and *D. sechellia* were identified: 101 genes on the X, 74 on 2L, 89 on 2R, 77 on 3L and 98 on 3R (Table 1, Fig. 2). We were unable to perform a similar analysis for the fourth chromosome due to the fact that only 19 genes of high confidence were identified, which precluded the use of percentiles.

Both Screening Strategies Identify a Large Number of Common Genes

When the results of the percentile search are combined with the results of the search of averages and standard deviations, we notice that a significant number of genes are represented in both searches. In particular, 73 (98.6%) genes common to both searches were found on 2L; 76 (85.4%) on 2R, 61(79.2%) on 3L; and 96 (98.0%) on the 3R. Interestingly, on the X chromosome only 14 (13.9%) genes common to both searches were found (Table 1). The relatively low percentage of common genes observed on the X chromosome stems from the fact that the average divergence plus the standard deviation on this chromosome is higher than in the autosomes (Fig. 2, note the different spacing between the dashed line and solid lines on the X with autosomes). This variation results in the exclusion of several genes found by the method of percentiles and at the same time, in the inclusion of others. Thus, when the results of both screening are combined, the number of genes found by the averages method on the X chromosome is inferior to those found by the percentiles method. The main conclusions that can be drawn from these results are that both searches identify a large number of common genes and that the X chromosome genes evolve slightly faster than autosomal genes.

Most Genes Identified are Orthologous

To test whether the 320 genes identified in *D. simulans* and *D. sechellia* by both methods correspond to true orthologues as opposed to paralogues, we blasted the *D. simulans* and *D. sechellia* genes separately against the genome of *D. melanogaster*. Out of these 320 genes, 307 genes from each species matched the chromosomal position and gene used in *D. melanogaster* at the beginning of the screening. Thus, 96% of these genes correspond to homologues, not paralogs. The remaining 13 (4%) were excluded since they correspond to paralogs.

Spatial Distribution of Ancestral Alleles within the Genome and Divergence Hotspots

We next tested whether the ancestral alleles are clustered in specific genomic locations or whether they appear evenly distributed across the genome. To address this issue we plotted their occurrence along the 20 divisions of the major chromosomes, using chromosomal coordinates of *D. melanogaster* [20] (Fig. 3). Since the method of averages and standard deviations produces a distortion that results in fewer common genes on the X chromosome, we used the results obtained from the percentile search to plot the position of these genes. Our data suggest that although these alleles can be found in almost every division of the three major chromosomes, some regions are hotspots for ancestral alleles. These regions were identified by searching for regions that have more ancestral alleles than the average plus 1 standard deviation. In particular, the X chromosome division 1 and division 9 have more ancestral alleles than most divisions on this chromosome. Similarly, three divisions on the left arm of the second chromosome (i.e. 22, 23 and 34) also harbor more ancestral alleles than the average plus one standard deviation for this arm. The right arm of the second chromosome also seems to have three hotspots in divisions 44, 54 and 59, while the distribution of ancestral alleles on the left arm of the third chromosome does not contain prominent hotspots, except perhaps by divisions 61, 64, 68 and 70. Finally, on the right arm of the third chromosome, one prominent hotspot appears at division 82. The significance of this clustering is not yet clear, but we note that some of these hotspots are located nearby known rearrangement breakpoints observed in *D. simulans* such as in divisions 1–2, divisions 21–22, 59–60 and 82 [21].

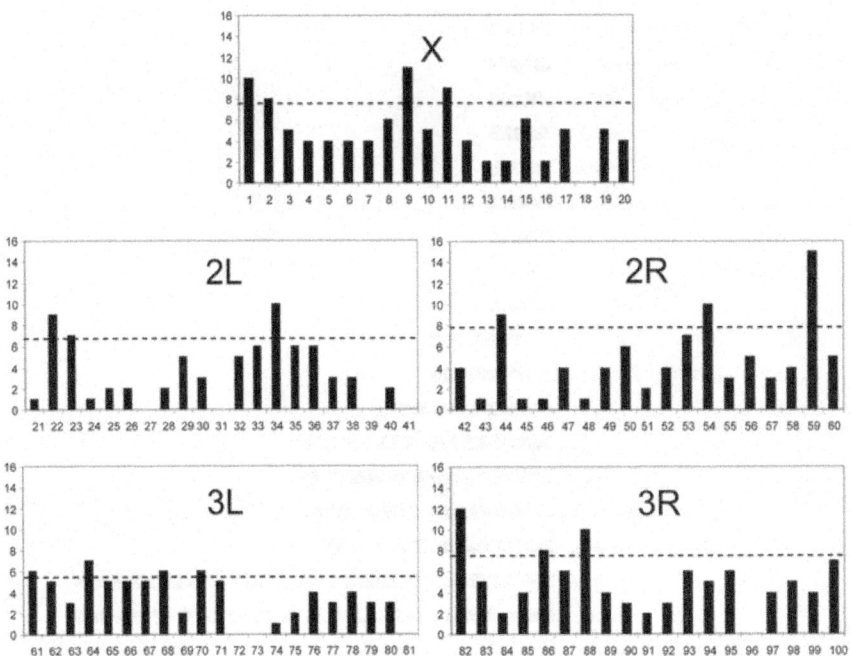

Figure 3: Distribution of ancestral alleles in the three major chromosomes.

The 20 division coordinates used are those of *D. melanogaster*. Almost all divisions have one or more ancestral alleles. The dashed lines indicate the average number of alleles plus one standard deviation. Note that some divisions have a higher density of ancestral alleles than others.

doi:10.1371/journal.pone.0010485.g003

Annotated Biological Functions of Ancestral Alleles

To test whether the genes identified have functions consistent with roles in species differentiation, we cross-referenced them to Gene Ontology (GO). If ancestral alleles participated in the segregation of these species, then we should expect to find biological functions consistent with pre and post-zygotic barriers, such as those that interfere with mating and cause interspecific lethality, sterility and mitotic defects in hybrids.

The GO referencing shows that despite the fact that less than 40% of these genes have either known molecular or biological functions, several ancestral alleles fall in discrete GO functional groups such as hybrid lethality, oogenesis, gamete generation, female meiosis, sperm competition and displacement, chemical perception of taste and olfaction, and immunity (Fig. 4).

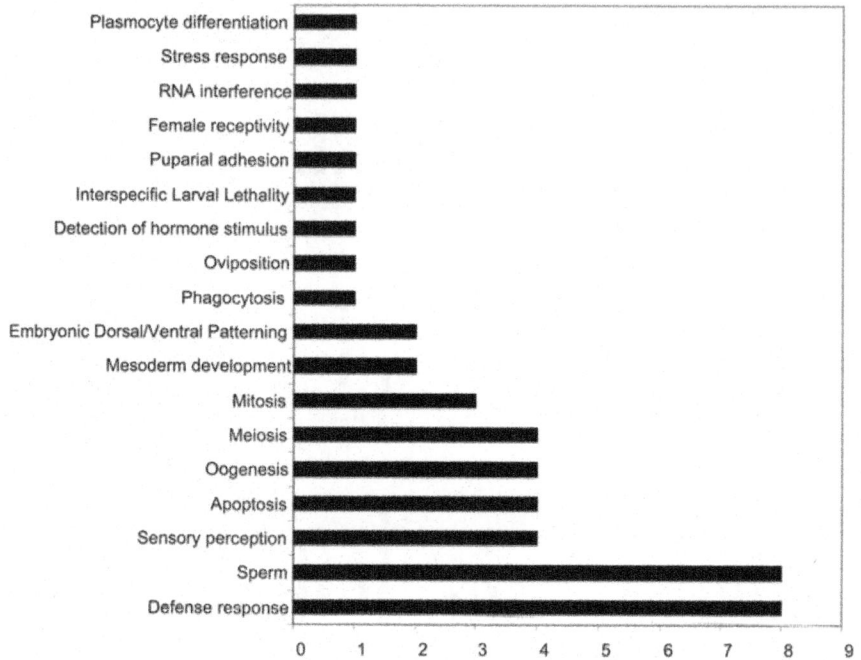

Figure 4: The biological functions of 48 ancestral alleles defined by Gene Ontology.

The graph shows only genes with biological functions assigned by assays, inferred by sequence similarity or phenotype and does not include general biochemical properties such as phosphorylation, transcription initiation, signal transduction, proteolysis among others.

doi:10.1371/journal.pone.0010485.g004

Among known genes that cause zygotic barriers, our search readily identified *Lethal hybrid rescue* (*Lhr*) as an ancestral allele. The wild type allele of *Lhr* in *D. simulans* is responsible for the larval lethality in the sons of *D. simulans* males and *D. melanogaster* females [15], [19]. Similarly, the gene *CG14781* appears as an ancestral allele. *CG14781* has been implicated in mitotic spindle elongation and recently shown to correspond to *mei-38* [22], [23]. Null mutants of *mei-38* cause abnormal meiotic non-disjunction in females, abnormal mitosis and consequently lethality due to aneuploidy [23]. Thus, *mei-38* could be potentially involved in the sex-specific offspring hybrid lethality in females.

Our search also identified a number of genes with functions consistent with the formation of pre-zygotic barriers. For instance, accessory gland proteins such as Acp29AB and Acp98AB appear as ancestral alleles and it has been suggested that Acp29AB confers a resistance to the sperm of one male to be displaced by the sperm of another male, while Acp98AB appears to negatively regulate female receptivity [24], [25]. We also found genes involved in perception of taste such as *Gr59d* and *Gr59f* and odors like the *Odorant binding proteins Obp19a, Obp22a*and *Obp47a* [26]. These genes have been implicated in the sensory perception of chemical stimuli [27] and could potentially participate in the formation of pre-mating barriers through species-specific mate recognition.

Finally, our search identified ancestral alleles that potentially have novel and previously unsuspected functions such as immunity (Fig. 4). Together, these results suggest that the segregation of an ancestral population into three distinct species involved changes in reproduction, embryonic development, nervous system development and physiology, and immunity.

DISCUSSION

Genomes and Footprints of Evolution

Extensive circumstantial evidences suggest that the genes that once created a sharp barrier between the ancestor of *D.melanogaster* and its sibling species might share an unusual conservation in *D. simulans* and *D. sechellia*. We tested this hypothesis by comparing their coding sequences and found 439 genes with little divergence in *D. simulans* and *D. sechellia*, but that diverge significantly from *D. melanogaster*.

The ancestral alleles identified in this work possibly record the earliest events in the differentiation of these *Drosophila* lineages that can be detected in extant species. The fact that these genes are very similar in *D. simulans/D. sechellia* but diverged from *D. melanogaster* more than most genes in the genome suggests two possible scenarios. In the first, the high divergence of ancestral alleles was acquired focally in time (i.e. this divergence is the result of one or few events that happened in short periods of time). The second possible scenario is that they were acquired over a longer period of time. (i.e. these genes are more prone to mutations and evolve faster than other genes). There are at least two evidences that favor the first hypothesis, but these are not yet conclusive. If these alleles were more prone to mutations, then we should expect that they would continue diverging at high rates after the separation of *D. simulans* from *D. sechellia*, but we did not observe such continuing divergence in the genome samples currently available. Also, if these alleles were more prone to mutations, then we should expect to observe high rates of polymorphism in *D. melanogaster*and *D. simulans*, which has not been reported in genomic results of different strains sequenced yet. In addition, we found a higher frequency of ancestral alleles near known chromosomal rearrangements, which raises the interesting possibility that these alleles could have been generated at the time those rearrangements appeared.

The High Divergence of X-Chromosome Genes, Recombination and Segregation Patterns

In our search, we analyzed each chromosome separately to test whether there were variations in the rate of divergence among distinct chromosomes. The existence of such differences might provide an insight into the mechanism involved in the generation of ancestral alleles. Our analysis reveals that the average divergence plus the standard deviation for the X chromosome genes is higher than that of autosomal genes. Conversely, the same analysis suggests that the fourth chromosome has a lower average divergence than the other autosomes and the X chromosome. Together, these results show that the chromosome X evolves faster than the other autosomes and suggest that the fourth chromosome evolves slower. Since the rates of mutation and recombination on the X chromosome and in the two large autosomes do not appear significantly different [28](http://flybase.org/maps/chromosomes/maps.html), the discrepancy between the divergences of the X and the remaining chromosomes is intriguing. However, this discrepancy possibly stems from the fact that the X chromosome is the only chromosome that exists in one or two copies (X/X and X/Y) in every generation. The existence of a hemizygote state allows recessive mutations on the X chromosome to be subject to the scrutiny of natural selection at least one generation before and in more individuals per generation, than a similar recessive mutation in an autosome. Thus, even with the same rate of mutation, recessive mutations on the X chromosome are subject to more rounds of selection than mutations in the autosomes, and consequently should have a better chance to become fixed.

The difference in the Mean plus Standard Deviation on the Fourth Chromosome Genes suggest a Possible Mechanism for the Generation of Ancestral Alleles

While analyzing the fourth chromosome, we detected an unusually low divergence in this chromosome. One possible explanation for this low divergence is that this is the only autosome that does not recombine during meiosis. Without recombination, errors acquired due to abnormal crossover are almost inexistent and the possibility of combining in a single chromosome different alleles floating in a population is equally low. Thus, the lack of errors during recombination and the combination of these mutations in a single chromosome could be accountable to some extent for the low levels of generation and accumulation of ancestral alleles on the fourth chromosome. However, since only 19 out of the 85 genes on this chromosome could be analyzed, this hypothesis needs to be more thoroughly tested as new high quality sequences become available for this chromosome.

The Advantages and Limits of the Analysis of Ancestral Alleles

The literature of speciation mechanisms has some examples of cleverly designed experiments to isolate genes required to block gene flow among closely related species. However, despite the fact that these screenings are of great significance and provide invaluable information about the approximate position of genes involved in speciation, researchers often face tremendous challenges to identify them molecularly. A typical example is *Lhr*, a gene identified genetically in 1979, which was only molecularly cloned almost 30 years later. This gene was readily identified as an ancestral allele in our search.

Our search can also potentially simplify the identification of other genes involved in speciation. For instance, Sawamura and cols. (2004) genetically mapped a female sterile mutant from *D. simulans*, presumably involved in the sterility observed in *D. melanogaster/D. simulans* hybrids, near the chromosomal division 32. Despite their efforts to narrow the region down to a 170 kb interval containing 20 coding sequences, they could not identify molecularly which of those 20 genes had a major effect on female fertility [29]. Our screening has identified 5 ancestral alleles on subdivision 32, and within the interval identified by Sawamura et al, there are only two ancestral alleles: Vm32E and CG14926. The GO of Vm32E suggests a role in the formation of embryonic vitelline membrane, which is consistent with female sterility, while CG14926 has no defined function but is expressed in male spermatocytes.

Although our analysis can potentially simplify the search and characterization of novel genes involved in speciation, there are some limits to its capabilities. The first one is the quality of the sequenced genomes. For instance, our search failed to identify *Hybrid male rescue*, since this gene does not have the same coverage in *D. simulans* and *D. sechellia* and for this reason was excluded from our analysis. Several other genes in the genome of *D. simulans* and *D. sechellia* also have a poor coverage. We expect that the search of evolutionary genes using the strategy outlined here will be greatly improved as more sequence gaps are filled.

The sequence comparison tools developed in this work can also be used in other types of screenings to identify genes involved in other biological processes unique to each sibling *Drosophila* species. For instance, since our screening was designed to identify ancestral alleles of *D. simulans* and *D. sechellia*, it eliminated genes required for particular specializations in each species. Our screening most likely missed genes that might be necessary for the feeding habits that make *D. simulans* and *D. melanogaster* cosmopolitan and *D. sechellia* restricted to *Morinda*. To identify the genes required for such differences, it would be necessary to screen for highly conserved genes in *D. simulans* and *D. melanogaster*that diverged in *D. sechellia*. Similarly, genes involved in the female choice for males would be expected to be missed by the current screening since females of the three species prefer males of their own species. To identify these genes the search should be directed to fast evolving genes (i.e. genes that are most divergent in the three species). Together, our results identify a relatively small number of genes that can be tested for speciation roles among *D. melanogaster* sibling species.

MATERIALS AND METHODS

Gene Extraction and Searches

Usually genome searches that aim to find variation in coding sequences focus on translations since non-synonym amino acid variation is generally believed to produce phenotypic variation. However, this approach eliminates synonym substitutions that result in protein variation (e.g. mutations in splicing enhancers). For this reason, here we took all nucleotide variation in consideration.

Annotated sequences from the *Drosophila* library NT corresponding to the X (AE014298), 2L (AE014134), 2R (AE013599), 3L (AE014296), 3R (AE014297) and 4 (AE014135) arms or chromosomes were downloaded from the NCBI website and the coding sequences (CDS) were extracted using Extractor, a software developed by us. The extracted genes were then Blasted against the Whole Genome Sequences of *D. melanogaster* (mel-mel), *D. simulans* (mel-sim) and *D. sechellia* (mel-sec) obtained from the Whole Genome Shotgun (WGS) NCBI›s library (14-Mar-2008). This library contains sequences from different strains of *D. simulans* and thus provides samples of gene variation in different populations. We developed another program, the Analyst, to automatically assemble the blast hits of the clone with best coverage using the*D. melanogaster* positions of the annotated coding sequences as a template. The Analyst also reported the coverage of all coding genes in *D. simulans* and *D. sechellia* and calculated the divergence between the pairs mel-sim, mel-sec and sim-sec by using their respective alignments.

Blast Settings and Controls

Several different Blast settings were used in control experiments where annotated coding sequences of Drosophila were blasted against Whole Genome

Sequences of *D. melanogaster*. These controls were used to define the Blast program that most consistently identifies the largest number of complete coding sequences in *D. melanogaster*. Discontinuous Megablast was chosen since it yielded the best reconstruction of the coding sequences in the control mel-mel.

Triangulation of Alleles with the same or Similar Nucleotide Composition in *D. Simulans* and *D. Sechellia* that Diverged from *D. Melanogaster*

Identity values generated by Blast alignments provide information about the percentage of substitutions within a DNA segment, but not about the position of these substitutions. Thus, if a query gene in one species has the same identity of the subjects in two other species and this identity is different from 0, the two subjects may or may not contain mutations in the same position. To identify mutations in the same position, we triangulated the position of these substitutions by coverage and identity in pair wise comparisons between mel-sim, mel-sec and sim-sec. Using this system, genes that diverged significantly from *D. melanogaster* and were inherited by *D. simulans* and *D. sechellia* lineages should appear with the same coverage and an identical or similar identity in the *D. simulans/D. sechellia* comparison, and with equally fewer identities in the *D. melanogaster/D. simulans* and *D. melanogater/D. sechellia*comparisons.

To minimize errors due to the incorrect automatic assembly of the coding sequences in *D. simulans* and *D. sechellia* that could interfere with the evaluation of the divergence (i.e. truncated coding sequences due to the inability of Blast to identify particular exons), the coding sequences of *D. melanogaster* were Blasted against the *D. melanogaster* WGS library and the predicted coding sequences assembled from the Blast hits. The identities of these predicted coding sequences were then compared to the actual coding sequences in the annotated genome and only genes with at least 99% of identity with the annotated coding sequences were included in the analysis. To avoid false-positives due to incomplete clone representation, the data was sorted by coverage in sim and sec, and only genes with the same coverage were selected.

Cross-Referencing to Gene Function

The functional cross-referencing of the genes identified was done using annotated biological functions from flybase (http://flybase.bio.indiana.edu/), as well as descriptions in the literature.

ACKNOWLEDGMENTS

We would like to thank Claudia Mieko Mizutani and Joseph Schinaman for comments on the manuscript.

AUTHOR CONTRIBUTIONS

Conceived and designed the experiments: RSN AR. Performed the experiments: RSN AR. Analyzed the data: RSN AR. Contributed reagents/materials/analysis tools: RSN AR. Wrote the paper: RSN AR.

REFERENCES

1.   Tamura K, Subramanian S, Kumar S (2004) Temporal patterns of fruit fly (Drosophila) evolution revealed by mutation clocks. Mol Biol Evol 21: 36–44.

2.   Acebes A, Cobb M, Ferveur JF (2003) Species-specific effects of single sensillum ablation on mating position in Drosophila. J Exp Biol 206: 3095–3100.

3.   Lachaise D, David JR, Lemeunier F, Tscas L, Ashburner M (1986) The Reproductive Relationships of Drosophila sechellia with D. mauritiana, D. simulans, and D. melanogaster from the Afrotropical Region. Evolution 40: 262–271.

4.   Lemeunier F, Ashburner MA (1976) Relationships within the melanogaster species subgroup of the genus Drosophila (Sophophora). II. Phylogenetic relationships between six species based upon polytene chromosome banding sequences. Proc R Soc Lond B Biol Sci 193: 275–294.

5.   Masly JP, Jones CD, Noor MA, Locke J, Orr HA (2006) Gene transposition as a cause of hybrid sterility in Drosophila. Science 313: 1448–1450.

6.   Podemski L, Ferrer C, Locke J (2001) Whole arm inversions of chromosome 4 in Drosophila species. Chromosoma 110: 305–312.

7.   Seiler T, Nöthiger R (1974) Somatic cell genetics applied to species hybrids of Drosophila. Experientia: 30: 709.

8.   Sanchez L, Düberdorfer A (1983) Development of Imaginal Discs from Lethal Hybrids between Drosophila melanogaster and Drosophila mauritiana. Roux's Arch Dev Biol 192: 48–50.

9.   Orr HA, Madden LD, Coyne JA, Goodwin R, Hawley RS (1997) The developmental genetics of hybrid inviability: a mitotic defect in Drosophila hybrids. Genetics 145: 1031–1040.

10. Sturtevant AH (1929) Contributions to the genetics of Drosophila simulans and Drosophila melanogaster. I. The genetics of Drosophila simulans. Publs Carnegie Instn 399: 1–62.

11. Hadorn E (1961) Zur Anatomie und Phasenspezifität der Letalität von Bastarden zwischen Drosophila melanogaster und D.simulans. Rev Swisse Zool 68: 197–207.

12. Watanabe TK, Kawanishi M (1979) Mating Preference and the Direction of Evolution in Drosophila. Science 205: 906–907.

13. Tomaru M, Doi M, Higuchi H, Oguma Y (2000) Courtship song recognition in the Drosophila melanogaster complex: heterospecific songs make females receptive in D. melanogaster, but not in D. sechellia. Evolution 54: 1286–1294.

14. Wirtz P (1999) Mother species-father species: unidirectional hybridization in animals with female choice. Anim Behav 58: 1–12.

15. Watanabe TK (1979) A gene that rescues the lethal hybris between Drosophila melanogaster and D.simulans. The Japanese Journal of Genetics 325–331.

16. Hutter P, Roote J, Ashburner M (1990) A genetic basis for the inviability of hybrids between sibling species of Drosophila. Genetics 124: 909–920.

17. Sawamura K, Yamamoto MT, Watanabe TK (1993) Hybrid lethal systems in the Drosophila melanogaster species complex. II. The Zygotic hybrid rescue (Zhr) gene of D. melanogaster. Genetics 133: 307–313.

18. Sawamura K, Taira T, Watanabe TK (1993) Hybrid lethal systems in the Drosophila melanogaster species complex. I. The maternal hybrid rescue (mhr) gene of Drosophila simulans. Genetics 133: 299–305.

19. Brideau NJ, Flores HA, Wang J, Maheshwari S, Wang X, et al. (2006) Two Dobzhansky-Muller genes interact to cause hybrid lethality in Drosophila. Science 314: 1292–1295.

20. Bridges CB (1935) Salivary chromosome maps with a key to the banding of the chromosomes of Drosophila melanogaster. J Hered 26: 60–64.

21. Horton IH (1939) A Comparison of the Salivary Gland Chromosomes of Drosophila Melanogaster and D. Simulans. Genetics 24: 234–243.

22. Baker BS, Carpenter AT (1972) Genetic analysis of sex chromosomal meiotic mutants in Drosophilia melanogaster. Genetics 71: 255–286.

23. Wu C, Singaram V, McKim KS (2008) mei-38 Is required for chromosome segregation during meiosis in drosophila females. Genetics 180: 61–72.

24. Wolfner MF, Harada HA, Bertram MJ, Stelick TJ, Kraus KW, et al. (1997) New genes for male accessory gland proteins in Drosophila melanogaster. Insect Biochem Mol Biol 27: 825–834.

25. Chapman T (2001) Seminal fluid-mediated fitness traits in Drosophila. Heredity 87: 511–521.

26. Clyne PJ, Warr CG, Carlson JR (2000) Candidate taste receptors in Drosophila. Science 287: 1830–1834.

27. Galindo K, Smith DP (2001) A large family of divergent Drosophila odorant-binding proteins expressed in gustatory and olfactory sensilla. Genetics 159: 1059–1072.

28. Haag-Liautard C, Dorris M, Maside X, Macaskill S, Halligan DL, et al. (2007) Direct estimation of per nucleotide and genomic deleterious mutation rates in Drosophila. Nature 445: 82–85.

29. Sawamura K, Karr TL, Yamamoto MT (2004) Genetics of hybrid inviability and sterility in Drosophila: dissection of introgression of D. simulans genes in D. melanogaster genome. Genetica 120: 253–260.

Chapter 7

GENETIC ANALYSIS OF POPULATION STRUCTURE AND REPRODUCTIVE MODE OF THE TERMITE RETICULITERMES CHINENSIS SNYDER

Qiuying Huang[1], Ganghua Li[1], Claudia Husseneder[2], Chaoliang Lei[1]

[1]Hubei Insect Resources Utilization and Sustainable Pest Management Key Laboratory, Huazhong Agricultural University, Wuhan, China

[2]Department of Entomology, Louisiana State University Agricultural Center, Baton Rouge, Louisiana, United States of America

ABSTRACT

The subterranean termite *Reticulitermes chinensis* Snyder is an important pest of trees and buildings in China. Here, we characterized genetic structure and reproductive modes of *R. chinensis* from China for the first time. A total of 1,875 workers from 75 collection sites in Huanggang, Changsha and Chongqing cities were genotyped at eight microsatellite loci. Analysis of genetic clusters showed two subpopulations in Chongqing city. The Huanggang population showed a uniform genetic pattern and was separated from the other populations by the largest genetic distances (F_{ST}: 0.17–0.20). In contrast, smaller genetic distances (F_{ST}: 0.05–0.12) separated Changsha, Chongqing-1 and Chongqing-2 populations. Chongqing-1 was the only population showing a genetic bottleneck. Isolation by distance among colonies in the Huanggang population indicated limited alate dispersal or colony budding. Lack of isolation by distance among colonies within the populations of Changsha, Chongqing-1 and Chongqing-2, suggested long-range dispersal by alates and/ or human-mediated transport. Overall, extended family colonies (73.91%) were predominant in all four populations, followed by simple (20.29%), and mixed family colonies (5.80%). Most simple families were headed by inbred related reproductive pairs in the Changsha population, while most simple families in the Chongqing-1 population were headed by outbred unrelated pairs. Simple families in the Huanggang population were a mixture of colonies headed by outbred or inbred reproductive pairs. The sample size of simple families in the Chongqing-2 population was too small to yield significant results. Extended families in all four populations were headed on the average by ≤10 neotenics. Mixed families likely originated from pleometrosis. Presence of heterozygote genotypes showed that all neotenic reproductives collected in addition from five field colonies in Wuhan city were sexually produced, suggesting that these colonies did not undergo parthenogenesis. This study contributes to better understanding of the variance of genetic structure and reproductive mode in the genus *Reticulitermes*.

INTRODUCTION

An increasing number of studies have employed molecular markers, such as microsatellites, to investigate population genetic structure and colony breeding system in termites, which has improved our understanding of evolutionary genetics, phylogeography, invasion biology, patterns and processes of dispersal, and social organization in termites [1], [2]. The analyses of genetic differentiation and patterns of gene flow within and among termite populations promoted understanding of population dynamics and dispersal patterns [3]–[5], and can help clarify the effect of ecological factors and human disturbance on the levels of genetic differentiation and gene flow [6]–[8]. Moreover, the studies on population genetic structure of invasive termites are helpful to identify potential source populations and to assess whether changes in genetic variability and/or breeding system are associated with invasiveness[9]–[11]. Previous studies on colony breeding structure showed that most populations of subterranean termites were composed of different percentages of simple and extended families[4], [12]–[16], while mixed families were generally less common [6], [17]. Additionally, the number of reproductives and the degree of relatedness among them in termite colonies can also be inferred using molecular markers [3],[18]–[22]. Recent studies on the modes of reproduction using microsatellite genotyping found that queens of *Reticulitermes speratus* and*R. virginicus* can use conditional parthenogenesis, where primary queens produce secondary queens by parthenogenesis (asexual queen succession AQS), but use sexual reproduction to produce workers [23], [24]. Studies on the model of conditional use of sexual and asexual reproduction are helpful to understand the advantages and disadvantages of different reproductive modes in termites [25]. However, it is still unknown how widespread AQS is among *Reticulitermes* spp.

The genus *Reticulitermes* has received much attention among the Isoptera when it comes to population and colony genetic studies, but almost all the species used in these studies were from America and Europe. In Asia, only colony genetic structure of *R. speratus* from Japan was studied using a small number of colonies on a limited spatial scale [26], but population genetic structure of this species has not yet been investigated. Since phylogeographic divisions were found among species of *Reticulitermes* termites across different countries even within Europe [27]–[30], it is necessary to thoroughly investigate and compare population genetic structure and colony breeding system of several representative termite species from different places in Asia. In a phylogenetic analysis of the family Rhinotermitidae, Austin *et al.*[31] found that *R. speratus* and *R. chinensis* were close relatives within the genus*Reticulitermes*, belonging to the same clade. Since *R. speratus* is one of the species featuring AQS, knowledge of the evolution of this unusual breeding strategy could be significantly advanced by studying how widespread AQS is within the genus *Reticulitermes* and if the development of AQS is congruent with phylogeny. Thus, we test the hypothesis that colonies of *R. chinensis* undergo AQS as well.

The subterranean termite *R. chinensis* is widely distributed in China, including Beijing, Tianjin, Shanxi and the Yangtze River drainage basin [32]. This termite species is an important pest of forest trees and urban buildings. To date, some basic information on the biology of *R. chinensis* has been obtained through field observations and laboratory experiments, including the growth of incipient colonies, the formation of neotenics, and the time of alate swarms [33],[34]. Since *R. chinensis* has cryptic foraging and nesting habits with hidden royal chambers, it is very difficult to find and collect reproductives. Thus, population genetic structures and colony breeding systems of *R. chinensis* are still poorly understood. The objectives of this study were to use microsatellite genotyping to (1) describe the population genetic structure of*R. chinensis*, (2) infer colony breeding system, (3) and determine whether colonies of *R. chinensis* undergo AQS as it has been suggested for other *Reticulitermes* species.

RESULTS

Colony Assignments and Population Genetic Structure

STRUCTURE analysis showed that the *R. chinensis* colonies from Huanggang, Changsha and Chonqing cities (Figure 1) belong to two major genetic clusters (K=2, DeltaK=1348.8) but the Evanno plot (Figure S1) also shows the presence of a secondary pattern consisting of four genetic clusters indicating that there might be subpopulations in one of the regions (DeltaK=417.5). The 25 samples collected from Huanggang (Figure 1c) belonged to 24 different colonies. No subpopulations were detected in Huanggang (Figure 2), but significant isolation by distance at 500 m spatial scale was found (Figure 3). The 25 samples collected from Changsha (Figure 1d) belonged to 23 colonies.

STRUCTURE indicated the potential presence of multiple subpopulations (2–4 clusters) among the samples from Changsha (Figure 2). However, assignment of most colonies to any one of the clusters was weak (<80%). Clusters were not congruent with the two geographic clusters of sample sites (Figure 1d) and no significant isolation by distance was detected (Figure 3). Therefore, we considered all Changsha colonies to belong to one population in the further analyses. The 25 samples collected from Chongqing (Figure 1e) belonged to 23 colonies. Results generated from STRUCTURE indicated the presence of two distinct genetic clusters in Chongqing, with 13 colonies assigned to the first subpopulation (Chongqing-1) and 9 colonies assigned to a second subpopulation (Chongqing-2). One colony could not be assigned to either of the two subpopulations of Chongqing (Figure 1e, open triangle) but belonged to the genetic cluster representing the Changsha population (Figure 2). The majority (83%) of colonies were strongly ($\geq$80%) assigned to one or the other subpopulation. Genetic separation was supported by 10 private alleles distinguishing subpopulation Chongqing-1 from Chongqing-2 and 15 private alleles distinguishing Chongqing-2 from Chongqing-1. Thus, colonies from the Chongqing-1 and Chongqing-2 subpopulations were treated as separate populations in the following analyses, thereby increasing the total number of populations to four. There was no significant isolation by distance within either subpopulation (Figure 3). Interestingly, there was no obvious relationship between the genetic clusters and geographical location of the colonies (Figure 1e). Evidence was found for a recent genetic bottleneck in the Chongqing-1 subpopulation, but not in Chongqing-2. The Chongqing-1 subpopulation had significant heterozygote excess under the infinite allele model and the two-phase model (P=0.002) and marginal heterozygote excess under the stepwise mutation model (P=0.097).

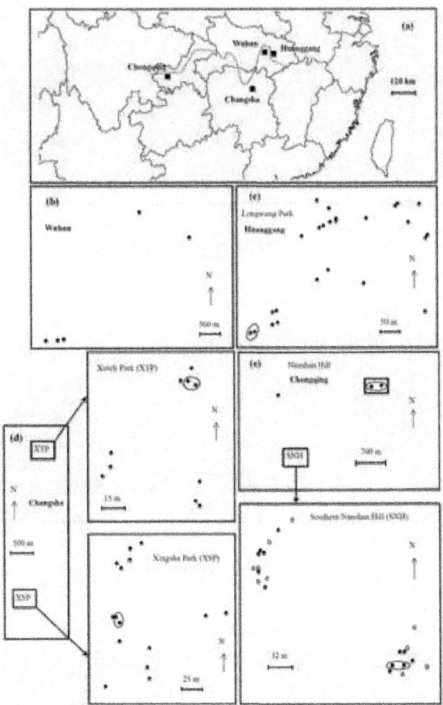

Figure 1: Map of the sample locations for the studies of population structure and reproductive mode.

(a) Overview over the locations of the cities in China where samples were obtained from (Huanggang, Changsha and Chongqing and Wuhan). (b) Sample sites of entire nests including neotenics from Wuhan City, Hubei Province. (c) Sample sites of workers from Longwang Park,Huanggang City, Hubei Province. (d) Sample sites of workers from Xuteli Park and Xingsha Park, Changsha City, Hunan Province. (e) Sample sites of workers from Nanshan Hill, Nanan District, Chongqing City. The dotted line represents the Yangtze River. Genetic analyses revealed two subpopulations in Chongqing city (*filled circle* subpopulation 1; *open circle* subpopulation 2; *Open triangle* representing one colony that could not be assigned to either subpopulation of Chongqing. The loops represent samples assigned to the same colony.

doi:10.1371/journal.pone.0069070.g001

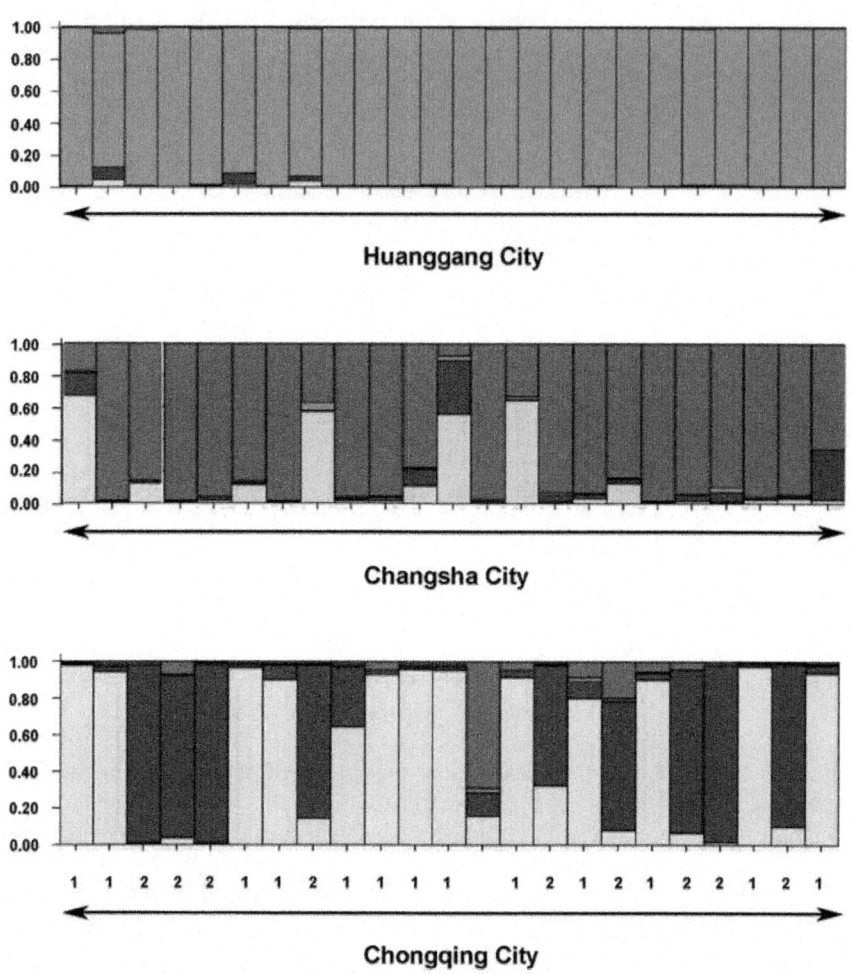

Figure 2: Assignment of individuals (each representing a colony) sampled from three cities to genetic clusters.

Columns represent colonies, each color represent a different genetic cluster defined by STRUCTURE (*K*=4). The colors in each column represent the likelihood with which a colony is assigned to each genetic cluster. The numbers (1 and 2) in this figure represent the colonies assigned to the Chongqing-1 population and the Chongqing-2 population, respectively.

doi:10.1371/journal.pone.0069070.g002

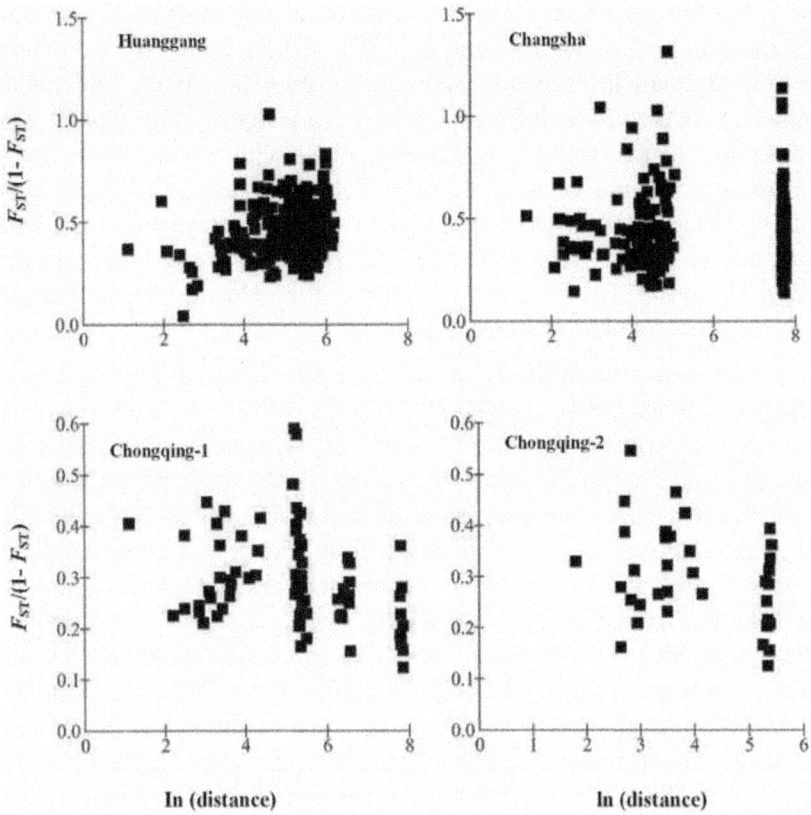

Figure 3: Isolation by distance analysis for the colonies within each population.

The correlation coefficient was significant for the relationship for Huanggang population (r=0.2245, P=0.0014, Mantel test). The correlation coefficients were not significant for either of the relationships (r=0.0130, −0.3284 and −0.3426, P=0.4462, 0.9654 and 0.9015, Mantel test, for Changsha, Chongqing-1 and Chongqing-2 population, respectively).

doi:10.1371/journal.pone.0069070.g003

Permutation tests showed significant genetic differentiation among all four populations (P=0.001, 6000 permutations) confirming the existence of four genetically differentiated groups indicated by STRUCTURE analysis. Genetic distances (F_{ST}) among the four populations of *R. chinensis* ranged from 0.05 to 0.2 (Table S1). The Huanggang population was clearly separated from the other populations by large genetic distances ranging from 0.17 to 0.20. Its genetic pattern was uniform showing little contributions to the genetic make-up of the other populations in the dataset. Although the remaining three populations were significantly differentiated from each other according to the permutation tests, the genetic distances ranging from 0.05 to 0.12 between the populations, were rather small in particular between the two Chongqing subpopulations (Table S1). Two colonies from Changsha were assigned to the genetic clusters representing the two Chongqing subpopulations with >80%. One colony from Chongqing could not be assigned to either subpopulation of Chongqing but could be assigned with 70% to the genetic cluster representing the Changsha population (Figure 2). The STRUCTURE results showing two main genetic clusters (Figure S1) confirm the separation of the Huangang population from the other three populations (Figure S2).

Table 1 shows the summary statistics for genetic variability and inbreeding in the populations. Between eight (*RS78*) and 21 (*Rf21-1*) alleles were detected per locus with a mean of 14.38 per locus. The expected heterozygosity ranged from 0.69 to 0.79, and the observed heterozygosity ranged from 0.55 to 0.84 (Table 1). The inbreeding coefficient ranged from −0.1 to 0.21. Overall, the four populations deviated significantly from Hardy–Weinberg equilibrium. There was a deficit of heterozygotes at three loci in the Huanggang population, and at two loci in the Changsha population. This heterozygote deficit at a subset of loci might have been caused by amplification failure of null alleles. Micro-Checker software confirmed deviations from Hardy-Weinberg-Equilibrium due to heterozygote deficit and suggested the occurrence of null alleles at locus *Rf21-1* (frequency: 0.31), *RS15* (frequency: 0.15) and *RS68* (frequency: 0.11) in the Huanggang population and at locus *Rf21-1* (frequency: 0.11) and *RS78* (frequency: 0.28) in the Changsha population. Heterozygote excess was detected across all loci in Chongqing-1 possibly due to a recent genetic bottleneck (see above). No evidence for linkage disequilibrium, large allele dropout or scoring error due to stuttering was detected in any of the four populations.

Table 1: Summary statistics for the four genetically differentiated populations

Population	N	N_A	H_E	H_O	F_{IS}
Huanggang	24	9	0.79	0.65	0.19
Changsha	23	7.5	0.69	0.55	0.21
Chongqing-1	13	5.88	0.76	0.84	−0.1
Chongqing-2	9	5.88	0.75	0.65	0.13

N, number of colonies screened per location; N_A, average number of alleles per locus; H_E, expected heterozygosity; H_O, observed heterozygosity; F_{IS}, inbreeding coefficient.
doi:10.1371/journal.pone.0069070.t001

doi:10.1371/journal.pone.0069070.t001

Colony Genetic Structure and Breeding System

Overall, extended family colonies were predominant (55–83%) in the four populations from Huanggang, Changsha and Chongqing, followed by simple family colonies (17–31%), and mixed family colonies (0–9%). The proportion of extended, simple and mixed family colonies varied among the four populations, with the highest proportion of extended families (83%) in the Huanggang population, the highest proportion of simple families (31%) in the Chongqing-1 population and the most mixed family colonies (9%) in the Chongqing-2 population (Table 2).

Table 2: F-statistics and relatedness coefficients for workers from *R. chinensis* colonies from four populations in China and expected for possible breeding systems of subterranean termite colonies previously derived from computer simulations [20], [21]

	F_{IT} (SE)	F_{CT} (SE)	F_{IC} (SE)	r (SE)
Huanggang				
Simple family colonies (n = 4)	0.14 (0.10)[a]	0.33 (0.05)	−0.29 (0.06)[a]	0.58 (0.05)[a]
Extended family colonies (n = 20)	0.18 (0.08)[A]	0.30 (0.02)	−0.19 (0.09)[A]	0.52 (0.02)[A]
Changsha				
Simple family colonies (n = 4)	0.23 (0.06)[b]	0.36 (0.05)	−0.20 (0.02)[b]	0.58 (0.05)[a]
Extended family colonies (n = 18)	0.14 (0.09)[C]	0.29 (0.02)	−0.21 (0.11)[AB]	0.50 (0.03)[A]
Mixed family colonies (n = 1)				
Chongqing-1				
Simple family colonies (n = 4)	−0.07 (0.06)[c]	0.19 (0.03)	−0.31 (0.07)[a]	0.40 (0.07)[a]
Extended family colonies (n = 8)	0.03 (0.04)[D]	0.25 (0.03)	−0.28 (0.04)[C]	0.48 (0.04)[A]
Mixed family colonies (n = 1)				
Chongqing-2				
Simple family colonies (n = 2)	0.03 (0.10)[a]	0.29 (0.07)	−0.35 (0.09)[a]	0.56 (0.10)[a]
Extended family colonies (n = 5)	0.11 (0.10)[A]	0.23 (0.02)	−0.16 (0.12)[AB]	0.42 (0.05)[A]
Mixed family colonies (n = 2)	−0.06 (0.04)[x]	0.11 (0.03)	−0.19 (0.04)[x]	0.24 (0.06)[y]
Simulated breeding systems				
(A) Simple family colonies with				
(1) outbred unrelated pairs	0[ac]	0.25	−0.33[a]	0.5[a]
(2) inbred related pairs	0.33[ab]	0.42	−0.14[ab]	0.62[x]
(B) Extended family colonies with inbreeding among multiple neotenics				
(1) $N_f = N_m = 1$, $X = 1$	0.26[AC]	0.65	−0.14[A]	0.55[A]
(2) $N_f = 2$, $N_m = 1$, $X = 3$	0.52[B]	0.59	−0.17[A]	0.78[b]
(3) $N_f = N_m = 10$, $X = 1$	0.33[A]	0.34	−0.01[A]	0.51[A]
(4) $N_f = 200$, $N_m = 100$, $X = 3$	0.33[A]	0.34	0[B]	0.5[A]
(C) Mixing between unrelated colonies, $N_f = N_m = 1$, $X = 3$, $p = 0.8$	0.57[b]	0.43	0.25[x]	0.55[y]
(D) Mixing between related colonies, $Nf = Nm = 1$, $X = 3$, $p = 0.9$	0.66[b]	0.64	0.04[y]	0.77[z]
(E) Pleometrosis				
(1) colonies headed by two queens and one king	0[a]	0.19	−0.23[x]	0.38[x]
(2) colonies headed by two queens and one king, then $N_f = N_m = 10$, $X = 3$	0.27[a]	0.29	−0.03[y]	0.45[y]
(3) colonies headed by five queens and five kings, then $N_f = N_m = 10$, $X = 3$	0.1[a]	0.12	−0.02[y]	0.22[x]

n, number of colonies; *SE*, standard error derived from jackknifing over colonies (or loci if sample size ≤3). One sample *t*-tests were performed using F_{IT}, F_{IC}, and *r* values across individual colonies. Significant differences between empirical values and expected values are indicated by different letters (uppercase for extended families, lowercase for simple families, Greek alphabet lower case for mixed families). For simulated breeding systems, *X* represents generation number of neotenics within a colony; N_f and N_m represent number of replacement females and males per generation respectively; *p* presents mixing proportion between workers from different colonies.
doi:10.1371/journal.pone.0069070.t002

doi:10.1371/journal.pone.0069070.t002

The F-statistics and relatedness values estimated from the worker genotypes in each population are shown in Table 2, along with values derived from simulations for different breeding systems computed in previous studies [20], [21], [35]. Mean observed F_{IT} and F_{IC} values of simple family colonies in Changsha were significantly higher than the expected values in outbred unrelated pairs (Table 2, case A1) (all $P<0.05$, one sample t test), but not significantly different from the expected F_{IT} and F_{IC} values for inbred related pairs (Table 2, case A2). Thus, most of simple-family colonies in Changsha were headed by monogamous pairs of inbred related reproductives. However, mean observed F_{IT} and F_{IC} values of simple family colonies in Chongqing-1 were significantly and marginally significantly lower than the expected values in inbred related pairs, respectively (Table 2, case A2) ($P<0.01$, one sample t test), but not significantly different from the expected F_{IT} and F_{IC} values for outbred unrelated pairs (Table 2, case A1). Therefore, the majority of simple-family colonies in Chongqing-1 were likely headed by monogamous pairs of outbred unrelated reproductives. The empirical values for simple families in Chongqing-2 also matched closest with the assumption of outbred unrelated founder pairs, however, no statistical significances could be achieved due to the small sample size ($n=2$). In Huanggang, the F-statistics and relatedness values of simple family colonies were not significantly different from either of the expected values and the standard errors were large (Table 2, cases A1 and 2). This indicates that some colonies in the Huanggang population might be headed by outbred unrelated pairs and others by inbred related pairs.

The values of F_{IT}, F_{IC}, and r of the extended family colonies in the Huanggang population were consistent with those expected for colonies having small numbers (≤ 10) of neotenic reproductives representing the first generation of replacement reproductives (Table 2, cases B1 and 3), but differed significantly ($P<0.05$, one sample t test) from breeding systems assuming three generations of replacement reproductives and/or hundreds of neotenics in at least one of the three genetic parameters (Table 2, cases B2 and 4). In Changsha, the extended family colonies showed F_{IT}, F_{IC}, and r values consistent with those expected for colonies headed by a pair of first generation neotenic reproductives (Table 2, case B1), but differed significantly ($P<0.05$, one sample t test) from the other breeding systems in at least one of the three genetic parameters (Table 2, cases B2, 3 and 4). In Chongqing-1, the values of F_{IT}, F_{IC}, and r for the extended family colonies differed significantly ($P<0.05$, one sample t test) from all the four simulated breeding systems in at least one of the three genetic parameters (Table 2, cases B1, 2, 3 and 4), thus no single predominant type (in regard to numbers and generations of neotenics) could be assigned to the extended families in this population. In Chongqing-2, the F_{IT}, F_{IC}, and r values in extended family colonies were consistent with the simulated values for colonies headed by the first generation of ≤ 10 neotenics (Table 2, cases B1, and 3), but differed significantly ($P<0.05$, one sample t test) from the other breeding systems with multiple generations and/or high numbers of neotenics in at least one of the three genetic parameters (Table 2, case B2 and 4).

The Changsha and Chongqing-1 populations contained only one mixed family colony each, so F-statistics could not be obtained. The F-statistics and relatedness estimates for the two mixed family colonies from Chongqing-2 were consistent with simulated values for pleometrosis (Table 2, case E1, 2 and 3), but F_{IT} differed significantly ($P<0.05$, one sample $t$test) from the breeding systems involving mixing of colonies (Table 2, cases C and D).

Inferring Reproductive Mode of Neotenics' Parents

Of the nine field colonies collected in Wuhan city (Figure 1a) and dissected to obtain the reproductives, four colonies contained single pairs of primary reproductives and were thus simple families. Five colonies (Figure 1b) contained multiple neotenics with a range of 1 to 51 secondary kings and 5 to 74 secondary queens (Table 3). Most of the secondary reproductives developed from nymphs, but there were one male and 5 female ergatoid neotenics (i.e. derived from workers) in colony 5, and two female ergatoid neotenics in colony 2 (Table S2). Neither primary kings nor primary queens were found in addition to the neotenics. We determined that all of the 59 secondary queens and 29 secondary kings genotyped were sexually produced because they were heterozygote at one to eight loci (Table S3). This allows us to test whether these neotenics originated from simple, extended or mixed family colonies. The results showed that the neotenics in colony 1, 3, 4 and 5 had more genotypes than possible for offspring of a monogamous pair or the observed frequencies of the genotypes deviated significantly from those expected in simple families ($P<0.05$, G-test). Thus, the neotenics from these four colonies were offspring of extended family colonies headed by multiple neotenics and as such arose at least two generations after colony foundation (Table 3). However, neotenics in colony 2 had six alleles at locus $Rf21-1$, which suggests that they originated from a mixed family colony (Table 3).

Table 3: Compositions of the five colonies headed by neotenics and reproductive mode of neotenics' parents

Colony	Location	Collection date	No. of secondary queens genotyped/ collected	No. of secondary kings genotyped/ collected	Reproductive mode of neotenics' parents	Breeding system of the colonies of the neotenics' parents
1	Shizi Hill, Wuhan City, China	1 Aug 2011	20/30	3/5	Sexual reproduction	Extended family
2	Shizi Hill, Wuhan City, China	15 Aug 2011	20/74	20/51	Sexual reproduction	Mixed family
3	Shizi Hill, Wuhan City, China	23 Apr 2011	8/8	4/6	Sexual reproduction	Extended family
4	Houshan Hill, Wuhan City, China	20 Apr 2011	6/6	1/1	Sexual reproduction	Extended family
5	Yujia Hill, Wuhan City, China	20 Apr 2011	5/5	1/1	Sexual reproduction	Extended family

doi:10.1371/journal.pone.0069070.t003

doi:10.1371/journal.pone.0069070.t003

DISCUSSION

Large-scale population genetic structure analysis revealed the existence of four populations with significant genetic differentiation in the dataset. The four populations were divided into two main genetic clusters. The population from Huanggang was separated from the other populations by comparatively large genetic distances and showed a uniform genetic pattern, suggesting that there was little contribution of alleles to or from the other populations. The sample sites from Huanggang lie north of the Yangtze River and are thus separated from the sample sites of Changsha and Chongqing, which are located south of the Yangtze River. Thus, geographical barriers, including the Yangtze River in addition to long distances, might limit gene flow between Huanggang and the other three populations. In contrast, the remaining three populations from Changsha, Chongqing-1 and -2 showed some level of gene flow among them. Surprisingly, colonies from Changsha were being assigned to the genetic clusters representing Chongqing populations and *vice versa*, which is indirect evidence for gene flow even across large distances (approx. 640 km). The most likely routes of migrants connecting Changsha and Chongqing are occasional human-mediated movement of infested material from one population to the other through roads and trade network, such as building timbers, ornamental plants, and barges travelling along rivers. Similarly, human transport was considered to be a factor in the gene flow for introduced *R. flavipes* populations between North American and France [9], for *R. urbis* between Balkans and Western Europe [8], and for *Coptotermes formosanus* between Fukue and Kyushu islands [7].

The existence of two populations with slightly different genetic characteristics in the same area in Chongqing was peculiar. The Chongqing-1 population showed signs of a genetic bottleneck, but the Chongqing-2 population was in mutation-drift equilibrium. This difference between the two populations colonizing the same area in Chongqing might be explained by the introduction history. The first introduction to Nanshan Hill in Chongqing took place in the early 1990s when a number of ornamental plants and landscape timbers were transported to Nanshan Hill for constructing Nanshan Arboretum [36]. The second introduction to Nanshan Hill in Chongqing happened in the early 2000s when a lot of building timbers and ornamental plants were transported to construct hotels in Nanshan Hill. To explain the difference between the two populations, we suggest that the earlier introduced population (Chongqing-2) might have contained a larger population size mitigating the initial bottleneck effects and/or enough time since the introduction has passed for the bottleneck to become undetectable, while the more recent introduction (Chongqing-1) still is experiencing the effects of a bottleneck [11]. Relative excess of heterozygosity is a transient effect and is expected to be detectable only for a limited number of generations [37]. Interestingly, bottlenecks could be detected as far back as the 1950's and 60's in another subterranean termite with a known introduction history, *C. formosanus* [11]. Either *R. chinensis* has a higher generation rate accelerating the return to mutation-drift equilibrium, the initial bottlenecks were not as severe as in *C. formosanus*'intercontinental journey (i.e. the initial founder population size was larger), and/or the small sample size prevented detection of weak bottleneck effects in the Chongqing-1 population [38]. Since there are no conceivable barriers preventing gene flow, we expect the differences between the sympatric populations of Chonqing-1 and -2 to disappear over time.

There was significant isolation by distance among colonies in the Huanggang population, indicating that dispersal by primary reproductives is relatively limited over the spatial scale studied (550 m), either because of short range mating flights by alates and/or frequent colony reproduction by budding. Similarly, significant isolation by distance has been found in two *R. flavipes* populations from France [9]. In contrast, there was no significant isolation by distance among colonies in the Chongqing-1 and Chongqing-2 populations, at the spatial scale studied (about 2.6 km). This lack of isolation by distance suggests that during mating flights reproductives disperse relatively far over the spatial scale studied. Simple-family colonies in Chongqing-1 were headed by monogamous pairs of outbred unrelated reproductives, which further supports the hypothesis of long-range dispersal flights of alates. In addition, the human-mediated introduction and movement of colonies via landscaping and building material and activities might have further masked any isolation by distance imposed by natural dispersal limits.

Similar to the Chongqing populations, the colonies in the Changsha population also did not show isolation by distance at the spatial scale studied (2.4 km). In contrast to the Chongqing colonies, however, simple family colonies in Changsha were headed by inbred related monogamous pairs. This result does not support the existence of long-range mating flights in this population. Alternatively, as suggested for the Chongqing population, human-mediated colony movement may play an important role in population expansion and reducing population viscosity, i.e. isolation by distance, in Changsha.

The four populations of *R. chinensis* in our study consisted of a combination of simple family and extended family colonies. A few mixed family colonies were found in all but the Huanggang population. Overall, extended families with a low number of neotenics and generation turnovers were the dominant breeding system. This finding is consistent with previous studies in *R. flavipes* [9], [13], [21], [39], *Zootermopsis nevadensis nuttingi* [40], *R.*

grassei [35], [41], *R. urbis* [8] and *C. formosanus* in Guangdong, Hunan, Kauai, Maui 1 and Charleston [10], [11], [16]. Moreover, small percentages of mixed family colonies were also detected for *R. flavipes* [13], [39] and *R. grassei* [41]. Proportions of termite colonies with different breeding systems vary across populations and depend on age structure of the colonies, dynamics of colony-colony interactions, food quantity and quality, soil characteristics and disturbance or treatment [4], [14], [20], [21], [42]. There were plenty of 20–30 years old pine trees and camphor trees in the parks and hills of sample collection in Huanggang, Changsha and Chongqing, suggesting that the *R. chinensis* colonies had enough time and food to grow and develop into extended families headed by neotenics [35]. Even more recently established populations, such as the Chongqing populations, showed predominance of extended family colonies, possibly due to disturbance and movement of colonies by human activities, which might lead to splitting colonies, death and replacement of colony founders and an overall increase in extended family colonies in the population as was previously shown in a population of another subterranean termite, *C. formosanus*, which was subjected to disturbance by building and landscaping activities [42].

Simple family colonies in Changsha were headed by inbred related monogamous pairs. This result suggested that dispersal by primary reproductives is limited in the Changsha population as discussed above. Similarly, simple family colonies headed by inbred related monogamous pairs were found in studies of *R. grassei* in southwestern France [35], *Z. n. nevadensis* in North Carolina [40], and *C. formosanus* in Kyushu, Fukue and Hawaii [7], [11], [16]. However, simple family colonies in Chongqing-1 were headed by outbred unrelated monogamous pairs, which suggested that colonies reproduce by relatively long-range mating flights within this population. Similarly, simple family colonies headed by outbred unrelated monogamous pairs were reported in studies of other termite species, including *R. flavipes* in Massachusetts [21]and North Carolina [3], [13], [43], [44], [45], *R. virginicus* in North Carolina [44], and *C. formosanus* in Charleston [10]. Moreover, our results indicated that simple family colonies in Huanggang were a mixture of colonies whose workers were offspring of outbred and unrelated or inbred and related monogamous pairs. A previous study in one *R. grassei* population in southwestern France has also reported that simple family colonies were a mixture of colonies headed by outbred or inbred monogamous pairs [35]. The sample size of only two simple families in the Chongqing-2 population was too small to reject either of the simulated breeding systems with sufficient statistical support.

The strongly negative F_{IC} values for the extended families in Huanggang and Changsha suggest that the mean number of neotenics in our colonies was low (≤ 10), and that all neotenics were inbred and ultimately the descendants from monogamous pairs of founders. Similarly, extended family colonies were found to be headed by low numbers of inbred neotenics in populations of *R. flavipes* [3], [43], [44], [45], *R. hageni* [45], *R. virginicus* [45], *C. formosanus* in South Carolina, Louisiana, North Carolina, Hawaii, Guangdong and Hunan [10], ,*Z. n. nevadensis* [40], *Z. angusticollis* and *Z. n. nuttingi* [5]. Surprisingly, the extended family colonies in Chongqing-1 were inconsistent with all of the simulated breeding systems. Similar results have been found in studies of other termite species and were suggested to arise from colonies with multiple reproductives representing a mixture of several breeding systems [35],[40], [43]. Although classified as extended family colonies, the breeding structure estimates for extended family colonies in Chongqing-1 were most similar to values expected for simple family colonies headed by outbred monogamous pairs. One possibility is that these colonies only recently made the transition from simple to extended family by developing neotenics and thus contained still a measurable proportion of offspring of the previous (i.e. simple family) generation [40], [42]. Since colonies of the Chongqing-1 population were probably recently moved by human activities (see above), it is plausible to assume that disturbance might have increased incidental death of the colony founders resulting in recent development of replacement reproductives and thus the transition to extended family colonies [42]. Overall, the results suggested that the extended family colonies in Chongqing-1 contained small numbers of neotenics as shown for the populations of Huanggang and Changsha. The values of F_{IT}, F_{IC}, and *r* for extended families in Chongqing-2 were not significantly different from those for the simulated breeding system based on high numbers of neotenics (Table 2, Case B4), which is possibly due to the low sample size of extended family colonies in this population. Nevertheless, the strongly negative F_{IC} values (-0.16) for the extended families in Chongqing-2 suggested that the number of neotenics in these colonies was low (≤ 10). Low numbers of neotenics were also found in the extended family colonies of the other three populations (Huanggang, Changsha and Chongqing-1).

Mixed family colonies in Chongqing-2 displayed breeding structure estimates consistent with pleometrosis. Evidence for mixed family colonies originating from pleometrosis was previously found in the lower termite *Z. nevadensis* [40] and the higher termites *Nasutitermes corniger*and *M. michaelseni* [46], [47]. In pleometrosis, multiple queens and kings (alates) cooperate in colony foundation and expansion during the primary stages of colony development [18], [47]. However, pleometrosis has never been found in any of the other *Reticulitermes* species reported from America and Europe [1]. Some other factors have been suggested to explain mixed family colonies, including colony fusion, invasion of mature colonies by other alates, or sharing foraging galleries by neighboring colonies [40], [43]. For example, mixed family colonies were the result of colony fusions in native and introduced populations of R. flavipes[43], [48]. Since the percentage of mixed families was low in R. chinensis populations, larger sample sizes are needed to unequivocally determine the origin of mixed family colonies.

In addition to information about breeding systems in R. chinensis colonies derived from worker genotypes, we also directly observed number and types of reproductives in nine field colonies collected from Wuhan. Four of the dissected field colonies were simple families headed by a pair of primary kings and queens, and five colonies contained multiple neotenics. Three of the five colonies had low numbers of reproductives (6–12). The range is largely consistent with the results from the genetic analyses of worker genotypes from the populations of Huanggang, Changsha and Chongqing. The percentage of our collected field colonies containing secondary kings in R. chinensis was higher than those in R. speratus and R. virginicus [23], [24]. In R. speratus, secondary kings have been found in only 6.67 percent of colonies (2 of 30 colonies;[23]). Secondary kings were not found at all in five colonies of R. virginicus [24]. In contrast, each colony of R. chinensis in the present study had at least one secondary king and up to 51 secondary kings in colony 2. Almost all the neotenics in R. speratus and R. virginicusdifferentiated from nymphs [23], [24], but there were eight ergatoid neotenics in R. chinensis in this study. Our results based on the inferred genotypes of the neotenics' parents in each of the five colonies showed that the preceding parental generation in the five colonies also consisted of multiple reproductives. The breeding system of the five colonies headed by the neotenics' parents was in four cases an extended family colony headed by neotenics and in one case a mixed family colony. These results suggest that extended family colonies headed by neotenics and mixed family colonies can continue the colony life cycle by producing the next generation of neotenics in R. chinensis.

Our results showed that all the neotenics in the five field colonies of R. chinensis were sexually produced, suggesting that these colonies did not undergo conditional parthenogenesis. Based on the classical model of AQS [23], we did not expect the ergatoid neotenics to be produced parthenogenetically since they are derived from workers, which are almost exclusively sexually produced in other Reticulitermes species that undergo AQS [24]. We also did not necessarily expect the secondary kings to be parthenogens. Queens usually increase their reproductive output and fitness by producing multiple replacement daughters via parthenogenesis, which at least in R. speratus can continue the queen's full genetic contribution throughout generations by producing parthenogenetic daughters themselves. However, none of the nymphoid secondary queens in *R. chinensis* was homozygote at all loci.

In contrast to our findings in *R. chinensis*, the seven colonies of *R. speratus* in Kyoto [23] and the five colonies of *R. virginicus* in North Carolina and Texas [24] have been shown to undergo AQS. A previous phylogenetic analysis of the family Rhinotermitidae found that *R. speratus*and *R. virginicus* belonged to distinctly different clades and were thus not closely related, which prompted the hypothesis that AQS might have arisen early in the evolution of the genus*Reticulitermes*. However, the same study showed that *R. chinensis* is a close relative of *R. speratus* with both species sharing the same clade [31]. Nevertheless their reproductive mode differs since *R. chinensis* apparently does not undergo AQS. Therefore, AQS has either evolved independently in *R. speratus* and *R. virginicus* [24] or *R. chinensis* secondarily lost the ability to undergo AQS.

Clinal variation of ecological factors, such as temperature, moisture, seasonality, availability of wood and soil composition have been shown to influence breeding structure in *Reticulitermes*species [49], however, geographic parthenogenesis has not yet been studied in termites. In facultative parthenogenic plants and animals in general, asexual populations tend to occur at higher latitudes and altitudes and in disturbed habitats [50]. Previous studies in populations of the stick insect, *Clitarchus hookeri* showed that the incidence of parthenogenesis increased at higher latitudes both in the Northern and Southern Hemisphere [51]. The colonies of *R. chinensis* in our study were collected from lower latitudes than the colonies of *R. speratus* and*R. virginicus* (>35°N) that exhibited parthenogenesis, except for one *R. virginicus* colony from South Houston (<30°N) [23], [24]. Thus, it would be interesting to study whether *R. chinensis*colonies located further north and at higher altitudes than the ones in this study display AQS or if parthenogenesis simply does not occur in this species regardless of the local ecological conditions.

MATERIALS AND METHODS

Sampling for Population Genetic Study

Between May and July 2010, samples of *R. chinensis* were collected from fallen trees and stumps from 75 collection sites in three geographic regions of China (Figure 1a): Longwang Park, Huanggang city, Hubei Province (Figure 1c); Xuteli Park and Xingsha Park, Changsha city, Hunan Province (Figure 1d); and Nanshan Hill, Nanan District, Chongqing city (Figure 1e). All the parks and hills of sample collection were inside cities. The geographic distance was 300 km, 650 and 800 km between Huanggang and Changsha, between Changsha and Chongqing, and between Chongqing and Huanggang, respectively. The latitudes ranging from 28°14′20.1″ to 30°27′34.5″ and longitudes ranging from 106°36′53.8″ to 114°52′00.3″ of all sample sites were determined using a GPS localizer (Table S4). More than 30 workers per sample were collected in the field and immediately preserved in 95% ethanol. A few soldiers per sample were also collected in order to identify the species.

None of the parks and hills where samples were collected are privately-owned or protected in any way, and R. chinensis is not endangered or protected. Thus, no specific permissions were required to access and sample at these locations.

Microsatellite Genotyping

DNA was extracted from whole worker bodies using the DNeasy Tissue Kit (Beijing Dingguo Changsheng Biotech Co. Ltd). Twenty-five workers from each collection site were genotyped at 8 microsatellite loci (Table S5), identified from R. flavipes [52], [53]. PCR amplifications were performed as described in Vargo [52]. PCR products were separated by electrophoresis on 6% polyacrylamide gels run on a LI-COR 4300 DNA analyzer. Alleles were scored using the computer program SAGA[GT] (LI-COR, Inc.). The software Micro-Checker [54] was used to test loci for null alleles and possible scoring errors derived from large allele dropout and the presence of microsatellite stutter bands.

Colony Identification

Samples from the 25 collection sites in each region were tested for significant genotypic differentiation using log-likelihood G-Statistics (FSTAT, [55]). P values were obtained by permutations of the multilocus genotypes between each pair of samples and standard Bonferroni corrections were used. If samples significantly differed from each other, they were assigned to different colonies [11], [14].

Population Genetic Structure

Since colony members are close kin, only a single randomly chosen individual per colony was used for population genetic analyses to avoid bias by nonindependent genotypes. The number of genetic clusters in the three geographic regions was assessed on the basis of a comparison of the penalized log likelihoods over independent simulation runs using STRUCTURE 2.3.3 with variable numbers of assumed genetic clusters (K) [56]. All simulations were done using the default settings, i.e. the admixture model and correlated allele frequencies, with 100,000 runs in the data collection phase following a burn-in period of 100,000 runs [11]. Three repeats of simulations were run for each model with different numbers of assumed genetic clusters (K=1 to 7). The number of genetic clusters was determined following the Delta K method of Evanno *et al.* [56] using STRUCTURE HARVESTER v.0.6 [57]. According to the methods of Pritchard *et al.* [58], the estimated membership coefficients in K clusters (degree of admixture) of each multilocus genotype (i.e. individual representing a colony) were determined and were graphically displayed in form of a histogram. Colonies from the three geographical regions were probabilistically assigned to the populations for which they showed dominant membership (>80%). As long as considerable structure was found, datasets would be separated into subsets and reanalyzed until the number of subpopulations was determined. Confirmed subpopulations were treated as genetically differentiated populations. Pairwise genetic differentiation among populations at the 1/1000 nominal level was confirmed via permutation tests using multilocus G-statistics with Bonferroni corrections as implemented in FSTAT [55]to obtain the final number of populations in the data set.

To estimate isolation by distance, pairwise F_{ST} values for all pairs of colonies within each population were obtained. The matrix correlation between $F_{ST}/(1-F_{ST})$ and the ln of geographic distance (m) for all pairs of colonies within each population was determined [10]. The significance of the correlation coefficient was estimated using Mantel tests (one sided p-values from 10,000 permutations, TFPGA v. 1.3, [59]. In addition, genetic differentiation (F_{ST}) between each pair of populations was computed based on a hierarchical analysis using the program GENETIC DATA ANALYSIS version 1.1 (GDA, [60]). Exact tests for Hardy–Weinberg equilibrium for each locus and linkage disequilibrium between all pairs of loci were performed using the program GDA (GDA, [60]) with 3200 shufflings [7]. To determine whether populations had experienced a recent genetic bottleneck, worker genotypes were tested for heterozygosity excess across loci based on a Wilcoxon sign-rank test under three mutation models (infinite allele model, two-phased model of mutation, and stepwise mutation model) using the program Bottleneck v. 1.2.02 [37].

Colony Genetic Structure and Breeding System

Referring to the previous methods [20]–[21], [43], colonies were classified into one of three family types: simple, extended or mixed families. Colonies would be regarded as simple families if the numbers of genotypes of the workers were compatible with those expected for offspring of a monogamous pair and if the frequencies of the observed genotypes were not different from the expected Mendelian ratios. Colonies would be regarded as extended families if there were more than four genotypes, or if genotype frequencies were significantly different from expected Mendelian ratios ($P<0.05$, G-test). Mixed family colonies would be characterized by the presence of more than four alleles at one or more loci.

Colony genetic structure was analyzed by F-statistics using the methods of Weir & Cockerham[61] as implemented in FSTAT [55]. Separate analyses were performed for each population to assure that F values (F_{IT}, F_{CT} and F_{IC}) used to infer breeding system were not confounded by higher-level genetic structure [7], [20], [21]. The coefficient of relatedness (r) was determined by the average nest mate relatedness within all colonies of the respective population using the program FSTAT [55]. These empirical values were then compared to expected values resulting from previously compiled computer simulations of possible breeding systems of termites via one-sample t-tests not assuming equal variances ($P<0.05$), to infer predominant breeding systems in the populations of this study [20], [21], [35].

Inferring Reproductive Mode of Neotenics' Parents

To infer reproductive mode of neotenics' parents, that is, parthenogenesis versus sexual reproduction in *R. chinensis*, we used the methods of Matsuura *et al.* [23] and Vargo *et al.* [24]to collect field colonies and conduct microsatellite analysis. Between July 2010 and September 2011, we collected four field colonies with primary reproductives (Figure S3a) and five field colonies with neotenics (Figure S3b) in Wuhan city (Figure 1a). All reproductives from each colony were immediately preserved in 95% ethanol in a vial for subsequent microsatellite analysis. We discriminated secondary neotenic reproductives from primary reproductives based on the lighter body color and absence of eyes in neotenics. Sex of all reproductives was identified based on the conformation of the caudal sternites under a stereoscope [62]. According to the presence or absence of wing pads, neotenics would be divided into nymphoid and ergatoid neotenics [23].

We genotyped up to 20 male and female neotenics (nymphoid and ergatoid) from five field colonies collected at Shizi Hill (colony 1, 2 and 3), Houshan Hill (colony 4), and Yujia Hill (colony 5) in Wuhan city using the eight microsatellite loci and protocols described above (Table S3). The latitudes ranging from 28°28′49.86″ to 30°32′15.3″ and longitudes ranging from 114°20′58.38″ to 114°24′9.66″ of the five field colonies were also determined using a GPS localizer (Figure 1b). If neotenics were homozygous at all loci, they would be considered to be produced by parthenogenesis. If neotenics were heterozygous at one or more loci, they would be considered to be sexually produced [23], [24]. For sexually produced neotenics, we tested whether these neotenics are offspring of simple, extended or mixed family colonies using the same methods as described for workers above [20]–[21], [43].

ACKNOWLEDGMENTS

We are thankful to Xintian Cai and Bing Zhou for their help in field collection. We thank the technical assistance for genotyping from Double Company. We thank Drs. Xianchun Li, Lane D. Foil and the anonymous reviewers for providing valuable comments on earlier drafts of this manuscript.

AUTHOR CONTRIBUTIONS

Conceived and designed the experiments: QH GL CH CL. Performed the experiments: QH GL. Analyzed the data: QH GL CH CL. Contributed reagents/materials/analysis tools: QH GL CH CL. Wrote the paper: QH GL CH CL. Designed the software used in analysis: QH CH GL CL.

REFERENCES

1.   Vargo EL, Husseneder C (2009) Biology of subterranean termites: insights from molecular studies of *Reticulitermes* and *Coptotermes*. Annu Rev Entomol 54: 379–403. doi: 10.1146/annurev.ento.54.110807.090443

2.   Vargo EL, Husseneder C (2011) Genetic structure of termite colonies and populations. In *Biology of Termites: A Modern Sythesis* (eds D. E. Bignell, Y. Roisin & N. Lo), 321–348. Dordrecht, The Netherlands: Springer Science+Business Media B.V.

3.   Vargo EL (2003) Hierarchical analysis of colony and population genetic structure of the eastern subterranean termite, *Reticulitermes flavipes*, using two classes of molecular markers. Evolution 57: 2805–2818. doi: 10.1111/j.0014-3820.2003.tb01522.x

4.   Husseneder C, Messenger MT, Su NY, Grace JK, Vargo EL (2005) Colony social organization and population genetic structure of an introduced population of formosan subterranean termite from New Orleans, Louisiana. J Econ Entomol 98: 1421–1434. doi: 10.1603/0022-0493-98.5.1421

5.   Booth W, Brent CS, Calleri DV, Rosengaus RB, Traniello JFA, et al. (2012) Population genetic structure and colony breeding system in dampwood termites (*Zootermopsis angusticollis* and *Z. nevadensis nuttingi*). Insectes Soc 59: 127–137. doi: 10.1007/s00040-011-0198-2

6.   Goodisman MAD, Crozier RH (2002) Population and colony genetic structure of the primitive termite *Mastotermes darwiniensis*. Evolution 56: 70–83. doi: 10.1554/0014-3820(2002)056[0070:pacgso]2.0.co;2

7. Vargo EL, Husseneder C, Grace JK (2003) Colony and population genetic structure of the Formosan subterranean termite, *Coptotermes formosanus*, in Japan. Mol Ecol 12: 2599–2608. doi: 10.1046/j.1365-294x.2003.01938.x

8. Leniaud L, Dedeine F, Pichon A, Dupont S, Bagnères AG (2010) Geographical distribution, genetic diversity and social organization of a new European termite,*Reticulitermes urbis* (Isoptera: Rhinotermitidae). Biol Invasions 12: 1389–1402. doi: 10.1007/s10530-009-9555-8

9. Dronnet S, Chapuisat M, Vargo EL, Lohou C, Bagnères AG (2005) Genetic analysis of the breeding system of an invasive subterranean termite, *Reticulitermes santonensis*, in urban and natural habitats. Mol Ecol 14: 1311–1320. doi: 10.1111/j.1365-294x.2005.02508.x

10. Vargo EL, Husseneder C, Woodson D, Waldvogel MG, Grace JK (2006) Genetic analysis of colony and population structure of three introduced populations of the Formosan subterranean termite (Isoptera: Rhinotermitidae) in the continental United States. Environ Entomol 35: 151–166. doi: 10.1603/0046-225x-35.1.151

11. Husseneder C, Simms DM, Delatte JR, Wang C, Grace JK, et al. (2012) Genetic diversity and colony breeding structure in native and introduced ranges of the Formosan subterranean termite, *Coptotermes formosanus*. Biol Invasions 14: 419–437. doi: 10.1007/s10530-011-0087-7

12. Copren KA (2004) The genetic and social structure of the western subterranean termite,*Reticulitermes hesperus*(Isoptera: Rhinotermitidae). PhD Thesis, University of California, Davis, CA.

13. Vargo EL, Carlson JC (2006) Comparative study of breeding systems of sympatric subterranean termites (*Reticulitermes flavipes* and *R.hageni*) in central North Carolina using two classes of molecular genetic markers. Environ Entomol 35: 173–187. doi: 10.1603/0046-225x-35.1.173

14. Husseneder C, Simms DM, Riegel C (2007) Evaluation of treatment success and patterns of reinfestation of the Formosan subterranean termite (Isoptera: Rhinotermitidae). J Econ Entomol 100: 1370–1380. doi: 10.1603/0022-0493(2007)100[1370:eotsap]2.0.co;2

15. Thompson GJ, Lenz M, Crozier RH, Crespi BJ (2007) Molecular-genetic analyses of dispersal and breeding behaviour in the Australian termite *Coptotermes lacteus*: evidence for non-random mating in a swarm-dispersal mating system. Aust J Zool 55: 219–227. doi: 10.1071/zo07023

16. Husseneder C, Powell JE, Grace JK, Vargo EL, Matsuura K (2008) Worker size in the Formosan subterranean termite in relation to colony

breeding structure as inferred from molecular markers. Environ Entomol 37: 400–408. doi: 10.1603/0046-225x(2008)37[400:wsitfs]2.0.co;2

17. DeHeer CJ, Vargo EL (2008) Strong mitochondrial DNA similarity but low relatedness at microsatellite loci among families within fused colonies of the termite *Reticulitermes flavipes*. Insectes Soc 55: 190–199. doi: 10.1007/s00040-008-0999-0

18. Atkinson L, Adams ES (1997) The origins and relatedness of multiple reproductives in colonies of the termite *Nasutitermes corniger*. Proc R Soc Lond B 264: 1131–1136. doi: 10.1098/rspb.1997.0156

19. Husseneder C, Brandl R, Epplen C, Epplen JT, Kaib M (1998) Variation between and within colonies in the termite: morphology, genomic DNA, and behaviour. Mol Ecol 7: 983–990. doi: 10.1046/j.1365-294x.1998.00416.x

20. Thorne BL, Traniello JFA, Adams ES, Bulmer M (1999) Reproductive dynamics and colony structure of subterranean termites of the genus *Reticulitermes* (Isoptera Rhinotermitidae): a review of the evidence from behavioral, ecological, and genetic studies. Ethol Ecol Evol 11: 149–169. doi: 10.1080/08927014.1999.9522833

21. Bulmer M S, Adams ES, Traniello JFA (2001) Variation in colony structure in the subterranean termite *Reticulitermes flavipes*. Behav Ecol Sociobiol 49: 236–243. doi: 10.1007/s002650000304

22. Ross KG (2001) Molecular ecology of social behaviour: analyses of breeding systems and genetic structure. Mol Ecol 10: 265–284. doi: 10.1046/j.1365-294x.2001.01191.x

23. Matsuura K, Vargo EL, Kawatsu K, Labadie PE, Nakano H, et al. (2009) Queen Succession Through Asexual Reproduction in Termites. Science 323: 1687. doi: 10.1126/science.1169702

24. Vargo EL, Labadie PE, Matsuura K (2012) Asexual queen succession in the subterranean termite *Reticulitermes virginicus*. Proc R Soc Lond B 279: 813–819. doi: 10.1098/rspb.2011.1030

25. Matsuura K (2011) Sexual and asexual reproduction in termites. In *Biology of Termites: A Modern Sythesis* (eds D. E. Bignell, Y. Roisin & N. Lo), 255–277. Dordrecht, The Netherlands: Springer Science+Business Media B.V.

26. Hayashi Y, Kitade O, Gonda M, Kondo T, Miyata H, et al. (2005) Diverse colony genetic structures in the Japanese subterranean termite *Reticulitermes speratus*(Isoptera: Rhinotermitidae). Sociobiology 46: 175–184. doi: 10.13102/sociobiology.v60i4.446-452

27. Clément JL, Bagnères AG, Uva P, Wilfert L, Quintana A, et al. (2001) Biosystematics of *Reticulitermes* termites in Europe: morphological, chemical and molecular data. Insectes Soc 48: 202–215. doi: 10.1007/pl00001768

28. Marini M, Mantovani B (2002) Molecular relationships among European samples of*Reticulitermes* (Isoptera, Rhinotermitidae). Mol Phylogenet Evol 22: 454–459. doi: 10.1006/mpev.2001.1068

29. Kutnik M, Uva P, Brinkworth L, Bagnères AG (2004) Phylogeography of two European*Reticulitermes* (Isoptera) species: the Iberian refugium. Mol Ecol 13: 3099–3113. doi: 10.1111/j.1365-294x.2004.02294.x

30. Luchetti A, Marini M, Mantovani B (2005) Mitochondrial evolutionary rate and speciation in termites: data on European Reticulitermes taxa (Isoptera, Rhinotermitidae). Insectes Soc 52: 218–221. doi: 10.1007/s00040-005-0806-5

31. Austin JW, Szalanski AL, Cabrera BJ (2004) Phylogenetic analysis of the subterranean termite family Rhinotermitidae (Isoptera) by using the mitochondrial cytochrome oxidase II gene. Ann Entomol Soc Am 97: 548–555. doi: 10.1603/0013-8746(2004)097[0548:paotst]2.0.co;2

32. Huang FS, Zhu SM, Ping ZM, He XS, Li GX, et al.. (2000) Fauna Sinica: Insecta, Vol. 17: Isoptera. Beijing: Science Press.

33. Pan YZ, Liu YZ, Tang GQ (1990) The establisment and development of colonies in*Reticulitermes chinensis* Snyder. Acta Entomol Sin 33: 200–206.

34. Liu YZ, Tang TY (1994) The colony development and growth by substitute reproductives of *Reticulitermes chinensis*.. Acta Entomol Sin 37: 38–43.

35. DeHeer CJ, Kutnik M, Vargo EL, Bagnères AG (2005) The breeding system and population structure of the termite *Reticulitermes grassei* in Southwestern France. Heredity 95: 408–415. doi: 10.1038/sj.hdy.6800744

36. Yang WT, Zhang JL (2011) Analysis Landscape of Plants in Nanshan Botanical Park. South China Agriculture 5: 1–5.

37. Piry S, Luikart G, Cornuet JM (1999) BOTTLENECK: a computer program for detecting recent reductions in the effective size using allele frequency data. J Hered 90: 502–503. doi: 10.1093/jhered/90.4.502

38. Peery ZM, Kirby R, Reid BN, Stoelting R, Doucet-Bëer E, Robinson S, et al. (2012) Reliability of genetic bottleneck tests for detecting recent population declines. Mol Ecol 21 3403–3418. doi: 10.1111/j.1365-294x.2012.05635.x

39. DeHeer CJ, Kamble ST (2008) Colony genetic organization, fusion and inbreeding in*Reticulitermes flavipes* from the Midwestern U.S. Sociobiology. 51: 307–325.

40. Aldrich BT, Kambhampati S (2007) Population structure and colony composition of two*Zootermopsis nevadensis* subspecies. Heredity 99: 443–451. doi: 10.1038/sj.hdy.6801022

41. Nobre T, Nunes L, Bignell DE (2008) Colony interactions in *Reticulitermes grassei*population assessed by molecular genetic methods. Insectes Soc 55: 66–73. doi: 10.1007/s00040-007-0971-4

42. Aluko GA, Husseneder C (2007) Colony dynamics of the Formosan subterranean termite in a frequently disturbed urban landscape. J Econ Entomol 100: 1037–1046. doi: 10.1603/0022-0493(2007)100[1037:cdo tfs]2.0.co;2

43. DeHeer CJ, Vargo EL (2004) Colony genetic organization and colony fusion in the termite *Reticulitermes flavipes* as revealed by foraging patterns over time and space. Mol Ecol 13: 431–441. doi: 10.1046/j.1365-294x.2003.2065.x

44. Vargo EL (2003) Genetic structure of *Reticulitermes flavipes* and *R. virginicus*(Isoptera, Rhinotermitidae) colonies in an urban habitat and tracking of colonies following treatment with hexaflumuron bait. Environ Entomol 32: 1271–1282. doi: 10.1603/0046-225x-32.5.1271

45. Parman V, Vargo EL (2008) Population density, species abundance, and breeding structure of subterranean termite colonies in and around infested houses in central North Carolina. J Econ Entomol 101: 1349–1359. doi: 10.1603/0022-0493(2008)101[1349:pdsaab]2.0.co;2

46. Thorne BL (1984) Polygyny in the Neotropical termite *Nasutitermes corniger*: life history consequences of queen mutualism. Behav Ecol Sociobiol 14: 117–136. doi: 10.1007/bf00291903

47. Hacker M, Kaib M, Bagine R, Epplen J, Brandl R (2005) Unrelated queens coexist in colonies of the termite *Macrotermes michaelseni*. Mol Ecol 14: 1527–1532. doi: 10.1111/j.1365-294x.2005.02507.x

48. Perdereau E, Bagnères AG, Dupont S, Dedeine F (2010) High occurrence of colony fusion in a European population of the American termite *Reticulitermes flavipes*. Insect Soc 57: 393–402. doi: 10.1007/s00040-010-0096-z

49. Vargo EL, Leniaud L, Swoboda LE, Diamond SE, Michael D, et al.. (2013) Clinal variation in colony breeding structure and level of inbreeding in the subterranean termites *Reticulitermes flavipes* and *R. grassei*. Mol Ecol: doi: 10.1111/mec.12166.

50. Glesener RR, Tilman D (1985) Sexuality and the components of environmental uncertainty: Clues from geographic parthenogenesis is terestrial animals. Am Nat 112: 659–673. doi: 10.1086/283308

51. Morgan-Richards M, Trewick SA, Stringer IAN (2010) Geographic parthenogenesis and the common tea-tree stick insect of New Zealand. Mol Ecol 19: 1227–1238. doi: 10.1111/j.1365-294x.2010.04542.x

52. Vargo EL (2000) Polymorphism at trinucleotide microsatellite loci in the subterranean termite *Reticulitermes flavipes*. Mol Ecol 9: 817–820. doi: 10.1046/j.1365-294x.2000.00915.x

53. Dronnet S, Bagnères AG, Juba TR, Vargo EL (2004) Polymorphic microsatellite loci in the European subterranean termite, *Reticulitermes santonensis* Feytaud. Mol Ecol Notes 4: 127–129. doi: 10.1111/j.1471-8286.2004.00600.x

54. Van Oosterhout C, Hutchingson WF, Wills DPM, Shipley P (2004) Micro-Checker: software for identifying and correcting microsatellite data. Mol Ecol Notes 4: 535–538. doi: 10.1111/j.1471-8286.2004.00684.x

55. Goudet J (2001) FSTAT, a program to estimate and test gene diversities and fixation indices (version 2.9.3). Unil website. Available:http://www.unil.ch/izea/softwares/fstat.html. Accessed 2012 Dec 23.

56. Evanno G, Regnaut S, Goudet J (2005) Detecting the number of clusters of individuals using the software structure: a simulation study. Mol Ecol 14: 2611–2620. doi: 10.1111/j.1365-294x.2005.02553.x

57. Earl DA, vonHoldt BM (2012) STRUCTURE HARVESTER: a website and program for visualizing STRUCTURE output and implementing the Evanno method. Conservation Genetics Resources 4: 359–361. doi: 10.1007/s12686-011-9548-7

58. Pritchard JK, Stephens M, Donnelly P (2000) Inference of population structure using multilocus genotype data. Genetics 155: 945–959.

59. Miller MP (1997) Tools for population genetic analyses (TFPGA) 1.3: A Windows program for the analysis of allozyme and molecular population genetic data. *February 2000.* Usu website. Available: http://bioweb.usu.edu/mpmbio. Accessed 2012 Nov 18.

60. Lewis PO, Zaykin D (2000) Genetic Data Analysis: Computer program for the analysis of allelic data, Version 1.0 (d12). Uconn website. Available:http://lewis.eeb.uconn.edu/lewishome/software.html. Accessed 2012 Jun 16.

61. Weir B, Cockerham CC (1984) Estimating F-statistics for the analysis of population structure. Evolution 38: 1358–1370. doi: 10.2307/2408641

62. Roonwal ML (1975) Sex ratio and sexual dimorphism in termites. J Sci Ind Res india 34: 402–416.

Chapter 8

GENETIC ANALYSIS OF COLD TOLERANCE AT THE GERMINATION AND BOOTING STAGES IN RICE BY ASSOCIATION MAPPING

Yinghua Pan[1,3], Hongliang Zhang[1], Dongling Zhang[2], Jinjie Li[1], Haiyan Xiong[1], Jianping Yu[1], Jilong Li[1], Muhammad Abdul Rehman Rashid[1], Gangling Li[1], Xiaoding Ma[2], Guilan Cao[2], Longzhi Han[2], Zichao Li[1]

[1]Key Laboratory of Crop Heterosis and Utilization, Ministry of Education, Beijing Key Laboratory of Crop Genetic Improvement, China Agricultural University, Beijing, 100193, China

[2]Institute of Crop Science, Chinese Academy of Agricultural Sciences, the National Key Facility for Crop Gene Resources and Genetic Improvement, Key Laboratory of Crop Germplasm Resources and Biotechnology, Ministry of Agriculture, Beijing, 100081, China

[3]Rice Research Institute, Guangxi Academy of Agricultural Sciences, Nanning, Guangxi, 530005, China

ABSTRACT

Low temperature affects the rice plants at all stages of growth. It can cause severe seedling injury and male sterility resulting in severe yield losses. Using a mini core collection of 174 Chinese rice accessions and 273 SSR markers we investigated cold tolerance at the germination and booting stages, as well as the underlying genetic bases, by association mapping. Two distinct populations, corresponding to subspecies *indica* and *japonica* showed evident differences in cold tolerance and its genetic basis. Both subspecies were sensitive to cold stress at both growth stages. However, *japonica* was more tolerant than *indica* at all stages as measured by seedling survival and seed setting. There was a low correlation in cold tolerance between the germination and booting stages. Fifty one quantitative trait loci (QTLs) for cold tolerance were dispersed across all 12 chromosomes; 22 detected at the germination stage and 33 at the booting stage. Eight QTLs were identified by at least two of four measures. About 46% of the QTLs represented new loci. The only QTL shared between *indica* and *japonica* for the same measure was *qLTSSvR6-2* for SSvR. This implied a complicated mechanism of old tolerance between the two subspecies. According to the relative genotypic effect (RGE) of each genotype for each QTL, we detected 18 positive genotypes and 21 negative genotypes in *indica*, and 19 positive genotypes and 24 negative genotypes in *japonica*. In general, the negative effects were much stronger than the positive effects in both subspecies. Markers for QTL with positive effects in one subspecies were shown to be effective for selection of cold tolerance in that subspecies, but not in the other subspecies. QTL with strong negative effects on cold tolerance should be avoided during MAS breeding so as to not cancel the effect of favorable QTL at other loci.

INTRODUCTION

Low temperature is one of the major abiotic stresses that threaten the adaptability of rice and its production. Due to the long growing season and frequent low temperature in North China, Korea, Japan and other countries at high latitudes, the growth duration of cultivars that cannot tolerate low temperature has to be shortened and this usually results in low yields. Cold stresses that often occur during flowering and grain filling all over the world [1–2] may also lower grain quality and yield. Cold tolerance of rice is a complex quantitative trait significantly influenced by environment. Thus research on the underlying genetic mechanisms and discovery of genes affecting to cold tolerance in rice will be helpful for developing elite cultivars with strong cold tolerance.

Segregating populations are usually used for mapping and cloning genes for cold tolerance. More than 30 quantitative trait loci (QTLs) for cold tolerance have been mapped on 12 chromosomes (according to www.gramene. com). Most of them were for cold tolerance at the germination stage, and a few were cloned. Using backcross inbred lines of Livorno/Haymasari, the QTL *qLTG3–1* was cloned and proven to contribute to the high germinability under low temperatures [3]. Mapping QTL for cold tolerance at the booting stage is more difficult because of difficulties in phenotyping and the underlying complex biological and genetic mechanisms involved. Eight QTLs related to spikelet fertility under cold stress were mapped on chromosomes 1, 4, 5, 7, 10 and 11 based on NILs derived from KMXBG as donor and Towada as the recipient [4–5]. Saito et al. mapped the QTL *Ctb1* encoding an F-box protein and contributing to normal anther development under cold stress [6].

In contrast to traditional QTL mapping, "association mapping" is an alternative approach for studying complex traits. Association mapping, also known as linkage disequilibrium mapping, utilizes allelic variation in natural populations, and is capable of simultaneously identifying many loci for multiple traits [7]. Association mapping has been employed in rice to identify QTLs or genes related to yield [8–10], heading date [11], disease resistance [12–13] and tolerance to abiotic stresses [14].

Using 174 rice accessions from the mini core collection of cultivated rice and 273 simple-sequence repeat (SSR) markers we detected QTLs associated with cold tolerance during the germination and booting stages. We subsequently screened the accessions with strong cold tolerance and QTL markers showing strong effects, which will be the potential parents and markers in developing cultivars with cold tolerance by molecular marker assisted selection.

MATERIALS AND METHODS

Plant Materials

The research material comprised 174 diverse accessions from the mini core collection of cultivated rice [15]. Among them, 118 were landraces, 56 were modern improved cultivars and 109 were subspecies *indica*, and 65 were *japonica*.

Evaluation of Cold Tolerance at the Germination Stage

Fifty seeds of each accession were placed in a drying oven at 50°C for 48 h to break dormancy, and then soaked in distilled water in a 15 cm Petri dish for 24 h. The pre-soaked seeds were germinated in a growth chamber at a constant temperature of 32°C for 36 h. Germinated seeds with 5 mm coleoptiles were stressed at 5°C for 10 days, and then moved to a greenhouse at 20°C for 10 days to allow seedlings to recover and resume normal growth. Seedling survival rates (SSvR) were then scored as a measure of cold tolerance at the germination stage.

Evaluation of Cold Tolerance at the Booting Stage under Naturally Low Temperature Conditions

One measure of cold tolerance at the booting stage is the seed setting rate under naturally low temperature conditions (SStR-NL). Suitable conditions are available at an experimental farm of Yunnan Academy of Agricultural Sciences, Kunming, where atmospheric temperatures are 15–19°C during the booting stage extending from June to August [16]. Seedlings of each cultivar were grown in a nursery seedbed planted 31 March 2006, and transplanted on 9 May with two replications. Twenty plants of each accession were transplanted in a single row with a row × plant distance of 25 × 15 cm. The field was fertilized with N (120 kg/ha^2) and P$_2$O$_5$ (80 kg/ha^2). SStR-NL was scored as the mean seed setting rate of 10 plants from the middle 18 plants in each row.

Evaluation of Cold Tolerance at the Booting Stage under Cold Water Stress

The other two measures of cold tolerance at the booting stage were seed setting rate under cold water (SStR-CW) and relative seed setting rate (RSStR-CW, i.e. ratio of seed setting rate under cold water stress to seed setting rate under normal conditions). Cold tolerance under cold water stress was evaluated in Gongzhuling, Jilin province. Seedlings were grown in a nursery seedbed after planting on 15 April 2006, and transplanted on 25 May with two replications. Twelve plants of each accession were transplanted in a single row with a line × plant distance 25 × 15 cm. The field was fertilized with N (120 kg/ha^2) and P$_2$O$_5$ (80 kg/ha^2). On 1 July at panicle initiation, the field was irrigated continuously for 40 days with cold water pumped from a deep well where the water temperature is a constant 19°C. The cold water was maintained at 20 cm depth at the early stage to 30 cm at the late growth stage, such that the young panicles were covered by cold water. Late accessions that could not be harvested from late August to mid-October were enclosed by a plastic greenhouse to avoid the effects of low atmospheric temperatures on grain filling. Another two replications were transplanted and managed under the same conditions, except that they were irrigated with surface water at higher temperatures (daily mean temperature of 22–26°C from 1 July to 10 August). Mean seed setting rates for all accessions under both cold and normal conditions were recorded on the middle 10 plants in each row.

SSR Polymorphism and Genotyping

DNA was extracted from the young leaves using the CTAB method. Two hundred and seventy three simple sequence repeat (SSR) markers randomly distributed across all 12 rice chromosomes were selected for genotyping. The average distance between markers was 835 kb. The minimum distance between adjacent markers was 0.06 kb and the maximum was 4,050 kb; 18% of distances between adjacent markers were less than 150 kb, more than 50% were less than 750 kb, and more than 90% were less than 1,800 kb. PCR conditions, gel electrophoresis of PCR products and genotype scoring methods were as described in Zhang et al. [15].

Statistical Analysis and Association Mapping

STRUCTURE 2.2 [17], a model-based program, was utilized to infer population structure with 60 unlinked SSRs from the 273 markers. Ten independent simulations were run for each K (the number of clusters, from 1 to 10) using a burn-in length of 10,000, run length of 100,000 and a model for admixture and correlated allelic frequencies. To determine the K value, the LnP(D) value in the STRUCTURE output and Evanno's K between successive K [18] were used.

Pair-wise linkage disequilibria between markers were evaluated by TASSEL V4.0 (http://www.maizegenetics.net/tassel) in a total of 174 accessions and in each of two clusters inferred by STRUCTURE (here, the clusters were subspecies, *indica* and *japonica*).

Association analyses between SSR markers and measures of cold tolerance were carried out by TASSEL V4.0. The Q-Q plot of GLM and MLM for four measures of cold tolerance indicated that MLM overestimated the effect of genetic relationship among individuals. The GLM model was therefore used to analyze the association between SSR markers and four measures of cold tolerance for *indica* and *japonica* separately. To control false discovery in association analyses, the significance level α was adjusted upward by α' in a rate $(1-α)^R$ for R-rejected hypotheses. Here, we set α = 0.05, then α' = 0.05(1–0.05) ^{R}. We denoted markers with p-values lower than α' as significantly associated ones.

To provide information about the markers and germplasm resources that may be used in breeding cultivars with strong cold tolerance, we investigated the genotypic effect of each genotype for each QTL. Due to the distinct difference in cold tolerance between the two subspecies, the relative genotypic effect for the ith genotype in the sth subspecies for the corresponding indicator was denoted as $RGE_{si} = (\bar{x}_{si} - \bar{x}_s)/\bar{x}_s$, where, s represents subspecies (*indica* or *japonica*), $\bar{x}_{si}$ is the average phenotype of ith genotype in the sth subspecies, and $\bar{x}_s$ is the average phenotype of the sth subspecies. Error MS for each marker (MS_E) estimated during association analysis was used to estimate the error MS of RGE as $MS_{RGEsi} = (MS_E/n_i)/\bar{x}_s^2$. Given that MS_E was estimated from a large sample, the z-test was used to test whether the RGE of each genotype was biased from zero. We denoted genotypes significantly larger than zero as positive or cold-tolerant genotypes, and those smaller than zero as negative or cold-sensitive genotypes.

RESULTS

Population Structure and Linkage Disequilibrium

When we ran the STRUCTURE simulation using 60 SSRs, the LnP(D) value increased with K from 1 to 10, but showed an evident knee and there was a sharp peak of Evanno's ΔK at $K = 2$ (Fig. 1). These results indicated that there were two distinctly divergent populations, corresponding to subspecies *indica* and *japonica*. To survey the influence of population structure on LD, we analyzed LD in the whole population and in each of the subgroups (i.e.*indica* and *japonica*) identified by structure analysis (Fig. 2). At the whole population level, the r^2 within the whole genome was 0.0624±0.0865. For SSR markers with physical distances less than 50 kb, the r^2 was 0.1443±0.2149; for those between 50 kb and 150 kb, $r^2$was 0.1332±0.1829; for those between 150 kb and 500 kb, r^2 was 0.0926±0.1353; for those between 500 kb and 1000 kb, r^2 was 0.0624±0.0765; and for those more than 1,000 kb, r^2 was 0.0605±0.0818. Thus LD for markers with physical distances shorter than 50 kb was not obviously different from those between 50 kb and 150 kb, but decreased dramatically for those further than 150 kb. Compared to the whole population, LDs within *indica* and *japonica* were obviously lower (Fig. 2). For SSR markers with physical distances shorter than 50 kb, mean r^2 were 0.1388±0.2135 and 0.1810±0.3203 for *indica* and *japonica*, respectively; for markers between 50 kb and 150 kb, 0.0761±0.0796 and 0.0760±0.0368; for markers between 150 kb and 500 kb, 0.0331±0.0573 and 0.0392±0.0616; for markers between 500 kb and 1000 kb, 0.0155±0.0148 and 0.0336±0.0505; and for markers more than 1000 kb, 0.0134±0.0165 and 0.0201±0.0261. These results indicated that LDs between close markers (such as physical distances less than 50 kb) in both *indica* and *japonica* did not decrease relative to the whole population, however, the LDs between distant markers especially those more than 150 kb decreased dramatically. Given the above results, we carried out the association analysis within *indica* and *japonica* independently in order to avoid false positive associations; and given that few non-linked markers showed strong LD, we used the GLM model that controls population structure but not kinship for the association analysis within*indica* and *japonica*.

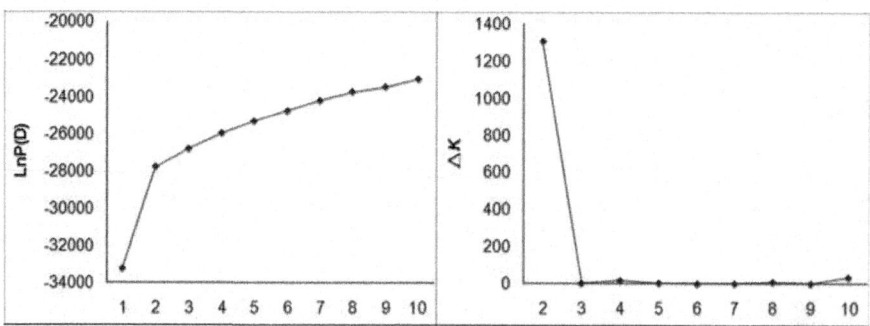

Figure 1: Average LnP(D) and Δ*K* **over 10 repeats of STRUCTURE simulation.**

doi:10.1371/journal.pone.0120590.g001

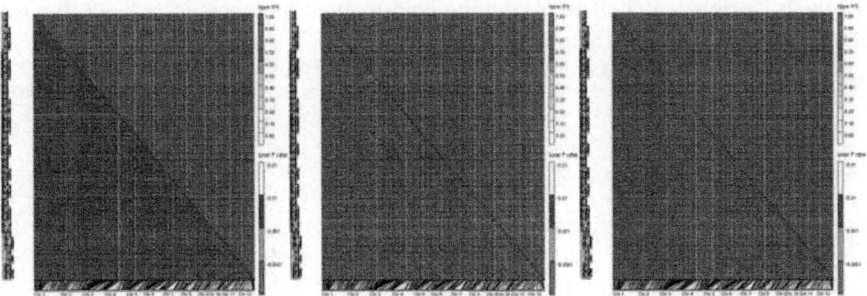

Figure 2: Distribution of LD across 273 SSR loci on 12 linkage groups in the total population (left), *indica* **(center) and** *japonica* **(right).**

doi:10.1371/journal.pone.0120590.g002

Cold Tolerance at the Germination and Booting Stages

All four measures of cold tolerance showed high variation among accessions at both the germination and booting stages, particularly the latter (Table 1). *Japonica* had apparently strong cold tolerance at the germination stage as measured by seedling survival rate (SSvR = 79.11%) under low temperature; this was more than three-fold stronger than *indica* (23.39%). Both *indica* and *japonica* were sensitive to cold stress at the booting stage, as measured by seed setting rate under natural low temperature conditions in Kunming(SStR-NL), seed setting rate under cold water irrigation in Gongzhuling (SStR-CW), and relative seed setting rate under cold water irrigation in Gongzhuling (RSStR-CW). However, *japonica* was more tolerant to cold stress (SStR-NL and RSStR-CW were higher than 25%) than *indica* (SStR-NL and RSStR-CW were lower than 25%). Correlation analyses among the four measures revealed low correlations between cold tolerance at the germination stage and at the booting stage, and low correlations between cold tolerances in different environments (Table 2). Higher correlations in cold tolerance among different stages and environments occurred in the case of *japonica*.

Table 1: Phenotypic variation in four cold-tolerant measures in *indica* and *japonica*

Measure	Indica				Japonica			
	Mean (%)	Range	SD (%)	CV %	Mean (%)	Range	SD (%)	CV %
SSvR	23.39	1.11–100.00	18.03	77.06	79.11**	1.11–100.00	28.67	36.23
SStR-NL	18.86	0–74.42	21.33	113.01	29.89**	0–80.43	26.43	88.35
SStR-CW	14.71	0–92.49	21.24	144.39	16.87	0–64.31	19.54	115.82
RSStR-CW	14.57	0–89.14	20.53	140.92	25.79*	0–90.47	28.84	100.16

* Significant at P = 0.05;
**: significant at P = 0.01;
SSvR, survival at low temperature expressed as percentage (%) of surviving plants;
SStR-NL, percentage (%) of filled grains per panicle under natural low temperature conditions in Kunming;
SStR-CW, percentage (%) of filled grains per panicle under cold water irrigation in Gongzhuling;
RSStR-CW, percentage (%) relative seed setting under cold water irrigation in Gongzhuling.

doi:10.1371/journal.pone.0120590.t001

doi:10.1371/journal.pone.0120590.t001

Table 2: Correlation coefficients among cold-tolerant measures in *indica* (above the diagonal) and *japonica* (below the diagonal)

Measure	SSvR	SStR-NL	SStR-CW	RSStR-CW
SSvR	1.00	0.16	0.11	0.03
SStR-NL	0.16	1.00	0.00	0.17
SStR-CW	0.17	0.23	1.00	0.98 **
RSStR-CW	0.27	0.25	0.91**	1.00

**: Significant at P = 0.01.

doi:10.1371/journal.pone.0120590.t002

doi:10.1371/journal.pone.0120590.t002

QTLs Associated With Cold Tolerance at the Germination and Booting Stages

Fifty one QTLs related to cold tolerance at the germination and booting stages were detected. They were distributed on all 12 chromosomes, with the maximum number on chromosome 1 and only one on chromosome 9 (Fig. 3). Among them, 22 were detected at the germination stage, 33 at booting, and eight were identified by at least two measures. The genetic contribution of each QTL was 29.20% on average, with a minimum 5.25% (*qLTSSvR3–1* on chromosome 3) and the maximum 59.28% (*qLTSSvR7–1* on chromosome 7). Of the reported QTLs related to cold tolerance in rice (www.gramene.com) more than 54% (21 of 39 QTLs) were detected in the current study (Fig. 3). We identified four loci associated with cold tolerance at the booting stage that were also detected by Cui *et al.* by association mapping [19].

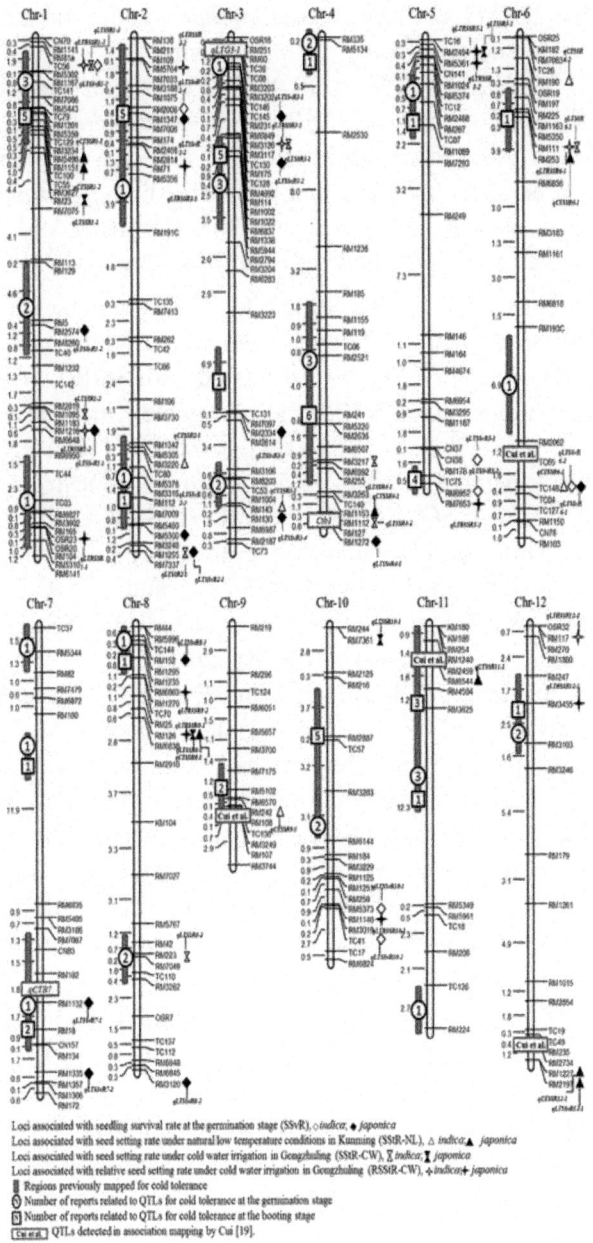

Loci associated with seedling survival rate at the germination stage (SSvR), ○ indica; ♦ japonica
Loci associated with seed setting rate under natural low temperature conditions in Kunming (SStR-NL), △ indica; ▲ japonica
Loci associated with seed setting rate under cold water irrigation in Gongzhuling (SStR-CW), ∑ indica; ∑ japonica
Loci associated with relative seed setting rate under cold water irrigation in Gongzhuling (RSStR-CW), ✦ indica; ✦ japonica
▌ Regions previously mapped for cold tolerance
⑤ Number of reports related to QTLs for cold tolerance at the germination stage
⑤ Number of reports related to QTLs for cold tolerance at the booting stage
▭ Cui et al ▭ QTLs detected in association mapping by Cui [19].

Figure 3: Chromosome maps of QLTs for cold tolerance at the germination and booting stages (the distances between markers are Mb).

doi:10.1371/journal.pone.0120590.g003

Among the 22 QTLs related to cold tolerance at germination as measured by SSvR, 6 were detected only in *indica* with low average contributions to phenotypic variation (CPV = 12.0%), 15 only in *japonica* with high average CPV (36.2%), and *qCTSSR6–3* in both *indica* and *japonica* with average CPV of 13.5%. The results showing that only one QTL was shared between *indica* and *japonica* and that CPV of QTLs in *japonica* were much higher than those in *indica*, implied that the mechanism of cold tolerance at the germination stage in *japonica* was different from that in *indica*.

Of 12 QTLs associated with SStR-NL, five were detected in *indica* with low average CPV (21.8%), and seven in *japonica* with higher average CPV (33.9%) (Fig. 3). Eight of 12 QTLs associated with SStR-CW were identified in *indica* with low average CPV (24.0%), compared with four in *japonica* with high average CPV (35.2%). Six of 15 QTLs associated with RSStR-CW, were in *indica* with low average CPV (24.3%), compared to nine in *japonica* with high average CPV (38.2%). No QTL related to SStR-NL, SStR-CW or RSStR-CW was shared between *indica* and *japonica*. Under the same stress and environmental conditions, three (*qLTRSSR6–1*, *qLTRSSR3–1* and *qLTRSSR1–3*) of eight QTLs identified by SStR-CW were also detected by RSStR-CW in *indica*, and similar findings were observed for two (*qLTRSSR8–1* and *qLTRSSR5–1*) of four QTLs in *japonica*. Although using the same measure (seed setting rate, SStR), only one QTL (*qLTRSSR8–1*) at the booting stage was detected under both naturally low air temperatures (the corresponding measure is SStR-NL) and cold water irrigation (the corresponding measure is SStR-CW) in *japonica* (Fig. 3). The results indicated that the mechanism of cold tolerance at the booting stage was complex. Moreover, only one QTL (*qLTRSSR8–1*) was detected by the same measure (seed setting rate, SStR) under different kinds of cold stress in *japonica*, i.e. naturally low air temperature in Kunming and cold water irrigation in Gongzhuling.

Among QTLs detected at both the germination and booting stages, the same two QTLs (*qLTSSR1–3* and *qCTSSR6–3* with CPV 9.1 and 10.9%, respectively) were detected at the germination and booting stages in *indica*, but there was no similar QTL in *japonica*. That is, *qLTSSR1–3* and *qCTSSR6–3* may provide a common mechanism of cold tolerance at the germination and booting stages to some degree in *indica*, although most of the cold tolerance at both stages may be attributed to different mechanisms. In *japonica* the mechanisms of cold tolerance at the germination and booting stages appear to be more different than in *indica*.

Markers and Germplasm Resources with Strong Cold Tolerance Potentially useful for Breeding

According to the relative genotypic effect (RGE) of each genotype at each QTL, we detected 18 positive genotypes and 21 negative genotypes in *indica*, and 19 positive genotypes and 24 negative genotypes in *japonica*. Generally, the negative effects were much stronger than the positive effects in both subspecies. Using these genotypes as selection markers we independently screened *indica* and *japonica* accessions containing different numbers of positive genotypes (Fig. 4 and 5). For positive genotypes, the cold tolerance as measured by the four measures increased with the increasing number of positive genotypes in *indica*, but not in *japonica* (Fig. 4). Likewise, for positive genotypes detected in *japonica*, the cold tolerance as measured by the four measures generally increased in *japonica*, but not in *indica*, apart from two exceptions for measures SStR_NL (Fig. 5B) and RSStR_CW (Fig. 5D). We checked the accessions with four genotypes positive for SStR_NL (Fig. 5B), among which one accession (Ye Li Cang Hua) had quite low seed setting (18%) and contained a strong negative genotype (RM5496_136), which reduced seed setting by up to 60%. Similarly, among the accessions with three genotypes positive for RSStR_CW (Fig. 5D), two cultivars (Ning Hui 21 and Dan Dong Lu Dao) had low seed setting (7% and 13.5%, respectively), and both contained a strong negative genotype (RM6863_155) that reduced seed setting by as much as 67%.

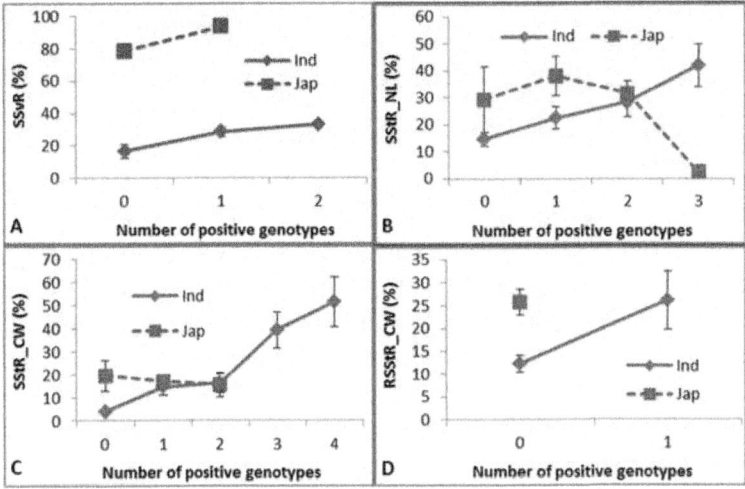

Figure 4: Phenotypes of four measures in *indica* **and** *japonica* **accessions with different numbers of positive QTLs detected in** *indica*.

doi:10.1371/journal.pone.0120590.g004

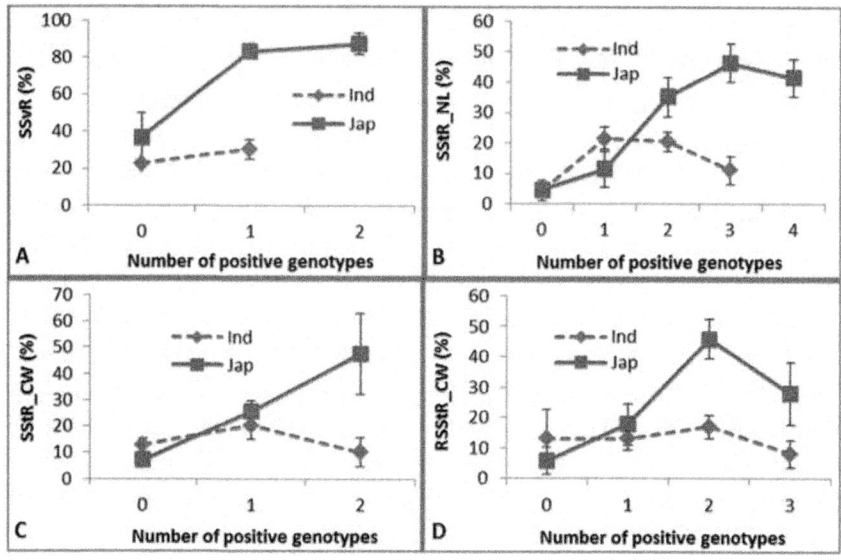

Figure 5: Phenotypes of four measures in *indica* and *japonica* accessions with different numbers of positive QTLs detected in *japonica*.

doi:10.1371/journal.pone.0120590.g005

In addition to screening accessions using positive genotypes we also screened the top five cultivars showing high cold tolerance according to each measure in *indica* and *japonica*, respectively, and subsequently investigated the distribution of the positive and negative genotypes among these accessions in each subspecies. There were three obvious characteristics for the distribution of the positive and negative genotypes for each measure within each subspecies. Firstly, the top cultivars contained at least one positive genotype of the corresponding measure in the corresponding subspecies, but not necessarily in the other subspecies, although exceptions existed in *indica* for SStR_NL (where accessions Z102, Z146, Z54 and Z64 had no positive genotypes) and RSStR_CW (where accessions Z126, Z5 and Z60 had no positive genotypes). Secondly, the top accessions excluded most of the negative genotypes, especially those with strong negative effects. Thirdly, both positive and negative genotypes in one subspecies may exist in the other subspecies but with no effect on cold tolerance in that subspecies, implying that *indica* and *japonica* do not share the same set of tolerance genes or alleles at most loci.

DISCUSSION

QTLs for Cold Tolerance in *Indica* and *Japonica*

Association analysis is an effective approach to dissect the genetic bases underlying complex traits in plants [9–10], animals [20] and humans [21]. In the present study we used association analysis to identify 51 QTLs related to cold tolerance at the germination and booting stages in rice. Among them, 46.17% were new QTLs, and 27 corresponded to QTLs reported previously. For example, *qLTSSR4–1* associated with SStR-CW was near the cloned *Ctb1* on chromosome 4; *qLTSSvR7–1* associated with SSvR is 1.4 cM from *qCTB7* on chromosome 7; and *qLTRSSR12–1* associated with RSStR-CW is near *qCTS12*. Four loci identified in the present study were detected at the booting stage by Cui *et al.* [17]. Of these, *qLTSSvR6–1* on Chr6 in the current study is near RM528 as reported by Cui *et al.*, *qCTSSR9–1* on Chr9 is near RM160, *qCTSSR11–1* on Chr11 is near RM4B, and *qCTSSR12–1* on Chr12 is near RM235.

Our results clearly indicated that *japonica* has a higher level of tolerance to cold stress than*indica* at both the germination and booting stages. This agrees with results published by others [22–25]. More QTLs were identified in *japonica* (36) than in *indica* (26). Five QTLs were shared among measures in *indica* or *japonica* at the booting stage. Only one QTL (*qCTSSR6–3*) associated with seedling survival rate (SSvR) under low temperatures was shared by *indica*and *japonica*. Reasons for this deserve further discussion.

During the thousands of years of cultivation and utilization under diverse environmental conditions tremendous genetic differentiation has occurred in rice produced under various agro-ecosystems. *Indica* is known to be adapted to tropical environments at low latitudes or altitudes typified by warm climatic conditions, whereas *japonica* cultivars were grown in temperate areas at high latitudes or altitudes with relatively cool conditions [26], at high altitudes in tropical and sub-tropical areas, and in areas where paddy fields are irrigated by cold water. In such areas, water and soil temperatures at sowing are below 15°C [27–28], and thus*japonica* cultivars were preferred due to tolerance to prevailing low temperatures. In double-cropped rice-growing regions, water and soil temperatures at sowing are higher than 15°C but cold air currents from Siberia often occurring in April may cause early rotting of seedlings, resulting in heavy seedling losses and a delayed growing period. *Indica* cultivars grown in such areas are usually characterized by tolerance to short periods of low temperature. Evident correlation has been found between cold stress and altitude or atmospheric temperature. Low temperatures (15–19°C) at booting cause sterile pollen which directly leads to spikelet sterility and ultimately causes serious yield losses [23,29–31]. In Heilongjiang province an increase of 1°C in negative accumulated temperature GDD17 at the inflorescence differentiation stage can result in a 3.51 kg/ha yield loss, and an increase of 1°C in negative accumulated temperature GDD19 at booting stage can cause a 4.95 kg/ha yield loss [32]. *Indica* is adapted to warm areas and is more sensitive to low temperatures. Pollen sterility is observed in *indica* when average temperatures remain below 22°C for three days [33]. The seasonal duration for rice gets progressively shorter due to decreased average temperature and accumulated temperature from southern to northern China. *Japonica* cultivars are mainly grown in the area north of the Yellow River, whereas most *indica* cultivars are grown to the south. This distribution implies that *indica-japonica* differentiation resulted from adaptation to different climatic and geographic environments.

The significant *indica–japonica* difference in cold tolerance was probably enhanced by rare genetic exchanges between the two subspecies. During cultivation and improvement over thousands of years, *indica* and *japonica* became genetically separated with adaptation to different agro-ecosystems. Differentiation between the subspecies was further enforced by fertility barriers that prevented genetic exchange between the two subspecies [34].

Observations of apparent differences in cold tolerance between *indica* and *japonica* may be partially attributed to the use of SSRs originating from noncoding regions. Only 15% of the SSRs used in the present study were from ESTs: this is higher than the proportion (1.5–7.5%) among all SSRs reported in rice [35–36]. Detection of marker-trait associations based on linkage disequilibria (LD) in genetically diverse materials can help to identify QTLs controlling target traits. However, this depends on the chromosome distance between marker and QTL and times between mutational changes in marker and the causal mutation of the target QTL. An association can be detected easily in a population when the mutation of the test marker is near in distance and time to the functional mutation in the target QTL [37]. However, 85% of SSRs in the study were nonfunctional markers and there are few genomic exchanges between the two subspecies as mentioned above. This may partially explain why *indica* and *japonica*share so few QTLs of cold tolerance.

Both Positive and Negative Genotypes should be considered in Breeding by MAS

Association mapping is a proven and efficient way to detect natural variation in a diverse population and provides many candidate genes or genomic regions for marker assisted selection in breeding. For example, Huang *et al.* used a worldwide panel of cultivated rice varieties to map 32 new loci associated with flowering time and 10 grain-related traits, allowing identification of candidate genes for 18 associated loci [10]. Marker assisted selection is an efficient breeding approach to pyramid elite genes in crop improvement; for example, Narayanan *et al.* developed a line combining blast and blight resistance genes Pi-1, Piz-5 and Xa2 using markers [38]. In the present study, we identified 18 and 19 positive genotypes (i.e. cold-tolerant genotypes) in *indica* and *japonica*, respectively, based on 51 QTLs. Our results indicated that the more positive genotypes the selected cultivars contain, the stronger cold tolerance they show. This proved that positive markers can be used to pyramid cold tolerance genes. In addition to those positive genotypes, however, we identified even more negative genotypes (sensitive to cold stress), numbering 21 and 24 in *indica* and *japonica*, respectively. Some cultivars (such as Ning Hui 21 and Dan Dong Lu Dao) are very sensitive to low temperatures due to a single strong negative genotype (such as RM6863_155) even though they may contain up to three positive genotypes. This indicates that certain strong negative genotypes may play the role of the shortest piece of wood as showed in the bucket theory and should be avoided by the breeders lest they cover up the advantage of the positive genotypes.

ACKNOWLEDGMENTS

We thank Professor Robert A McIntosh, University of Sydney, for suggested revisions to the manuscript.

AUTHOR CONTRIBUTIONS

Conceived and designed the experiments: YHP HLZ ZCL. Performed the experiments: XDM GLC YHP HLZ DLZ. Analyzed the data: YHP HLZ DLZ JJL HYX JPY JLL MARR GGL. Contributed reagents/materials/analysis tools: YHP HLZ DLZ JJL XDM GLC LZH. Wrote the paper: YHP HLZ ZCL MARR.

REFERENCES

1. Guirguis K, Gershunov A, Schwartz R, Bennet S. Recent warm and cold daily winter temperature extremes in the Northern Hemisphere. Geophysical Research Letters. 2011 Sep 2. pii: S0094–8276. doi: 10.1029/2011GL0487622.

2. Kodra E, Steinhaeuser K, Auroop R. Persisting cold extremes under 21st-century warming scenarios. Geophysical Research Letters. 2011 Apr 28. pii: S0094–8276. doi: 10.1029/2011GL047103.

3. Fujino K, Sekiguchi H, Matsuda Y, Sugimoto K, Ono K, Yano M. Molecular identification of a major quantitative trait locus, *qLTG3–1*, controlling low-temperature germinability in rice. Proc Natl Acad Sci USA. 2008; 105: 12623–12628. doi: 10.1073/pnas.0805303105. pmid:18719107

4. Xu LM, Zhou L, Zeng YW, Feng MW, Zhang HL, Shen SQ, et al. Identification and mapping of quantitative trait loci for cold tolerance at the booting stage in a *japonica*rice near-isogenic line. Plant Sci. 2008; 174: 340–347. doi: 10.1016/j.plantsci.2007.12.003

5. Zhou L, Zeng YW, Zheng WW, Tang B, SM Yang, Zhang HL, et al. Fine mapping a QTL, qCTB7, for cold tolerance at the booting stage on rice chromosome 7 using a near-isogenic line. Theor Appl Genet. 2010; 121: 893–905. doi: 10.1007/s00122-010-1358-x

6. Saito K, Hayano-Saito Y, Kuroki M, Sato Y. Map-based cloning of the rice cold tolerance gene *Ctb1*. Plant Sci. 2010; 179: 97–102. doi: 10.1016/j.plantsci.2010.04.004

7. Flint-Garcia S, Thuillet AC, Yu JM, Gael-Pressoir G, Romero S, Mitchell SE, et al. Maize association population: a high-resolution platform for quantitative trait locus dissection. Plant J. 2005; 44: 1054–1064. pmid:16359397 doi: 10.1111/j.1365-313x.2005.02591.x

8. Iwata H, Uga Y, Yoshioka Y, Ebana K, Hayashi T, Hayashi T. Bayesian association mapping of multiple quantitative trait loci and its application to the analysis of genetic variation among *Oryza sativa* L. germplasms. Theor Appl Genet. 2007; 114: 1437–1449. pmid:17356864 doi: 10.1007/s00122-007-0529-x

9. Agrama HA, Eizenga GC, Yan W. Association mapping of yield and its components in rice cultivars. Mol Breeding. 2007; 19: 341–356. doi: 10.1007/s11032-006-9066-6

10. Huang X, Wei X, Sang T, Zhao Q, Feng Q, Zhao Y, et al. Genome-wide association studies of 14 agronomic traits in rice landraces. Nature Genetics. 2010; 42: 961–970. doi: 10.1038/ng.695. pmid:20972439

11. Huang XH, Yan Z, Xing HW, Li CY, Wang A, Zhao Q, et al. Genome-wide association study of flowering time and grain yield traits in a worldwide collection of rice germplasm. Nature Genetics. 2012; 44: 32–39. doi: 10.1093/jncimonographs/lgs007. pmid:22623593

12. Li XB, Yan WG, Agrama H, Jia LM, Jackson A, Moldenhauer K, et al. Unraveling the complex trait of harvest index with association mapping in rice (*Oryza sativa* L.). PLoS ONE. 2012; 7: e29350. doi: 10.1371/journal.pone.0029350. pmid:22291889

13. Jia LM, Yan WG, Zhu CS, Agrama HA, Jackson A, Yeater K, et al. Allelic analysis of sheath blight resistance with association mapping in rice. PLoS ONE. 2012; 7: e32703. doi: 10.1371/journal.pone.0032703. pmid:22427867

14. Zhang YM, Zou DT. Association analysis of rice cold tolerance at tillering stage with SSR markers in *japonica* cultivars in Northeast China. Chin J Rice Sci. 2012; 26: 423–430.

15. Zhang DL, Zhang HL, Wang MX, Sun JL, Qi YW, Wang FM, et al. Genetic structure and differentiation of *Oryza sativa* L. in China revealed by microsatellites. Theor Appl Genet. 2009; 199: 1105–1117. doi: 10.1007/s00122-009-1112-4

16. Dai LY, Lin XH, Ye CR, Ise K, Saito K, Kato A, et al. Identification of quantitative trait loci controlling cold tolerance at the reproductive stage in Yunnan landrace of rice, Kunmingxiaobaigu. Breed Sci. 2004; 54: 253–258. doi: 10.1270/jsbbs.54.253

17. Falush D, Stephens M, Pritchard JK. Inference of population structure using multilocus genotype data: linked loci and correlated allele frequencies. Genetics. 2007; 164: 1567–1587.

18. Evanno G, Regnaut S, Goudet J. Detecting the number of clusters of individuals using the software structure: a simulation study. Molecular Ecology. 2005; 14: 2611–2620. pmid:15969739 doi: 10.1111/j.1365-294x.2005.02553.x

19. Cui D, Xu CY, Tang CF, Yang CG, Yu TQ, A XX, et al. Genetic structure and association mapping of cold tolerance in improved *japonica* rice germplasm at the booting stage. Euphytica. 2013; 193: 369–382 doi: 10.1007/s10681-013-0935-x

20. Bolormaa S, Hayes BJ, Savin K, Hawken R, Barendse W, Arthur PF, et al. Genome-wide association studies for feedlot and growth traits in cattle. J Anim Sci. 2011; 6: 1684–1697. doi: 10.2527/jas.2010-3079. pmid:21239664

21. Barrett JC, Hansoul S, Nicolae DL, Cho JH, Duerr RH, Rioux JD, et al. Genome-wide association defines more than 30 distinct susceptibility loci for Crohn›s disease. Nat Genet. 2008; 40: 955–962. doi: 10.1038/ng.175. pmid:18587394

22. Fujino K, Sekiguchi H, Sato T, Kiuchi H, Nonoue Y, Takeuchi Y, et al. Mapping of quantitative trait loci controlling low-temperature germinability in rice (*Oryza sativa* L.). Theor Appl Genet. 2004; 108: 794–799. pmid:14624339 doi: 10.1007/s00122-003-1509-4

23. Andaya VC, Mackill DJ. QTLs conferring cold tolerance at the booting stage of rice using recombinant inbred lines from a *japonica* × *indica* cross. Theor Appl Genet. 2003; 106: 1084–1090. pmid:12671757

24. Tai LM, Du J, Zeng YW. The cluster analysis of peculiarity of *indica* and *japonica* in Simao. Southwest China J Agaric Sci. 2006; 19: 49–53.

25. Zeng YW, Li ZC, Yang ZY. Geographical distribution and cline classification of*indica*/*Japonica* subspecies of Yunnan local rice resources. Acta Agronomica Sin. 2001; 1: 15–20.

26. Chang TT, Oka HI. Genetic variousness in the climatic adaption of rice cultivars. In: Proceedings of the Symposium on Climate and Rice. International Rice Research Institute, Los Banŏs, The Philippines. 1976; 87–111.

27. Fujino K, Sekiguchi H, Sato T, Kiuchi H, Nonoue Y, Takeuchi Y, et al. Mapping of quantitative trait loci controlling low-temperature germinability in rice (*Oryza sativa* L.). Theor Appl Genet. 2004; 108: 794–799. pmid:14624339 doi: 10.1007/s00122-003-1509-4

28. Sthapit BR, Witcombe JR. Inheritance of tolerance to chilling stress in rice during germination and plumule greening. Crop Sci. 1998; 8: 660–665. doi: 10.2135/cropsci1998.0011183x003800030007x

29. Nishiyama I. Male sterility caused by cooling treatment at the young microspore stage in rice plants. XXIII. Another length, pollen number and the difference in susceptibility to coolness among spikelets on the panicle. Jpn J Crop Sci. 1982; 514: 462–469. doi: 10.1626/jcs.51.462

30. Ji SL, Jiang L, Wang YH, Liu SJ, Liu X, Zhai HQ, et al. QTL and epistasis for low temperature germinability in rice. Acta Agron Sin. 2008; 34: 551–556. doi: 10.1016/s1875-2780(08)60021-8

31. Ye C, Fukai S, Godwin I, Reinke R, Snell P, Schiller J, et al. Cold tolerance in rice varieties at different growth stages. Crop Pasture Sci. 2009; 60: 328–338. doi: 10.1071/cp09006

32. Liu XF, Zhang Z, Shuai JB, Wang P, Shi WJ, Chen Y, et al. Effect of chilling injury on rice yield in Heilongjiang province. Acta Geographica Sin. 2010; 67(9): 1223–1232.

33. Wen S, Yao H. Global warming over the period 1961–2008 did not increase high-temperature stress but did reduce low-temperature stress in irrigated rice across China. Agricultural and Forest Meteorology. 2011; 9: 1193–1201. doi: 10.1016/j.agrformet.2011.04.009

34. Kato S, Kosaka H, Hara S. On the affinity of rice varieties as shown by the fertility of hybrid plants. Bull Sci Fac Agric. 1928; 3: 132–147.

35. Thiel T, Michalek W, Varshney RK, Graner A. Exploiting EST databases for the development of cDNA derived microsatellite markers in barley (*Hordeum vulgare* L.). Theor Appl Genet, 2003; 106: 411–422. pmid:12589540

36. Lawson MJ, Zhang L. Distinct patterns of SSR distribution in the Arabidopsis thaliana and rice genomes. Genome Biol. 2006; 7: R14. pmid:16507170

37. Meuwissen THE, Hayes BJ, Goddard ME. Prediction of total genetic value using genome-wide dense marker maps. Genetics. 2001; 157: 1819–1829. pmid:11290733

38. Narayanan NN, Niranjan B, Norman PO, Casiana MV, Gnanamanickam SS, Datta K, et al. Molecular breeding: marker-assisted selection combined with biolistic transformation for blast and bacterial blight resistance in *indica* rice (cv. CO39). Mol Breeding. 2004; 14: 61–71. doi: 10.1023/b:molb.0000037995.63856.2d

Chapter 9

GENETIC ANALYSIS OF LOW BMI PHENOTYPE IN THE UTAH POPULATION DATABASE

William R. Yates[1,2], Craig Johnson[3], Patrick McKee[1,4], Lisa A. Cannon-Albright[5,6]

[1]Laureate Institute for Brain Research, Tulsa, Oklahoma, United States of America

[2]University of Oklahoma College of Medicine, Tulsa, Tulsa, Oklahoma, United States of America

[3]Eating Recovery Center, Denver, Colorado, United States of America

[4]University of Oklahoma College of Medicine, Oklahoma City, Oklahoma, United States of America

[5]Department of Medicine (Genetic Epidemiology), University of Utah School of Medicine, Salt Lake City, Utah, United States of America

[6]George E. Wahlen Department of Veterans Affairs Medical Center, Salt Lake City, Utah, United States of America

ABSTRACT

The low body mass index (BMI) phenotype of less than 18.5 has been linked to medical and psychological morbidity as well as increased mortality risk. Although genetic factors have been shown to influence BMI across the entire BMI, the contribution of genetic factors to the low BMI phenotype is unclear. We hypothesized genetic factors would contribute to risk of a low BMI phenotype. To test this hypothesis, we conducted a genealogy data analysis using height and weight measurements from driver's license data from the Utah Population Data Base. The Genealogical Index of Familiality (GIF) test and relative risk in relatives were used to examine evidence for excess relatedness among individuals with the low BMI phenotype. The overall GIF test for excess relatedness in the low BMI phenotype showed a significant excess over expected (GIF 4.47 for all cases versus 4.10 for controls, overall empirical p-value<0.001). The significant excess relatedness was still observed when close relationships were ignored, supporting a specific genetic contribution rather than only a family environmental effect. This study supports a specific genetic contribution in the risk for the low BMI phenotype. Better understanding of the genetic contribution to low BMI holds promise for weight regulation and potentially for novel strategies in the treatment of leanness and obesity.

INTRODUCTION

Genetic factors increase the risk for a high body mass index (BMI), overweight and obesity[1]. The role of genetic factors in low BMI is less well understood.

Family studies have found family clustering for low BMI [2]–[4]. However, family studies cannot distinguish between genetic and familial factors. Twin studies have estimated the heritability of BMI across the entire BMI range at between 50 and 74% [5]–[6].

Molecular genetic studies have identified a series of candidate genes for low BMI including a thyrotropin-releasing hormone (TRH) receptor polymorphism [7], the Ser23 allele of the serotonin 2C receptor [8], a genetic variant on chromosome 16p11.2 [9] and a copy number variant identified as gremlin1 [10]. Allelic variants of the *FTO* gene linked to obesity risk are infrequently found in thin individuals [11].

Understanding the genetic and environmental contributions to low BMI are important because low BMI has been linked to medical and psychiatric illnesses as well as increased mortality. In females, low BMI during childhood and adolescence increases women›s risk for later endometriosis [12], preterm birth [13], low infant birth weight [14] and increased risk for placental abruption [15]. Infants born to mothers with low BMI have an increased risk for atrial septal defect, genital abnormalities including hypospadias [16].

The psychiatric illness *anorexia nervosa* is defined by a low BMI in association with extreme fear of becoming fat [17].

Mortality rates across BMI categories in many studies display a U-shaped curve with increased death rates for the low BMI as well as those with a high BMI. Mortality rates are higher by an estimated 73% in those with BMI below 18.5 [18]. A Japanese study estimated the mortality risk increased by 78% in those with a BMI<18.5 and increased by 155% in those with a BMI<16 [19]. The exact mechanism for the mortality increase in low BMI populations is unclear. Some of the increase may be due to higher rates of death in severe illness and surgical procedures such as lung transplantation [20].

A study of excess deaths related to being low BMI and high BMI in the United States provides a reference for the relative contribution of low BMI to mortality [21]. High BMI was estimated to contribute to 111,909 deaths in the U.S. in 2000 while low BMI was estimated to be associated with 33,746 deaths. Thus, low BMI is estimated to contribute about three deaths for every ten deaths related to being high BMI.

To further understand the prevalence and genetic contributions of leanness, we examined the low BMI phenotype in the Utah Population Data Base (UPDB). The UPDB provides a strategy to examine the possibility of both environmental and genetic contributions to a phenotype by estimating risk in both close and distant relatives and by testing for excess familial clustering. An observation of excess close relationships alone would not have allowed discrimination between shared genes and shared environment, but the UPDB allows us to consider more distant relationships that are unlikely to represent lifestyle sharing beyond what is expected in the Utah population. We hypothesized that low BMI individuals would demonstrate near and distant familial clustering consistent with a genetic contribution.

METHODS

Ethics Approval

The protocol for this study was approved by the University of Utah Institutional Review Board and the Resource for Genetic and Epidemiologic Research. The Resource for Genetic and Epidemiologic Research is the oversight board for the Utah Population Data Base. All data used in this research study contains no individual identifiers. A waiver of consent was approved for this study due to the lack of individual identifiers for all subjects. Consent requirements were waived for this study since obtaining consent would have unnecessarily identified individuals in the anonymous database. Review of this protocol by the University of Utah Institutional Review Board and the Resource for Genetic and Epidemiologic Research includes a review and approval of consent issues and other ethical aspects of the research.

Utah Population Data Base (UPDB)

The UPDB is a unique computerized database primarily representing the pioneer founders of Utah and their modern day descendants. It includes up to 15 of genealogy data dating back to the original Utah founding pioneers [22], as well as current generations. The genealogy data has been linked to statewide data including driving license (DL) data, births, deaths, the Utah Cancer Registry, and Utah Hospital Discharge Data, among other data sets (www.huntsmancancer.org/groups/ppr). The DL data includes height and weight and is available for over three million Utah drivers.

For the genetic analyses performed here we selected only from those 1,192,768 individuals in the UPDB who have genealogy data for both parents, all four grandparents, and six of their eight great grandparents and whose genealogy connects to the original Utah genealogy, and the 593,704 of these individuals who have Utah Drivers License data. These strict criteria allow for appropriate matching of cases and controls in terms of quality and quantity of genealogical data.

The oversight board for the UPDB encourages collaboration with outside investigators and institutions. Researchers with interest in using the UPDB to test hypotheses may contact the board or one of the authors for information about methods to apply for access.

The UPDB has been successfully used to define familial clustering and genetic influences in a variety of disorders including cancer [23]–[27], coronary artery disease [28], diabetes[29] rotator cuff disease [30], and deaths due to influenza [31] and asthma [32]. The methods used to identify phenotypes, assess familial and genetic effects and identify pedigrees using UPDB data have been described in detail in these studies. The study of high-risk pedigrees identified in the UPDB has led to multiple gene identifications, including*BRCA1* [33], *BRCA2* [34], *CDKN2A* (melanoma) [35]–[36] and *HPC2/ELAC2* [37].

Low BMI Phenotype

The phenotype of adult leanness was established using Utah DL data available for 593,704 individuals (with acceptable genealogy data as described) included in the UPDB. We identified all male and female drivers whose most recent calculated BMI (from height and weight provided) was <18.5. Rates for the low BMI phenotype were calculated by age group and are shown in Table 1, which includes the age group, the number of lean individuals in the age group, the total number of individuals with DL data in the age group, the leanness prevalence and the 95% confidence interval for prevalence by age group, estimated by the method of Clopper and Pearson [38].

Table 1: Prevalence Rates for Low BMI (<18.5) in the UPDB

Age	BMI<18.5	N	Prevalence (%)	95% CI
15–19	6,259	57,010	11.0	10.7, 11.2
20–24	2,847	62,070	4.6	4.4, 4.8
25–29	1,629	63,548	2.5	2.4, 2.7
30–34	922	49.069	1.9	1.8, 2.0
35–39	544	39,353	1.4	1.3, 1.5
40–44	400	36,097	1.1	1.0, 1.2
45–49	311	40,719	0.8	0.7, 0.9
50–54	258	42,998	0.6	0.5, 0.7
55–59	176	38,040	0.4	0.3, 0.5
60–64	135	26,947	0.5	0.4, 0.6
65–69	191	27,076	0.7	0.6, 0.8
70–74	255	26,207	1.0	0.9, 1.1
75–79	354	25,122	1.4	1.3, 1.6
80 or older	569	26,979	2.1	1.9, 2.3

doi:10.1371/journal.pone.0080287.t001

doi:10.1371/journal.pone.0080287.t001

Statistical Analysis

The Genealogical Index of Familiality (GIF) statistic was used to test the hypothesis of excess relatedness among individuals in the low BMI phenotype. The GIF was developed specifically for the UPDB [39]–[40]. Briefly, the GIF measures the average pair-wise relatedness of a set of individuals and compares that measurement to the average pair-wise relatedness expected in the Utah population. The GIF test differs from relative risk (RR) in that it includes analysis of all genetic relationships, both close and distant. The GIF utilizes the Malecot coefficient of kinship to measure pair-wise relatedness. The coefficient is defined as the probability that randomly selected homologous genes from two individuals are identical by descent from a common ancestor [41]. The coefficient is 0.50 for parent/offspring, 0.25 for a sibling pair, 0.125 for an uncle/nephew pair, 0.0625 for a first cousin pair, and so forth. The contribution to the GIF statistic is therefore smaller for individual pairs with greater genetic distance between them; more closely related pairs contribute more.

To evaluate the significance of the GIF test, we estimated the average pair-wise relatedness for 1,000 sets of controls matched to the cases by birth year, sex, and birthplace (Utah or not). These controls are chosen from among the 593,704 individuals with acceptable quality genealogy data who also have DL data. The empirical significance of the GIF test is measured by comparing the case GIF to the distribution of 1,000 control GIF values.

The GIF statistic measures familial clustering, which can be due to genetic (genes related to low BMI phenotype), or to shared familial environmental effects (i.e. familial preference for low calorie diet or rigorous physical exercise), or to a combination of both. In order to better distinguish these effects, we recalculate the case GIF and the control GIFs while ignoring close relationships (first and second degree). If this distant GIF (dGif) test is significant, it provides strong evidence that there is significant distant excess relatedness that is unlikely to be due to shared environment.

The calculation of RR in relatives provides the more traditional mechanism for identifying genetic effects. A genetic contribution to a phenotype is supported when both close and distant relatives have elevated risk. RRs for the low BMI phenotype were estimated for first-, second- and third degree relatives of low BMI individuals as follows. First-degree relatives include parents, siblings and offspring; second-degree relatives are the first-degree relatives of the first-degree relatives (e.g. uncle, grandmother); third-degree relatives are the first-degree relatives of the second-degree relatives (e.g. first cousin, great grandchild), All 593,704 individuals in the UPDB with acceptable quality and quantity genealogy data as described and with DL data were assigned to one of 132 cohorts based on birth year (in five year groups), sex, and birthplace (Utah or not). Cohort-specific rates of low BMI were estimated by dividing the total number of low BMI individuals per cohort by the total number of individuals with DL data per cohort. Expected numbers of low BMI first-degree relatives were estimated by counting the number of first-degree relatives with DL data and genealogy data by cohort (without duplication), multiplying by the rate of low BMI in each cohort, and summing over all cohorts. Observed numbers of low BMI individuals (BMI<18.5) among relatives were counted without duplication. RRs were estimated for each degree of relationship as observed/ expected number of low BMI individuals; 95% confidence intervals for the RR were calculated using the method of Agresti [42]

High-risk Pedigree Identification

It is possible to identify pedigrees in the UPDB with a significant excess of low BMI using the same tools listed above. We first identify all possible related clusters of individuals with low BMI; no cluster is a subset of any other cluster, but individuals can be identified in more than 1 cluster. These clusters represent all sets of related individuals with low BMI descending from a common founder (a pedigree), but they are not necessarily high-risk for low BMI, they can represent chance clusters. To identify which of the clusters are high-risk for low BMI, we apply the internal cohort-specific rates for low BMI we estimated from the UPDB (see above methods for Relative Risks) to all of the descendants of each cluster. We compare the observed number of low BMI cases among the descendants to the expected number of low BMI cases among the descendants to identify those pedigrees with a significant excess of low BMI cases.

RESULTS

Prevalence Rates and Proband Selection

The prevalence of BMI<18.5 in the UPDB individuals with acceptable quality and quantity genealogy are summarized by age group in Table 1. The prevalence of BMI<18.5 ranged from .4% in the 55 to 59 year old age group to 11.0% in those aged 15 to 19 years. The BMI in the sample ranged from a minimum of 12.10 to a maximum of 62.93.

Table 1 illustrates some developmental factors associated with weight: i) stable adult weight often is not reached until midlife; ii) younger individuals have different ranges in BMI as they progress through adolescence and young adulthood, and iii) in older geriatric age groups, rates of low BMI increase with the loss of lean body mass with aging. To minimize phenotypic heterogeneity due to these other factors, we elected to include individuals between the ages of 25 and 64. This age period showed relatively stable low BMI population prevalence rates and occurred outside the periods of early and late adult development. We identified 4,375 individuals between the ages of 25 and 64 years with BMI<18.5 with acceptable quality genealogy data.

Genealogy Index of Famililiality (GIF)

The GIF test for excess relatedness in low BMI was performed on all of the 14,867 low BMI individuals and the results were compared to the average relatedness observed among 1,000 sets of matched controls selected from all individuals who had a Utah driver's license.Table 2 shows the number of cases, the average relatedness of the cases, the mean control relatedness, the empirical p value for the test for excess overall relatedness, and the empirical p value for the test for excess distant relatedness. The overall GIF test for excess relatedness for everyone with low BMI shows a significant excess over expected (p-value<0.001). The distant GIF test, ignoring close relatives (genetic distance<4) also showed a significant excess (p-value=0.031).

Table 2: GIF Test for Low BMI (<18.5) in the UPDB

Group	N	Case GIF	Control GIF	GIF p-value	dGIF p-value
All BMI<18.5	14,867	4.47	4.10	<0.001	0.031
BMI<18.5, ages 25–64	4,375	4.84	4.19	<0.001	<0.001

doi:10.1371/journal.pone.0080287.t002

doi:10.1371/journal.pone.0080287.t002

Table 2 also shows the results of the GIF test for the 4,375 low BMI adults aged 25–64 years of age. Similar results were obtained for this subgroup of lean individuals from which individuals at the extremes of the age distribution were removed to reduce bias. Figure 1shows the contribution to the GIF statistic, separately for cases and controls, by pairwise genetic distance where genetic distance 1=parent/offspring, 2=siblings, 3=avunculars, 4=first cousins, and so forth for the low BMI individuals aged 25–64 years. As seen, the distribution of relatedness for cases is in excess up to genetic distance=4 (e.g. first cousins) and beyond genetic distance=9 (e.g. second-cousins once-removed. Although the pairwise relatedness distributions for cases and matched controls cross at some points, as seen inFigure 1, the pairwise relatedness for cases is significantly elevated over that for matched controls when all genetic distances are considered as determined by distant Gif test (dGif p=0.031).

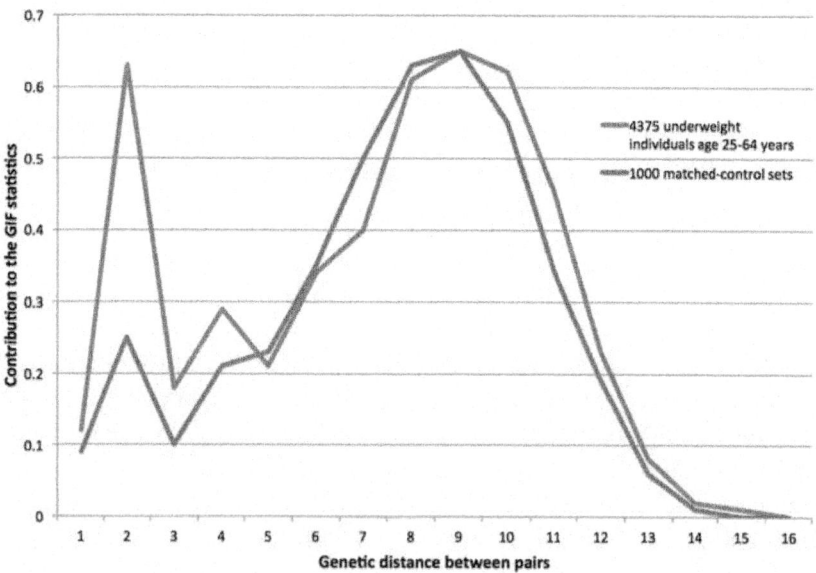

Figure 1: The contribution to the GIF statistic by genetic distance for 4,375 low BMI cases aged 25–64 years old compared to 1,000 sets of matched UPDB controls with BMI data.

Genetic distance between pairs is shown on the x-axis and represents an increasing measure of relatedness (1=parent/offspring; 2=siblings, e.g.; 3=uncle/niece, e.g.; 4=first cousins, e.g.) from close to distant; the most distant relationships noted (genetic distance=16) could represent, for example, two individuals who have a common ancestor 8 generations past. The cumulative contribution to the GIF statistic for each relatedness (as measured by genetic distance) for all pairs identified at that genetic distance is represented on the y-axis. The contribution to the GIF statistic for each larger genetic distance is one-half as large; the contribution for genetic distance 1=½, for genetic distance 2=¼, and so forth. The distribution for controls represents the expected relatedness of a group of individuals just like the cases (ignoring BMI) and is smoother because it is averaged over 1,000 different sets of controls tested. The distribution for cases represents only the analysis of the single set of cases and is more irregular. The peak at genetic distance=2 (e.g. siblings) in comparison with genetic distance=1 (parent/offspring) is seen for both cases and controls and represents that we observe more sib pairs than parent/offspring pairs in our data. A similar peak for cases at genetic distance 4 also indicates that we observed more cousins (same generation) than avunculars, for example.

doi:10.1371/journal.pone.0080287.g001

Relative Risks (RRs)

Estimates of relative risks in relatives of lean adults are shown in Table 3. The table shows the total number of each type of relative with BMI data (# relatives), the observed number of those relatives with BMI<18.5 (obs), and the expected number of relatives with BMI<18.5 (expected) based on birth year, sex, and birthplace cohort-specific rates for low BMI among all UPDB individuals with BMI data.RRs for adult leanness were significantly elevated among first-, second-, and third-degree relatives of lean adults. The smaller number of second degree lean adults observed is not unexpected, given that second degree relatives are primarily in different generations (avunculars, grandparent/child), while first and third-degree relatives occur in the same generation (siblings and cousins, respectively) as well as in different generations (parent/offspring). Since the DL data exist only after 1980, there is a very narrow window that limits observations across generations; however, our results support the GIF results, where the contribution from first- (genetic distance 1 and 2), second- (genetic distance=3) and third-degree relatives (genetic distance=4) was in excess for cases compared to controls. Table 4 shows the RR for leanness among the relatives of those lean Utah adults who were aged 25–65 years. Although the RR estimates are slightly larger, the conclusions are not different.

Table 3: Relative Risk of Low BMI (<18.5) in Relatives of All Lean Utah Adults

Relatives	N	Observed	Expected	p-value	Relative Risk	95% CI
First-degree	54,324	3,565	1,615.5	<0.00001	2.21	2.13, 2.28
Second-degree	96,151	2,236	1,799.7	2.1 e-23	1.24	1.19, 1.30
Third-degree	175,286	5,204	4,546.0	7.7 e-22	1.14	1.11, 1.18
Fourth-degree	317,807	6,897	6,773.4	0.133	1.02	.99, 1.04
Fifth-degree	501,420	12,417	12,271.5	0.191	1.01	.99, 1.03
Sixth-degree	571,643	14,180	14,147.9	0.788	1.00	.99, 1.02
Seventh-degree	588,223	14,796	14,772.6	0.849	1.00	.99, 1.02

doi:10.1371/journal.pone.0080287.t003

doi:10.1371/journal.pone.0080287.t003

Table 4: Relative Risk of Low BMI (<18.5) in Relatives of Lean Utah Adults Ages 25–65 Years

Relatives	N	Observed	Expected	p-value	Relative Risk	95% CI
First-degree	20,551	1,177	458.0	<0.00001	2.57	2.42, 2.72
Second-degree	32,849	1,056	775.4	<0.00001	1.36	1.28, 1.45
Third-degree	63,856	1,668	1,363.8	7.0 e-16	1.22	1.17, 1.28
Fourth-degree	153,915	3,597	3,427.9	0.0039	1.05	1.02, 1.08
Fifth-degree	309,372	6,899	6,774.1	.130	1.02	0.99, 1.04
Sixth-degree	474,053	11,512	11,376.2	.204	1.01	0.99, 1.03
Seventh-degree	562,037	12,984	13,955.3	.809	1.00	0.99, 1.02

doi:10.1371/journal.pone.0080287.t004

doi:10.1371/journal.pone.0080287.t004

Pedigree Identification

We identified all possible clusters of the 4,375 low BMI individuals between 25 and 65 years of age. These clusters merely represent all related sets of individuals with low BMI, they are not necessarily high-risk for BMI. We further evaluate each cluster by testing for an excess of low BMI among all the descendants of the founding pair of the cluster using the low BMI rates estimated from the UPDB. We identified over 4,000 clusters of low BMI relatives, ranging in size from 2 related cases (n=789 clusters) to size 168 (n=1 cluster).

We were able to identify thousands of individual pedigrees that may assist in future molecular genetic studies of the low BMI phenotype. As an example, we have identified 63 pedigrees with a significant excess of individuals with low BMI (p<0.0001) with at least 10 cases. An example of one of these pedigrees is shown in Figure 2. As can be observed, Utah driver's license data is only available for the most recent two or three generations of the Utah genealogy; earlier generations remain unknown for the phenotype of interest.

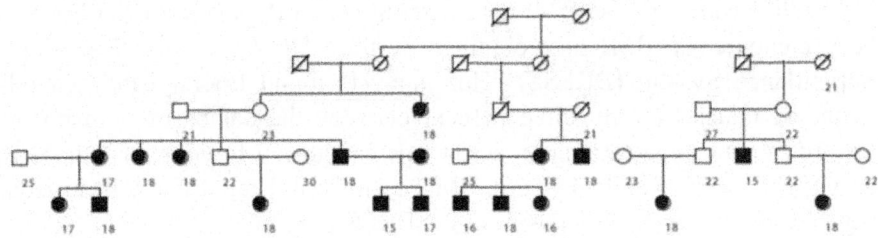

Figure 2: Example UPDB pedigree with a statitstical excess of low bmi (<18.5) individuals.

Individuals with bmi <18.5 are fully shaded and BMI is shown beneath subjects where available.

doi:10.1371/journal.pone.0080287.g002

DISCUSSION

Familial clustering of the low BMI phenotype in the UPDB is consistent with both genetic as well as environmental contributions to low BMI humans. Increased relative risks for the low BMI phenotype in first-, second-, and third-degree relatives suggests that genetic factors contribute to the familial clustering pattern.

The GIF analysis confirms a genetic contribution to low BMI in relatives with an even more distant degree of relatedness. Excess relatedness among close relatives could represent either shared environment or shared genetics, or a combination,. However, the finding of excess relatedness in distant relatives is much more likely to result from shared genes than shared environment.

Our study represents the largest genealogical population genetics study of low BMI to date. There are no comparable low BMI studies using a similar methodology. However, our identification of a genetic contribution to low BMI is consistent with findings in family studies[2]–[4] and in a single adoption study [43].

A primary limitation of this study is the reliability and validity of Department of Motor Vehicle height and weight self-reported measures. There is limited study of the accuracy of self-reported height and weight in DL data. Self-reported weights in overweight and obese individuals might be significantly underestimated due to the social stigma of obesity. In contrast, social stigma issues in reporting an accurate weight in the low BMI may be less than in obesity. Nevertheless, it remains possible that there is some social pressure to overestimate weight among the low BMI.

Utah driver›s license BMI data from the Utah Population Data Base has been compared to BMI data obtained by the CDC Behavioral Risk Factor Surveillance System (BRFSS). This analysis found BMI means generally varied in males by only three percent between the databases with no bias toward over or underestimation across age categories. In younger female age groups (between 25 and 34 years), BMI means from Utah driver›s license data was 5 to 8% lower than means from the BRFSS [44].

We compared rates of low BMI in the UPDB to the National Health and Nutrition Examination Survey (NHANES) to look for evidence of a self-report bias. [45]–[46]. The NHANES includes data from direct measurement of weight and height in a representative sample of individuals in the United States. The overall NHANES estimated prevalence of low BMI (BMI<18.5) is 1.8% in the U.S. adult population.

Prevalence rate estimates of low BMI in the UPDB generally are in close agreement with estimates from the NHANES. Rate estimates for BMI<18.5 by age group for the UPDB and (NHANES) were:20 to 39 years of age, 3.3%, (2.6%), 40 to 59 years of age, 0.9%, (1.2%), 60 to 74 years of age, .9%, (.9%), 75 years of age and older, 2.0%, (1.7%). This agreement supports the validity of the low BMI phenotype in the UPDB. Nevertheless, there is no easy method to directly measure the reliability and validity of the drivers license self-reported weight and height in the UPDB.

Some degree of inaccurate self-report of weight may be present and contribute to variance in our relative risk estimates. However, weight self-report bias would likely lead to underestimation of the relative risk.

Identification of familial relationships in the UPDB is also based on self-report. It is possible, self-reported relationships may differ from biological relationships. However, thousands of UPDB high-risk pedigrees have been genotyped with a very high rate of accuracy between self-reported relationships and genotype.

The UPDB cohorts cross several generations. Environmental dietary and physical exercise patterns likely change across generations and may have influenced a portion of BMI data.

Despite these potential limitations, this study finds excess clustering of the low BMI phenotype among both close and distant relatives supporting a significant genetic contribution. Environmental factors are also likely to contribute to familial clustering of low BMI. Families often share diet, exercise and other lifestyle patterns that contribute to body weight.

Pedigrees in the UPDB with a significant excess of low BMI individuals among the descendants of founder couples have been identified. These pedigrees could prove valuable for additional molecular genetic studies of low BMI. Figure 2 shows an example pedigree with a significant excess of individuals with low BMI. The pedigree founder has 257 descendants with BMI data. Eighteen of these individuals have a BMI<18.5. This is significantly greater than 5.3, the number expected in the pedigree (p=1.13 e^5).

Further studies to confirm our results and to explore the molecular genetics of the low BMI phenotype are needed. Such studies may uncover the mechanisms for increased morbidity and mortality in low BMI populations. Additionally, further understanding of the genetic and environmental contributions to low BMI may provide insight for prevention and treatment of both low BMI as well as obesity.

AUTHOR CONTRIBUTIONS

Conceived and designed the experiments: WRY CJ PM LACA. Performed the experiments: LACA. Analyzed the data: LACA. Contributed reagents/materials/analysis tools: LACA. Wrote the paper: WRY.

REFERENCES

1.   Day FR, Loos RJ (2011) Developments in obesity genetics in the era of genome-wide association studies. J Nutrigenet Nutrigenomics 24: 222–38. doi: 10.1159/000332158

2.   Lazkarzewski PM, Khoury P, Morrison JA, Kelly , Mellies MJ, et al. (1983) Familial obesity and leanness. Int J Obesity 1983 7: 505–27.

3.   Magnusson PKE, Rasmussen F (2002) Familial resemblance of body mass index and familial risk of high and low body mass index. A study of young men in Sweden. Int J Obesity 26: 1225–1231. doi: 10.1038/sj.ijo.0802041

4.   Borecki IB, Higgins M, Schreiner PJ, Arnett DK, Mayer-Davis E (1998) Evidence for multiple determinants of the body mass index: The National Heart, Lung, and Blood Institute Family Heart Study. Obes Res 6: 107–14. doi: 10.1002/j.1550-8528.1998.tb00323.x

5.   Stunkard AJ, Harris JR, Pedersen NL, McClearn GE (1990) The body mass index of twins who have been reared apart. N Engl J Med 322: 1483–7. doi: 10.1056/nejm199005243222102

6.   Dubois L, Ohm Kyvik K, Girard M, Tatone-Tokuda F, Perusse D (2012) Genetic and environmental contributions to weight, height, and BMI from birth to19 years of age: an international study of over 12,000 twin pairs. PLoS One 7 (2) e30153. doi: 10.1371/journal.pone.0030153

7. Liu XG, Tan LJ, Lei SF, Liu YJ, Shen H, et al. (2009) Genome-wide association and replication studies identified TRHR as an important gene for lean body mass. Am J Hum Genet 84: 418–23. doi: 10.1016/j.ajhg.2009.02.004

8. Bah J, Westberg L, Baghaei F, Henningsson S, Rosmond R, et al. (2010) Further exploration of the possible influence of polymorphisms in HTR2C and 5HTT on body weight. Metabolism 59: 1156–63. doi: 10.1016/j.metabol.2009.11.007

9. Jacquemont S, Reymond A, Zufferey F, Harewood L, Walters RG, et al. (2011) Mirror extreme phenotypes associated with gene dosage at the chromosome 16p11.2 locus. Nature 478: 97–102.

10. Hai R, Pei YF, Shen H, Zhang L, Liu XG, et al. (2012) Genome-wide association study of copy number variation identified gremlin1 as a candidate gene for lean body mass. J Hum Genet 57: 33–7. doi: 10.1038/jhg.2011.125

11. Hunt SC, Stone S, Xin Y, Scherer CA, Magness CL, et al. (2008) Association of the FTO gene with BMI. Obesity (Silver Spring) 16: 902–4. doi: 10.1038/oby.2007.126

12. Vitonis AF, Baer HJ, Hankinson SE, Laufer MR, Missmer SA (2010) A prospective study of body size during childhood and early adulthood and the incidence of endometriosis. Hum Reprod 5: 1325–34. doi: 10.1093/humrep/deq039

13. Khashan AS, Kenny LC (2009) The effects of maternal body mass index on pregnancy outcome. Eur J Epidemiol 24: 697–705. doi: 10.1007/s10654-009-9375-2

14. Kalk P, Guthmann F, Krause K, Relle K, Godes , et al. (2009) Impact of maternal body mass index on neonatal outcome. Eur J Med Res 14: 216–22.

15. Deutsch AB, Lynch O, Alio AP, Salihu HM, Spellacy WN (2010) Increased risk of placental abruption in low BMI women. Am J Perinatol 27: 235–40. doi: 10.1055/s-0029-1239490

16. Salihu HM, Lynch O, Alio AP, Mbah AK, Kornosky JL, et al. (2009) Extreme maternal low BMI and feto-infant morbidity outcomes: a population-based study. J Matern Fetal Neonatal Med 22: 428–34. doi: 10.1080/14767050802385764

17. American Psychiatric Association. (2000) Diagnostic and Statistical Manual of Mental Disorders, 4th ed, text rev. Washington, DC: American Psychiatric Association.

18. Orpana HM, Berthelot JM, Kaplan MS, Feeny DH, McFarland B, et al. (2010) BMI and mortality: results from a national longitudinal study of Canadian adults. Obesity (Silver Spring) 18: 214–8. doi: 10.1038/oby.2009.191

19. Tamakoshi A, Yatsuya H, Lin Y, Tamakoshi K, Kondo T, et al. (2010) JACC Study Group. BMI and all-cause mortality among Japanese older adults: findings from the Japan collaborative cohort study. Obesity (Silver Spring) 18: 362–9. doi: 10.1038/oby.2009.190

20. Lederer DJ, Wilt JS, D'Ovidio F, Bacchetta MD, Shah L, et al. (2009) Obesity and low BMI are associated with an increased risk of death after lung transplantation. Am J Respir Crit Care Med 180: 887–95. doi: 10.1164/rccm.200903-0425oc

21. Flegal KM, Graubard BI, Williamson DF, Gail MH (2005) Excess deaths associated with low BMI, overweight, and obesity. JAMA 293: 1861–7. doi: 10.1001/jama.298.17.2028

22. Skolnick M (1980) The Utah genealogical database: a resource for genetic epidemiology. In: Cairns J, Lyon JL, Skolnick M, editors. Cancer incidence in defined populations. Cold Spring Harbor, NY: Cold Spring Harbor Laboratories pp 285–97

23. Cannon LA, Bishop DT, Skolnick MH (1986) Segregation and linkage analysis of breast cancer in the Dutch and Utah families. Genet Epidemiol Suppl 1: 43–8. doi: 10.1002/gepi.1370030707

24. Cannon-Albright LA, Goldgar DE, Neuhausen S, Gruis NA, Anderson DE, et al. (1994) Localization of the 9p melanoma susceptibility locus (MLM) to a 2-cM region between D9S736 and D9S171. Genomics 23: 265–8. doi: 10.1006/geno.1994.1491

25. Shirts BH, Burt RW, Mulvihill SJ, Cannon-Albright LA (2010) A population-based description of familial clustering of pancreatic cancer. Clin Gastroenterol Hepatol 8: 812–6. doi: 10.1016/j.cgh.2010.05.012

26. Maul JS, Burt RW, Cannon-Albright LA (2007) A familial component to human rectal cancer, independent of colon cancer risk. Clin Gastroenterol Hepatol 5: 1080–4. doi: 10.1016/j.cgh.2007.04.025

27. Teerlink C, Farnham J, Allen-Brady K, Camp NJ, Thomas A, et al. (2012) A unique genome-wide association analysis in extended Utah high-risk pedigrees identifies a novel melanoma risk variant on chromosome arm 10q. Hum Genet 131: 77–85. doi: 10.1007/s00439-011-1048-z

28. Horne BD, Camp NJ, Muhlestein JB, Cannon-Albright LA (2006) Identification of excess clustering of coronary heart diseases among

extended pedigrees in a genealogical population database. Am Heart J 152: 305–11. doi: 10.1016/j.ahj.2005.12.028

29. Weires MB, Tausch B, Haug PJ, Edwards CQ, Wetter T, et al. (2007) Familiality of diabetes mellitus. Exp Clin Endocrinol Diabetes 115: 634–40. doi: 10.1055/s-2007-984443

30. Tashjian RZ, Farnham JM, Albright FS, Teerlink CC, Cannon-Albright LA (2009) Evidence for an inherited predisposition contributing to the risk for rotator cuff disease. J Bone Joint Surg Am 91: 1136–42. doi: 10.2106/jbjs.h.00831

31. Albright FS, Orlando P, Pavia AT, Jackson GG, Cannon Albright LA (2008) Evidence for a heritable predisposition to death due to influenza. J Infect Dis 197: 18–24. doi: 10.1086/524064

32. Teerlink CC, Hegewald MJ, Cannon-Albright LA (2007) A genealogical assessment of heritable predisposition to asthma mortality. Am J Respir Crit Care Med 176: 865–70. doi: 10.1164/rccm.200703-448oc

33. Miki Y, Swensen J, Shattuck-Eidens D, Futreal PA, Harshman K, et al. (1994) A strong candidate for the breast and ovarian cancer susceptibility gene BRCA1. Science 266: 66–71. doi: 10.1126/science.7545954

34. Tavtigian SV, Simard J, Rommens J, Couch F, Shattuck-Eidens D, et al. (1996) The complete BRCA2 gene and mutations in chromosome 13q-linked kindreds. Nat Genet 12: 333–7. doi: 10.1038/ng0396-333

35. Cannon-Albright LA, Goldgar DE, Meyer LJ, Lewis CM, Anderson DE, et al. (1992) Assignment of a locus for familial melanoma, MLM, to chromosome 9p13–p22. Science 258: 1148–52. doi: 10.1126/science.1439824

36. Kamb A, Shattuck-Eidens D, Eeles R, Liu Q, Gruis NA, et al. (1994) Analysis of the p16 gene (*CDKN2*) as a candidate for the chromosome 9p melanoma susceptibility locus. Nat Genet 8: 23–6. doi: 10.1038/ng0994-22

37. Vesprini D, Nam RK, Trachtenberg J, Jewett MA, Tavtigian SV, et al. (2001) HPC2 variants and screen-detected prostate cancer. Am J Hum Genet 68: 912–7. doi: 10.1086/319502

38. Clopper CJ, Pearson ES (1934) The use of confidence or fiducial limits illustrated in the case of the binomial. Biometrika 26: 404–413. doi: 10.1093/biomet/26.4.404

39. Hill JR (1980) A survey of cancer sites by kinship in the Utah Mormon population. In: Cairns J, Lyon JL, Skolnick M, editors. Cancer incidence in defined populations. Cold Spring Harbor, NY: Cold Spring Harbor Laboratories pp 299–318.

40.   Cannon-Albright LA (2008) Utah family-based analysis: past, present and future. Hum Hered 65: 209–20. doi: 10.1159/000112368

41.   Malecot G (1948) Les mathematiques de l'heredite. Paris: Masson & Cie.

42.   Agresti A (1990) Categorical Data Analysis. New York: Wiley.

43.   Costanzo PR, Schiffman SS (1989) Thinness—not obesity—has a genetic component. Neurosci Biobehav Rev 13: 55–58. doi: 10.1016/s0149-7634(89)80052-1

44.   Centers for Disease Control and Prevention (CDC) (2008) Behavioral Risk Factor Surveillance System Survey Data. Atlanta, Georgia: U.S. Department of Health and Human Services, Centers for Disease Control and Prevention.

45.   Centers for Disease Control and Prevention (CDC) (2011) National Center for Health Statistics (NCHS). National Health and Nutrition Examination Survey Data. Hyattsville, MD: U.S. Department of Health and Human Services, Centers for Disease Control and Prevention. Available:http://www.cdc.gov/nchs/nhanes/about_nhanes.htm

46.   Center for Disease Control and Prevention (CDC) National Center for Health Statistics (NCHS). Health E-Stat. Prevalence of Low BMI Among Adults: United States, 2003–2006. Available: http://www.cdc.gov/nchs/data/hestat/low BMI/low BMI_adults.htm

Chapter 10

GENETIC ANALYSIS OF INFLORESCENCE AND PLANT HEIGHT COMPONENTS IN SORGHUM (PANICOIDAE) AND COMPARATIVE GENETICS WITH RICE (ORYZOIDAE)

Dong Zhang[1,2], Wenqian Kong[1,3], Jon Robertson[1], Valorie H Goff[1], Ethan Epps[1], Alexandra Kerr[1], Gabriel Mills[1], Jay Cromwell[1], Yelena Lugin[1], Christine Phillips[1] and Andrew H Paterson[1,2,3,4,5]

[1]Plant Genome Mapping Laboratory, University of Georgia, Athens, GA 30602, USA

[2]Institute of Bioinformatics, University of Georgia, Athens, GA 30602, USA

[3]Department of Crop and Soil Sciences, University of Georgia

[4]Department of Plant Biology, University of Georgia

[5]Department of Genetics, University of Georgia

ABSTRACT

Background

Domestication has played an important role in shaping characteristics of the inflorescence and plant height in cultivated cereals. Taking advantage of meta-analysis of QTLs, phylogenetic analyses in 502 diverse sorghum accessions, GWAS in a sorghum association panel (n = 354) and comparative data, we provide insight into the genetic basis of the domestication traits in sorghum and rice.

Results

We performed genome-wide association studies (GWAS) on 6 traits related to inflorescence morphology and 6 traits related to plant height in sorghum, comparing the genomic regions implicated in these traits by GWAS and QTL mapping, respectively. In a search for signatures of selection, we identify genomic regions that may contribute to sorghum domestication regarding plant height, flowering time and pericarp color. Comparative studies across taxa show functionally conserved 'hotspots' in sorghum and rice for awn presence and pericarp color that do not appear to reflect corresponding single genes but may indicate co-regulated clusters of genes. We also reveal homoeologous regions retaining similar functions for plant height and flowering time since genome duplication an estimated 70 million years ago or more in a common ancestor of cereals. In most such homoeologous QTL pairs, only one QTL interval exhibits strong selection signals in modern sorghum.

Conclusions

Intersections among QTL, GWAS and comparative data advance knowledge of genetic determinants of inflorescence and plant height components in sorghum, and add new dimensions to comparisons between sorghum and rice.

BACKGROUND

The Sorghum genus has recently become an important botanical model for Andropogoneae grasses, by virtue of its relatively small and largely-sequenced genome, a minimum of gene duplication thanks to 70 million years of abstinence from polyploidy, and its close relationship to grasses such as maize, sugarcane and Miscanthus that have much more complex genomes [1]. Cultivated sorghum (*Sorghum bicolor*) ranks fifth in importance among the world's grain crops, is a versatile source of food, fodder, and fuel, and possesses a great diversity of cultivated forms that may reflect its wide range of adaptation [2-4].

The ~30 year history of using linked molecular markers to dissect complex traits in plants has broadly used two complementary approaches. Conventional biparental QTL mapping [5] has been widely used and has provided foundational information that led to some successes in the identification of causal genes in many organisms. However, biparental QTL mapping generally offers relatively coarse resolution that is not sufficient to determine causative genes. Highly saturated recombination maps, multiparent advanced generation intercrosses (MAGIC) [6] or nested association mapping (NAM) [7], offer options to enhance mapping resolution of QTLs. Dramatic increases in genomic data provide rich resources with which to investigate genes and gene functions on a much finer scale than QTL mapping by taking advantage of historical accumulation of recombination events in a gene pool using 'association genetics' [8]. However, association mapping can require extremely high DNA marker densities to thoroughly scan a genome for genes influencing a trait, and complex measures to distinguish between artifacts such as relatedness among genotypes (especially in improved germplasm) and true evidence of functional association between a mutation and a phenotype [8-11]. Although GWAS is able to explore for causative loci on a genome-wide scale, population structure and genetic relatedness may confound associations at causative loci. Some GWAS conducted in rice and *A. thaliana* have suggested that known causative loci showed weaker signals than nearby markers [12]. In contrast, carefully designed crossing schemes in QTL mapping may be more targeted to locate relevant QTLs. Identifying intersections between results from biparental QTL mapping and association genetic data is a potentially powerful means to mitigate constraints associated with each approach, accelerating progress toward identifying specific genes that function in biological processes of relevance to agriculture.

The grass inflorescence, the primary food source for humanity, has been repeatedly selected during domestication [13]. Some well-characterized domestication traits related to the inflorescence include pericarp color of seed, seed shattering, awn length/presence and seed size/yield. Of similarly high and recurring importance in plant domestication and crop improvement are flowering time and plant height, which often show significant genetic correlation with one another [13,14].

In this study, we use GWAS to investigate 6 components of sorghum inflorescence morphology and 6 traits related to plant height, then compare GWAS-based associations to positional evidence from meta-analysis of QTL likelihood intervals. QTL meta-analysis, the comparison of multiple independent QTL studies in different germplasm and environments, is used here to provide a more comprehensive picture of the true genetic control of a trait than analysis of any single population [15], for example revealing plant height and flowering of sorghum to be genetically more complex than had been realized after more than 70 years of investigation [16]. We note that a few classically-identified loci (*dw1-dw4*) had already been compared to GWAS data [4], one of which (*dw3*) had been cloned [17], however here we address additional loci identified only by meta-analysis of QTL likelihood intervals.

Inbreeding organisms have limits to the precision in the association mapping studies [12]. Sorghum is largely inbreeding, which can result in strong LD patterns, and may lead to low genetic resolution in specific local regions along the genome. Hence, we reported the hotspots underlying 12 traits in sorghum, instead of gene candidates. The identified 'hotspots' provide a valuable advance toward the goal of uncovering causative variants for the trait of interest.

It is accepted widely that QTL intervals controlling common traits have non-random correspondence across and within cereals, and GWAS adds a new dimension to the ability to compare the genetic control of common traits in different cereals (and other taxa). For example, in an early comparison, the observed probabilities that seed mass (size) QTLs in sorghum, rice, and maize would correspond so frequently by chance was conservatively estimated as 0.1 to 0.8% [18]. Another early study [14] indicated that 8 of 25 regions affecting flowering of maize fall into 4 homoelogous regions. Numerous studies have shown that some orthologs across taxa have similar functions underlying common phenotypes, but other causative genes have no obvious counterparts that contribute to similar traits even in their close relatives. Thus, genetic correspondence may reflect functionally conserved 'hotspots' existing across taxa, but may or may not be conserved 'genes'. Genes in a pathway exhibit significantly higher genomic clustering than expected by chance in eukaryotes [19]; for example, co-regulated clusters of genes have been implicated in QTLs affecting cotton fiber traits [20].

The common ancestor of rice, maize and sorghum experienced a whole-genome duplication (WGD; named rho) that is still readily discernible in their genomes [21], making it possible to test the hypothesis of convergent evolution across an estimated 70 million years, with the possibility of subfunctionalization among homoeologous regions. A hypothesis worthy of further exploration is that a co-regulated cluster of genes in the cereal common ancestor may have experienced gain/loss and functional divergence of some members in the subsequent 70 my of divergence, with independent domestications conferring additional functional changes in similar locations of different taxa but which are not strictly orthologous.

METHODS

Genotype

The imputed version of ~265,000 published SNPs characterized in 971 worldwide accessions based on genotyping-by-sequencing (GBS) was employed [4]. About 72% of annotated genes contain ≥ 1 SNP site. A total of 228 of 354 accessions from a US sorghum association panel (SAP) [22] are converted tropical lines that are photoperiod insensitive, early maturing, and short stature phenotypes, produced via crossing exotic lines and modern U.S. cultivars. It has been demonstrated [4] that the population has sufficient power to dissect a trait, such as inflorescence architecture, that was not a target of selection in the sorghum conversion program. We used the 354 accessions from the SAP to perform GWAS. A total of 502 accessions, each characterized to a unique morphological type, were used to analyze the population structure. Wild sorghums [*Sorghum bicolor* ssp. *verticilliflorum* (L.) Moench] (n = 31) of four races, namely *aethiopicum*, *arundinaceum*, *verticilliflorum* and *virgatum*, were used to calculate expected heterozygosity for the wild population, and to compare with the heterozygosity in the 502 accessions of cultivated sorghum.

Phenotype for GWAS

Phenotypic data for the 354 accessions in the SAP from three different growouts was utilized. We used a completely randomized design for all the 354 accessions (unreplicated trial with two observations per plot). On seed grown during 2008 in Lubbock, TX, the average RGB values for pericarp color for each genotype were determined from images of the surfaces of 5 seeds. Using the conversion formula of RGB→CIE-L*ab, RGB was transformed to CIE-L*ab color space. RGB is summarized by three values: R, G and B, and CIE-L*ab is summarized by: L, a and b. From growouts in 2009 (seeds sowed on May 19th) and 2010 (seeds sowed on May 26th), near Watkinsville GA, we took two representative samples (from each plot) per genotype in each year for the following measurements. Awn presence: presence or absence of awn, with 2 for abundant, 1 for occasional, and 0 for absent; base-flag length: measured in cm, from the plant base to the flag leaf; flag-rachis length: measured in cm, from the flag leaf to the top node, with positive values indicating the flag leaf below the rachis and negative values indicating the flag leaf above the rachis; inflorescence length: measured in cm, from the rachis to the top of inflorescence; inflorescence width: measured in cm, at the widest point; nodes: the number of nodes; whorls: the number of whorls; dry inflorescence weight; dry stalk weight; total plant height; and flowering time: the average number of days for the first five heads to flower, from the planting date. For traits measured on two samples, we used the average values per genotype to assess heritability and phenotypic correlations. Since the raw data in all the quantitative traits in our study is distributed in near-normal fashion, we conducted GWAS by using the data without transformation. Considering that data combined across years may weaken association levels for those traits with low heritability, we performed GWAS on data for each year individually.

QTL Mapping

QTL confidence intervals were from two resources. (1) We compiled 1-LOD likelihood intervals, which have been identified to underlie any of the 12 traits from published literature [13,14,18,23-29]. (2) We used a recombinant inbred line (RIL) from an interspecific cross between S. bicolor and S. propinquum (BTxSP), the widest cross that can be made with S. bicolor using conventional techniques, containing 141 loci on 10 linkage groups collectively spanning 773.1 cM to map the confidence intervals for base-flag length, flowering time, nodes and total plant height [29] for 2010 and 2011 data. Phenotypic data were collected with two replications per genotype in both 2010 and 2011. A LOD score of 2.5 was used for QTL detection. The methods for anchoring QTL intervals to the reference genome have been discussed in Zhang et al., 2013 [16]. Briefly, based on colinearity between genetic and physical positions of markers, a QTL region is delineated by two flanking markers nearest to the likelihood peak that have alignment information (BLASTN hits). Genomic positions for intervals are available from Additonal file 2: Table S2.

GWAS

The Compressed Mixed Linear Model (CMLM) involves genetic marker-based kinship matrix modeling of random effects, used jointly with population structure estimated by principal components analysis (PCA) to model fixed effects [30-32]. Since more extensive genetic variations may confer a phylogeny closer to the true one, the total of 265,487 SNPs in the SAP were used to analyze population structure. The compression level and optimal number of principal components that adequately explain population structure were previously determined by the Genomic Association and Prediction Integrated Tool [30-32]. Log quantile–quantile (QQ) P-value plots for 265,487 single-SNP tests of association implied that there were few systematic sources of spurious association using CMLM, noting the close adherence of P values to the null hypothesis over the most of the range.

Significance Threshold

We performed Bonferroni-like multiple testing correction [33] to determine significance thresholds for GWAS. Instead of 265,487 independent tests assumed in the Bonferroni method, the total number of tests was estimated by using the average extent of LD across the genome. On average, LD decays to background levels ($r^2 < 0.1$) within 150 kb in the current GBS data [4]. The effective number of independent tests was defined as LD bins [reference genome size (730 Mb)/average LD extent (150 kb)]. Given 0.05 as the desired experiment wide probability of type I error, a significance cutoff within about an order of magnitude of 10^{-5} was estimated.

Overlap between QTL and Heterozygosity Reduction

Since the hypergeometric probability distribution (sampling without replacement) $p = \dfrac{\binom{1}{m}\binom{n-1}{s-m}}{\binom{n}{s}}$ can assess the correspondence between QTLs [14,23], we used the hypergeometric probability distribution to evaluate genetic overlap between plant height/flowering time QTLs and significant heterozygosity reduction in wild sorghum. n is the total number of intervals (defined as 30 cM, approximating a QTL likelihood interval) along the whole genome; l is the number of intervals having significant heterozygosity reduction; s is the number of intervals having plant height/flowering time QTL; m is the number of intervals having both features (overlapping intervals). The purpose of this test is to show that biparental QTL mapping may capture genetic variations for domestication traits that were evolved from wild sorghum and of importance in the history of sorghum selection. Thus, we used QTL intervals for plant height/flowering time that were only determined by biparental QTL mapping. The regions with the largest 1% of heterozygosity reduction values were selected for testing. We utilized alignment between a high-density genetic recombination map [34] and the sorghum reference genome [1] to unify the genetic positions of QTLs and heterozygosity reduction regions (based on windows of 500 consecutive SNPs).

Reference Genomes

The gene annotations refer to JGI annotation release Sbi1.4 [1] and Michigan State University Rice Genome Annotation Project (MSU-RGAP release 7) [35].

RESULTS AND DISCUSSION

Phylogenetic Relationships of Five Main Sorghum Races

Morris et al used a genome-wide SNP map to explore the population structure of 971 sorghum accessions, and illustrated the differentiation of their geographic origins [4]. However, several intriguing questions remain, for example: (1) what is the most primitive sorghum type?, and (2) how many independent domestications has sorghum experienced?

It is generally accepted that the domestication of sorghum started in Africa. Bicolor, guinea, caudatum, durra and kafir are five main morphological types that are well recognized to represent genetic diversity in the cultivated sorghum. Wild sorghum [*Sorghum bicolor* ssp. *verticilliflorum* (L.) Moench] including four races (namely *aethiopicum, arundinaceum,verticilliflorum* and *virgatum*), is the progenitor of the cultivated sorghum [*Sorghum bicolor* (L.) Moench] [36,37]. The phylogenetic relatedness in sorghum races has been discussed in several studies [38-41], but ambiguous clustering patterns have often been found, part of which may be attributable to the limitations of either low-density markers or small population size. To re-investigate inferences drawn before, we focused on a subset of 502 accessions, in which 471 are each characterized uniquely to one and only one of the five primary cultivated races and 31 are wild types. Both the phylogenetic tree and the PCA plots indicate that bicolor is the most primitive race, based on having close phylogenetic relationship with wild types (Figure 1a and b). The level of population differentiation, fixation index (F_{ST}), was measured between wild types and each of five primary types. Race bicolor ($F_{ST} = 0.04$) exhibits closer genetic relationship with wild sorghums than any of the other 4 primary races (F_{ST} (guinea-wild) = 0.11, (durra-wild) = 0.20, (kafir-wild) = 0.33, (caudatum-wild) = 0.14), with guinea and caudatum apparently representing early derivatives. A large block of bicolor accessions are intermediate among durra types, potentially consistent with an ancestral relationship (Figure 1b), noting that F_{ST} (0.14) supports small population differentiation between bicolor and durra. Likewise, a large block of guinea accessions clustering within caudatum may suggest another derivation in the history of sorghum selection, also supported by evidence of minimal population differentiation ($F_{ST} = 0.17$). Races caudatum, durra and kafir show clustering patterns that are substantially distinct from one another, save for occasional single accessions that could be misclassifications (Figure 1b). Pairs of these 3 sorghum races show relatively high levels of population differentiation [F_{ST} (durra-caudatum) = 0.26, (durra-kafir) = 0.46, (caudatum-kafir) = 0.33]. Although the first three components of PCA are able to explain 89.1% of the total genetic variance, ambiguous clustering patterns are still occasionally observed, especially for the two primitive races bicolor and guinea. Both the PCA plot and neighbor-joining tree show quite similar clustering patterns by using the complete SNP set (265,487 SNPs) (Figure 1) and a set including more informative SNPs (missing rate ≤ 50% and MAF ≥ 0.02). The complex history of diffusion and selection in sorghum may confound phylogenetic inference using the clustering patterns. The phylogenetic relationships of sorghum races are still open to questions that may be more accurately addressed with growing genotype data.

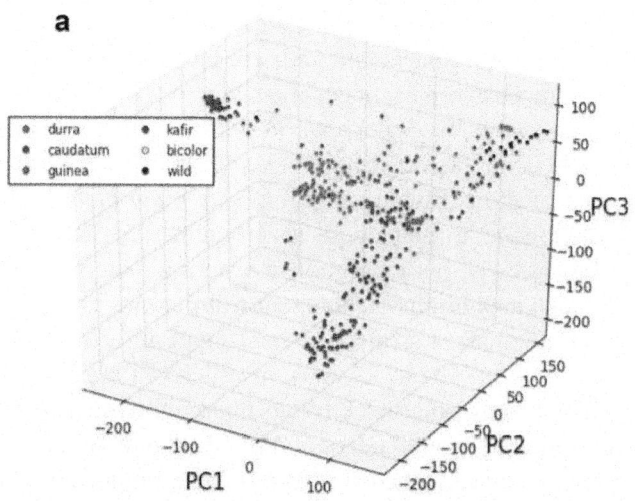

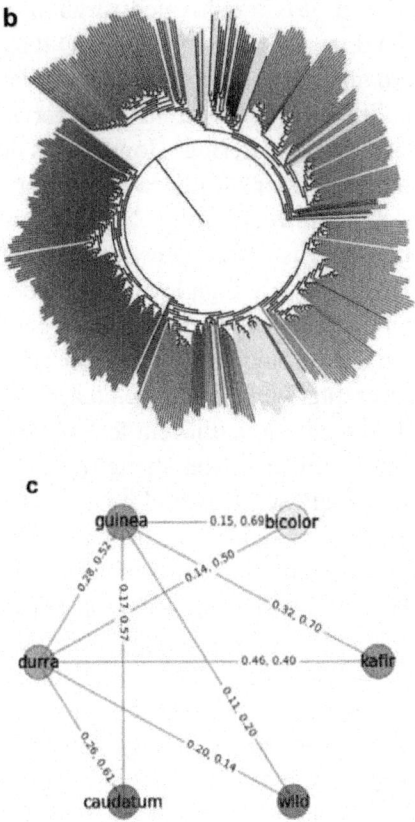

Figure 1: Population structure of 502 worldwide sorghum accessions. 476 belong to the five main cultivated races and 26 are wild types. (a)PCA plots of the first three components for 265,487 SNPs. The five main cultivated races and the wild type are color-coded. (b)Neighbor-joining tree of 502 sorghum accessions. (c) Population differentiations and frequencies of common two-locus haplotypes for 100 SNPs adjacent to the *Sh1* gene [2] for pairs of populations. All the connections for guinea and durra are shown. The F_{ST} values paired with frequencies of common two-locus haplotypes are indicated.

Using representative SNPs across the *Shattering1* (*Sh1*) gene, a key domestication locus [2], a recent discovery revealed four haplotypes, in which three represent non-shattering forms. Specifically, guinea and durra share a common haplotype (SC265); a second haplotype (Tx623) is prevalent in kafir; and a third haplotype (Tx430) is dominant in caudatum. Three non-shattering haplotypes strongly suggest at least three domestication episodes for reduced shattering in the five main sorghum races [2,3]. In general, domestication loci can exhibit very strong LD, because of reduction in genetic divergence. Unusually high genetic divergence at *Sh1* in cultivated sorghum races may explain why the surrounding region does not show stronger LD than the background level. The clustering pattern in the neighbor-joining tree and F_{ST} values support the inference that kafir and caudatum have experienced two independent domestication events for non-shattering. However, is the non-shattering allele derived from a common ancestor of guinea and durra? Generally, we did not observe a close phylogenetic relationship between guinea and durra on the basis of the clustering patterns of the genetic tree (Figure 1b), and the large value of F_{ST} (0.28). This relatively high level of genetic differentiation implies that guinea and durra may have experienced convergent domestication events. The types of haplotypes resulting from selection could be restricted by the limited number of representative SNPs detected across *Sh1*. To overcome this limitation, we examined the region adjacent to *Sh1* to provide a pool enriched by SNPs that are tightly linked to the shattering locus. Since *Sh1* was not genotyped in the current data set, it would be challenging to know the exact number of SNP sites that are linked to *Sh1*. Instead, using 100 SNPs in a 200 kb region adjacent to *Sh1*, we examined haplotypes in consecutive pairs of loci for each sorghum population. The ratio of common haplotypes/total haplotypes was calculated for each pair of sorghum types. Among the 9 possible pairwise comparisons of populations (Figure 1c), guinea-durra only shows a modest frequency (0.52) of common two-locus haplotypes, that is lower than guinea-bicolor (0.69), guinea-caudatum (0.57), guinea-kafir (0.70) and durra-caudatum (0.61). Both guinea and durra have extremely low frequencies (0.20 and 0.14) of common haplotypes with wild sorghum. Below, using two additional domestication genes mapped herein, we further investigate the hypothesis that guinea and durra may have experienced independent domestication events but achieved non-shattering by convergence.

Meta-Analysis of Sorghum QTLs

Biparental QTL mapping is based on the principle that genes and linked DNA markers largely co-segregate during meiosis save for occasional recombination events, thus allowing their analysis in the progeny. The limited number of recombination events captured in progeny of recent crosses may result in QTL likelihood intervals that contain dozens or even hundreds of genes. Further, the environment and parental lines used in a cross can limit the power to accurately estimate the number of QTLs and magnitude of their effects. Using a database that we have recently described [16], we compiled 1-LOD QTL likelihood intervals resulting from 11 independent biparental QTL mapping studies to yield a more complete picture of the genetic control of a trait than could be obtained from any individual study.

The pericentromeric region of sorghum chromosome 6 (Sb06) has repeatedly shown evidence of genetic control of plant height, and was thought to harbor two classic dwarfing genes (*dw2* and *dw4*) [4]. The general lack of recombination in this region allowed QTL confidence intervals to cross centromeres and cover broad genomic areas. Two other regions repeatedly associated with plant height are in the euchromatin of Sb07 and Sb09, which were considered to contain *dw3* and *dw1* respectively [4]. An additional 9 nonoverlapping regions in the sorghum genome containing height QTLs show that genetic control of sorghum plant height involves substantially more than the four genes reported in classical studies [42]. Likewise, 14 flowering QTL likelihood intervals published in six studies fall into at least 11 non-overlapping regions (Figure 2), strongly indicating far more than the classically suggested six genes, *Maturity1* (*Ma1*) to *Ma6*, in genetic control of sorghum flowering time [16,42,43]. Flowering time and plant height show significant genetic correlations on chromosomes Sb01, Sb03, Sb04, Sb06, and Sb09, indicating that their inheritance is linked either functionally (pleiotropy) or physically (linkage disequilibrium) [13].

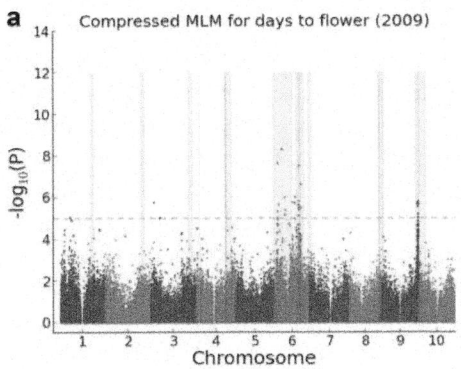

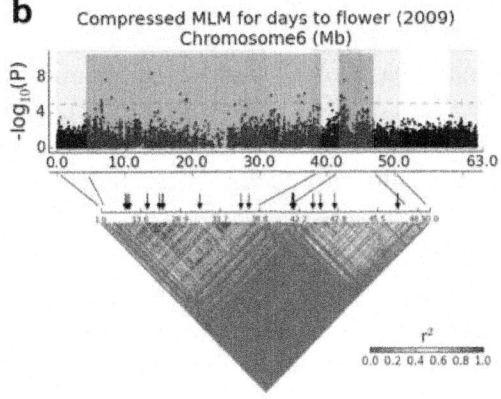

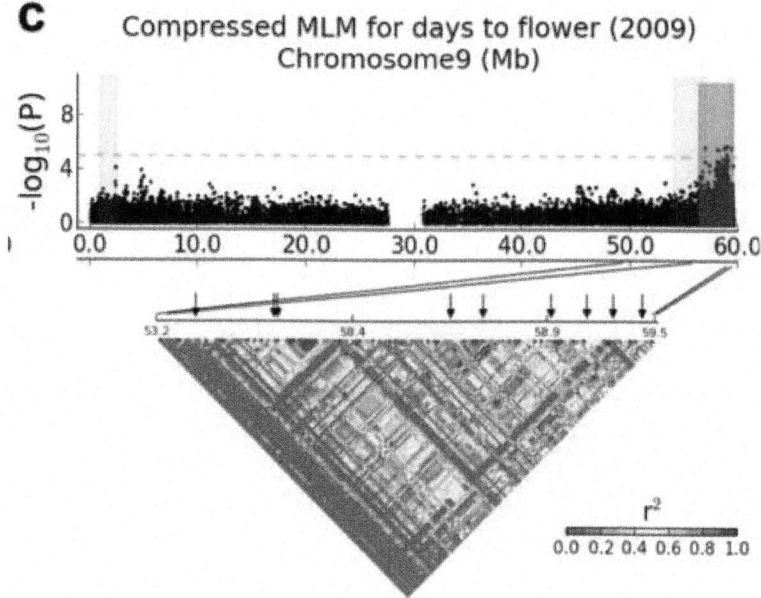

Figure 2: GWAS for flowering time in 2009. (a) Genome-wide Manhattan plot of CMLM. Significance threshold is denoted by the gray dashed line. The 10 sorghum chromosomes are plotted against the negative base-10 logarithm of the association P value. Areas highlighted in green indicate confidence intervals for flowering time determined by QTL mapping. (b) Chromosome Sb06 Manhattan plot of CMLM (top). Red areas show hotspots for 2009 flowering time identified by association mapping. Linkage disequilibrium (r^2 were calculated from SNPs with association $p \leq 0.05$ and missing data $\leq 50\%$) matrices (bottom) are plotted for regions denoted by anchoring lines. Regions of strong LD are shown in red. Significant association markers are denoted by black arrows. (c) Chromosome Sb09 Manhattan plot of CMLM.

GWAS for Sorghum

To dissect the genetic basis of 12 traits in sorghum by GWAS, we used a compressed mixed linear model (CMLM) [30,32] to assess evidence of phenotype-genotype association. Three steps were taken into consideration: (1) determination of significance thresholds for association, (2) identification of linkage disequilibrium (LD) regions for significant association signal and (3) positive control of association.

A major issue with genome scans, which involve many thousands of independent statistical tests, is multiple testing. The Bonferroni method approximates the significance cutoff for an overall (i.e. genome-wide) 5% probability of type I error as $0.05/265,487 = 1.89 \times 10^{-7}$ in our studies. However, this method has been criticized for its stringency [44] owing to the fact that genotype at some SNP loci are correlated thus are not independent hypotheses. Sorghum is largely inbreeding, which can result in strong LD patterns along the genome, so that an appropriate significance threshold may be larger than 1.89×10^{-7}. Here, we used the quantified average LD information in sorghum to adjust the Bonferroni correction [33] (See details in Methods). A significance cutoff within about an order of magnitude of 10^{-5} is inferred to balance an acceptable false positive rate with sufficient power to detect true associations.

It is also important to determine LD with single SNP association, especially when causative variants are not genotyped (or at least not known). On the basis of pairwise measures of LD (r^2), 'block-like' structures can be visually apparent. It is now well understood that the extent of LD in the pericentromeric region, which experiences relatively little recombination, is greater than in the euchromatin, which experiences more frequent recombination. A long LD block with association signals is most likely to contribute striking features to the 'skyline' of a genome-wide Manhattan plot.

Known genes and biparental QTL intervals that have been identified previously are useful to assess association validity. If knowledge of such candidate genes/intervals is limited in the species of interest, information from closely related species might be utilized, using synteny-based approaches to deduce orthology. In addition to the compilation of QTL likelihood intervals and *Dwarf1(Dw1)-Dw4* loci [4] in sorghum, sorghum maturity (*Ma*) genes *Ma1* [14] and *Ma6* [45], and genes *yellow seed1* (*y1*) [46,47] and *Tannin1* (*Tan1*) [48] for sorghum pericarp color provide positive controls for GWAS.

Traits related to the Sorghum Inflorescence

We investigated 6 properties of the sorghum inflorescence, including awn presence, pericarp color, dry inflorescence weight, inflorescence length and width, and whorl number.

Using a *S. bicolor* intraspecific map (BTxIS), Hart et al., [24] identified an interval controlling the presence of awns in euchromatin near the 3' end of chromosome Sb03. The most striking association based on GWAS of awn presence for two years was in the genetically mapped interval (Figure 3a, Figure 4b). Both mapping strategies achieve similar genetic resolution, with intervals spanning ~4.7 megabases (Mb). We found 10 additional significant association hotspots in 2009 and 7 in 2010, none of which are consistent in the two years, which could represent modifiers affected by environment, or false positive associations.

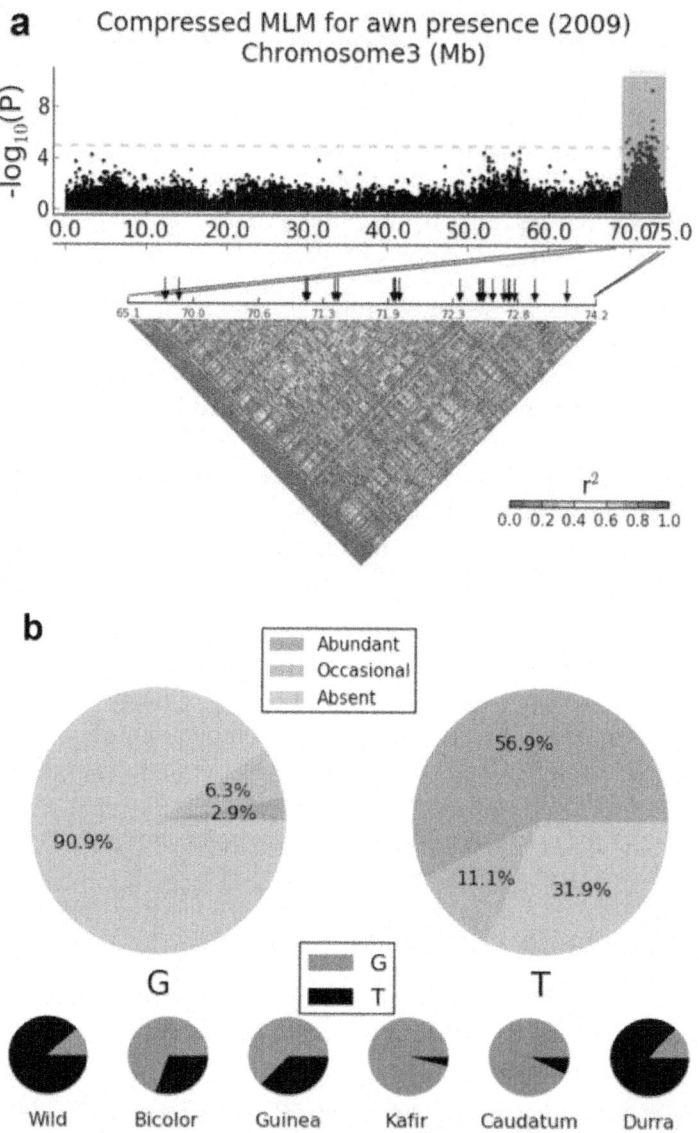

Figure 3: The spectrum of awn presence variation and allele frequencies at locus "S3_72702502" on chromosome Sb03. (a) Chromosome Sb03 Manhattan plot for 2009 awn presence in sorghum is plotted with the hotspot (red area) identified by GWAS, with the prior interval (green area) determined by QTL mapping [24], and with the LD pattern determined by r^2. Significant association markers are denoted by black arrows. (b) Three types of awn classification, which are "abundant", "occasional" and "absent" are color-coded in their frequencies plots for alleles "G" and "T". The allele frequencies are plotted for the five main cultivated races and wild sorghum.

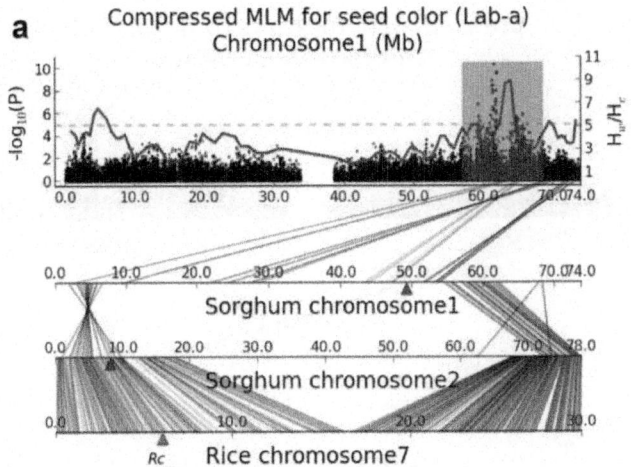

a Compressed MLM for seed color (Lab-a) Chromosome1 (Mb)

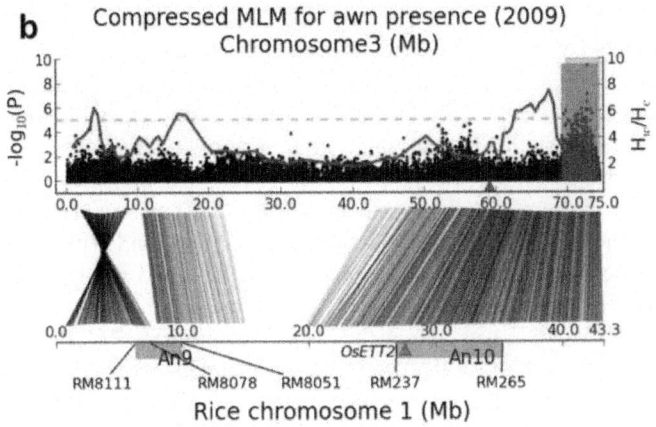

b Compressed MLM for awn presence (2009) Chromosome3 (Mb)

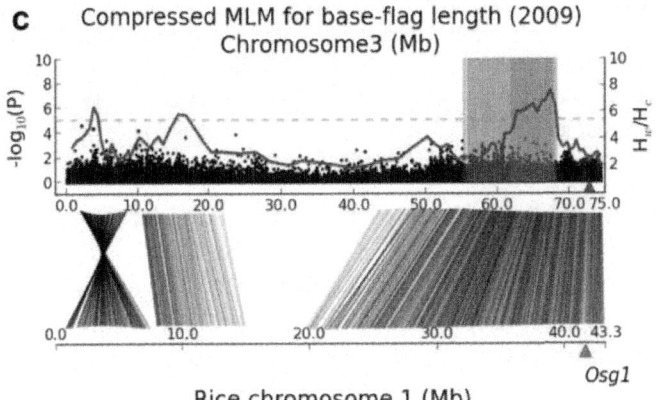

c Compressed MLM for base-flag length (2009) Chromosome3 (Mb)

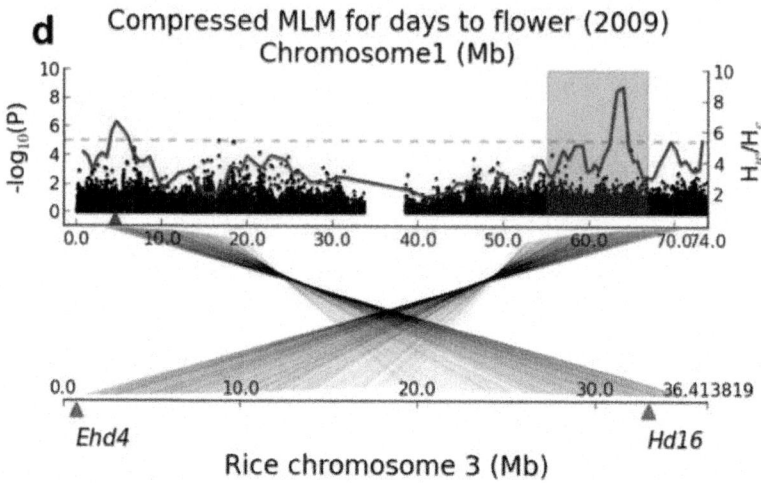

Figure 4: Genetic correspondence across taxa. (a) Chromosome Sb01 Manhattan plot for pericarp color (a value of L*ab model) of sorghum is plotted with the hotspot (red area) identified by GWAS, and with the curve of H_w/H_c ratios. The *Rc* [52] gene is denoted by a red triangle. Two sorghum homologs of *Rc* on chromosomes Sb01 and Sb02 are denoted by blue triangles. Gray connecting lines indicate pairs of duplicated genes. (b) Chromosome Sb03 Manhattan plot for 2010 awn presence in sorghum is plotted with the hotspot (red area) identified by GWAS, and with the prior interval (green area) determined by QTL mapping [24]. Rice awns co-segregated with SSR marker RM8078 tightly linked to An9 on chromosome Os01 [63]. The interval An10 for rice awn [63] was associated with SSR markers RM265 and RM237 on chromosome Os01. The *OsETT2* [64] gene is denoted by a red triangle. Sorghum ortholog of *OsETT2* is indicated by a blue triangle. (c) The genomic interval (~55 Mb-67 Mb) on chromosome Sb03 is implicated by three linkage studies [14,26,28] to affect plant height and flowering time, but doesn't harbor association signal in our GWAS. The *Osg1* [66] gene and its sorghum ortholog are indicated by a red and a blue triangle individually. (d) The genomic interval (~58 Mb-64 Mb) on chromosome Sb01 is implicated by two linkage studies [24,26] to control plant height, but doesn't harbor association signal in our GWAS. The genes *Ehd4* [67] and *Hd16* [68] are indicated by red triangles, and sorghum ortholog of*Hd16* is denoted by a blue triangle.

Inheritance studies of endosperm color in sorghum proposed that the trait was oligogenic [49,50]. To date, a single gene (*Rc*) has been verified to be responsible for pericarp pigmentation of rice grains (Figure 4a), while gene *Rd* is assumed to play a role in spreading pigment [51,52]. In view of the similar levels of gene duplication in rice and sorghum, it is plausible that pericarp color is also an oligogenic trait in sorghum. In sorghum, it is known that the gene *yellow seed1* (*y1*) is required for the production of red phlobaphene pigments in the grain pericarp [46,47], and the *Tan1* gene controls tannin biosynthesis to affect pericarp color [48]. In order to minimize the limitations of artificial descriptions of colors, we applied two commonly used color models, RGB and CIE-L*ab (See the Methods also for the details), to phenotype pericarp color of sorghum seeds. On chromosome Sb01, an association peak (~61.45 Mb) in a hotspot (58 Mb-67 Mb) for red, green and blue in RGB measurement, and the values of 'L' and 'a' in the Lab model, is close to *y1* (~61 Mb). An additional association peak (~61.86 Mb) near *Tannin1* (*Tan1*) (~61.6 Mb), is localized in a hotspot 57.5 Mb-62 Mb on chromosome Sb04 to be associated with red and green in the RGB and 'L' in the Lab.

Dry inflorescence weight, inflorescence length/width, and whorl number, were each very sensitive to the environment, with no colocalized hotspots between the two years for which we have data. The most striking 'skyline' determined by GWAS for dry inflorescence weight is centered at ~57 Mb on chromosome Sb09, which is also associated with plant height and flowering time. For inflorescence length the most striking association hotspot centered at ~45 Mb on chromosome Sb06, a location also connected with plant height and flowering.

To further investigate the inference that guinea and durra may have experienced independent domestication events but achieved non-shattering by convergence, we examined two-locus haplotypes for loci associated with pericarp color and awn presence respectively, both considered to be traits that may be subject to selection. For pericarp color, the association peak (S1_61453639) and 7 linked ($r^2 \geq 0.6$) SNPs in the hotspot (chromosome Sb01: 57,017,032 bp-68,413,103 bp) are used to calculate the ratio of common haplotypes/total haplotypes for each pair of sorghum types (See the "Phylogenetic relationships of five main sorghum races" also for the details). Similarly, we used an association peak (S3_72702502) and 18 linked SNPs in the hotspot (chromosome Sb03: 68,912,931 bp-74,241,979 bp) to calculate two-locus haplotype ratios for awn presence. Compared to other sorghum race pairs, guinea-durra only shows modest frequencies of common two-locus haplotypes for pericarp color (freq=0.22) and awn presence (freq=0.26). Our findings further support the hypothesis that independent domestication of guinea and durra involved convergent selection for non-shattering.

Race-Specific Patterns in awn Presence Variation

Geographic origins and domestication history can result in patterns of phenotypic variation among genotypes within a gene pool. We investigated whether awn presence exhibits variation patterns correlated with race-specific alleles for sorghum. The hotspot (chromosome Sb03: 68,912,931 bp-74,241,979 bp) for sorghum awn presence, with the association peak (S3_72702502), is detected by both GWAS and QTL mapping (See "Traits related to the sorghum inflorescence" also). The association peak is located at an intergenic locus, which is flanked by two annotated genes [*Sb03g045420*(chromosome Sb03: 72,681,274 bp-72,687,688 bp) (similar to Hexokinase-3) and *Sb03g045430* (chromosome Sb03: 72,703,668 bp-72,704,913 bp) (similar to Putative uncharacterized protein)], based on published reduced representation sequence [4]. In this scenario, it is probable that the causative genes/loci were not genotyped, and that S3_72702502 and the causative loci shared history of mutation and recombination.

An early study [53] described morphology of panicles for two of the five main sorghum races. We (Figure 3b) show allele distribution at the locus S3_72702502, reflecting alleles 'T' and 'G' associated with awned and awnless sorghum panicles individually, finding correlation between the alleles and the race-specific morphology of awns. Allele 'G' with the dominant frequency in race kafir corresponds to awnless cylindrical-shaped panicles of kafir [53]. Indeed, we found 84% of kafir accessions to have awnless panicles in our phenotypic data. Similarly, allele 'T' with the dominant frequency in race durra, corresponds to bearded and hairy panicles of durra [53]. The major allele in race caudatum is 'G', consistent with the finding that 92% of caudatum accessions for which we have data are awnless. Our findings also suggest that the two most primitive sorghum types, bicolor and guinea, derived more allele 'G' from wild sorghum which is dominated by more ancestral allele 'T'.

Traits related to Plant Height

We conducted GWAS for 6 traits related to sorghum plant height, including total plant height, distance from base to flag leaf (base-flag), distance from rachis to flag leaf (rachis-flag), number of nodes on the main stalk (nodes), dry stalk weight and days to flowering. Strong positive correlation are observed among the phenotypic data of total plant height, base-flag, nodes, dry stalk weight and days to flowering. It is important to remember that 228 accessions in the sorghum association panel (SAP), are converted tropical lines that are photoperiod insensitive, early maturing, and short statured phenotypes, developed by crossing exotic lines and U.S. cultivars [14,22]. Three independent studies [4,14,54] revealed three consistent genomic regions in sorghum that contribute to the introgression of cultivar-specific alleles to exotic sorghum lines. It is also clear that *Dwarf* (*Dw*)*4* and *Dw2* are associated with the introgression region in the heterochromatin of chromosome Sb06, and *Dw3* and *Dw1* are associated with the introgression regions in the 3' terminal euchromatin of chromosome Sb07 and Sb09 respectively.

In our studies, the *Dw1-Dw4* regions reported by Morris et al. [4] show the most striking signals associated with base-flag leaf length for two years. Likewise, the association results of total plant height are consistent with that of Morris et al [4]. This suggests that environmental variables have relatively little effect on *Dw1-Dw4*.

In addition to *Dw1-Dw4*, we found noteworthy hotspot(s) for plant height located on chromosome Sb04. Our BTxSP mapping population [29] and Shiringani et al., [28] each suggested two overlapping QTL likelihood intervals on Sb04, 57.98 Mb-64.93 Mb and 48.80 Mb-58.58 Mb, to contribute to plant height. Within both intervals, we found multiple SNPs significantly associated with base-flag and dry stalk weight. Because the terminal region of Sb04 holding these QTLs exhibited weak LD, we could not set clear boundaries for the hotspot(s). GWAS of another plant height related trait, dry stalk weight in 2009 data, also shows a clear association 'skyline' in the region of 48.80 Mb-64.93 Mb on Sb04.

Plant height and flowering time show significant genetic correlation. In further agreement, three association loci for flowering time show strong LD with dwarfing genes, and are distributed in the introgression regions. There is some confusion concerning the identities of *Ma1* and *Ma6*, with one group suggesting that *Ma1* is a sorghum ortholog of a Triticeae flowering gene, *PRR37* [55], but which is located very near the published position of *Ma6* [45]. Many of the same authors who report that *PRR37* is *Ma1* have stated under separate cover that *PRR37* is *Ma6* [56]. The LD pattern suggests two haplotype blocks in the region of 6 Mb-46 Mb on chromosome Sb06 (Figure 2b), suggesting that *Ma1* (or *Ma6*) and *Dw4*[4] are tightly linked in the region of 6 Mb-39 Mb, and *Dw2* [4] is linked with *Ma6* (or *Ma1*) in the region of 40 Mb-46 Mb on Sb06. We also identified a haplotype block of 56 Mb-59.5 Mb on Sb09 (Figure 2c), which is strongly associated with plant height and flowering time. Greater mapping resolution is required to pinpoint the causative genetic mutation(s) that affect each trait.

Selective Signatures and Phenotype-Genotype Association

A phenotype-genotype association is no guarantee that the trait or its candidate gene has been historically important or is an adaptation [57]. Reduction in genetic diversity, which can be assessed by heterozygosity, can bear the signature of domestication [4,58,59]. Using quantified genome-wide heterozygosity, low heterozygosity regions in sorghum have been reported [4] but have not shown significant correspondence with identified loci associated with plant height, one of the key domestication traits. Here, we refined analysis of selective signatures from domestication and explored the genomic regions that have been important in sorghum adaptation for inflorescence morphology and plant height. Genetic diversity of the sorghum population was assessed by the ratio of expected heterozygosity in wild sorghum to that in cultivated types (H_w/H_c) across the sorghum genome (Figure 5). Overall, the refined pattern of reduction in genetic diversity is consistent with that measured previously [4]. Among the hotspots determined by association mapping for the 12 traits, the genomic region 58 Mb-64 Mb on chromosome Sb01 for pericarp color shows a strong selection signal based on reduction in heterozygosity ratio (Figure 4a). Such correspondence was also discovered in rice [60], consistent with the observation that pericarp color has profoundly influenced the popularity of cultivated cereal varieties [61].

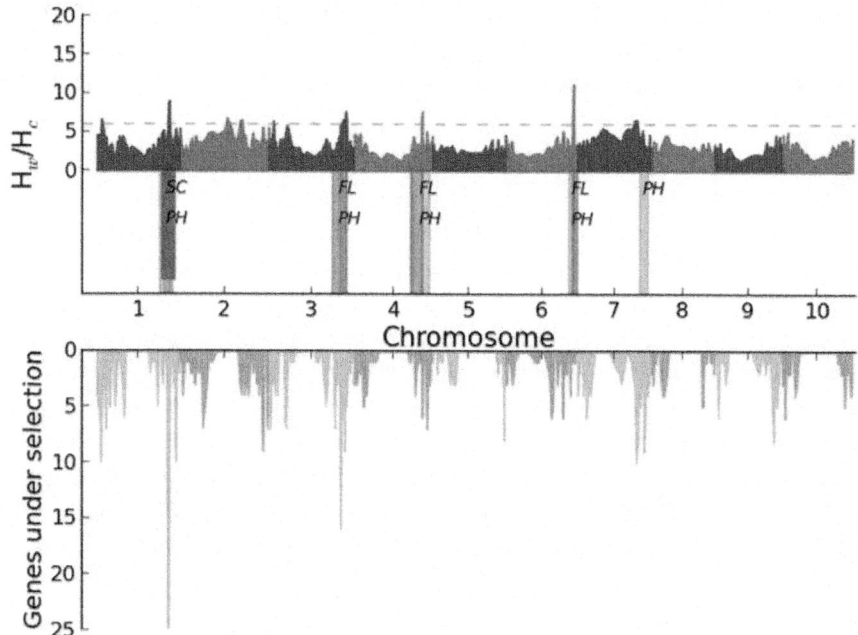

Figure 5: Genome-wide selection signatures. We calculated average ratios (expected heterozygosity of wild type sorghum (H_w) / expected heterozygosity of cultivated sorghum (H_c)) for windows of 500 consecutive SNPs throughout the genome. The gray dashed line is the cutoff for the top 5% of heterozygosity ratios. Green areas indicate confidence intervals determined by QTL mapping to contribute to either flowering time (FL) or plant height (PH), that also have strong selection signals. The hotspot on chromosome Sb01 identified by GWAS for pericarp color (SC) is highlighted by red. The distribution of 725 candidate genes implicated in domestication and/or improvement via gene-based population summary statistics [41] is shown below the heterozygosity ratios.

Some confidence intervals identified by QTL mapping for plant height and flowering time [14,24-28] colocalized non-randomly ($p = 0.231 \times 10^{-6}$) (See the overlapping testing in the "Methods") with regions having striking reductions in heterozygosity ratio (Figure 5). For example, Hart et al., [24] and Ritter et al., [26] each identified a region (55.1 Mb-66.7 Mb) for plant height on chromosome Sb01 which showed strong selection signals. A region of 55 Mb-62 Mb on chromosome Sb06, showing strong selection signals, was identified for plant height by Kebede et al., [25], Srinivas et al., [27] and Shiringani et al., [28], and for flowering time by Shiringani et al., [28]. Additional heterozygosity ratio peaks overlapping with plant height and flowering time intervals are found on Sb03 and Sb04. Most such colocalizing intervals affect plant height and flowering time simultaneously, and do not colocalize with hotspots determined by association mapping.

Using 8 M SNPs characterized in 44 sorghum lines, a recent study [41] implicated 725 candidate genes in sorghum domestication and/or improvement via gene-based population summary statistics. We show (Figure 5) at least two regions having striking reductions in heterozygosity ratio, to also be enriched for candidate genes under selection. One striking case is the 58 Mb-64 Mb interval on chromosome Sb01, in genetic control of pericarp color and plant height [24,26]. The other case is the 62 Mb-64 Mb interval on chromosome Sb03, in genetic control of plant height and flowering time [14,26,28].

Much of the 'missing heritability' from GWAS has been speculated to be attributed to the low/rare frequency of causative mutations, which are unlikely to be detected by most association studies [62]. Some genetic variations with large effect may be hidden by neutral genetic diversity in a natural population. By contrast, biparental QTL mapping uses genetic markers (e.g. SSR and RFLP), and is often derived from parental lines that differ strongly in phenotype such as our *S. bicolor* × *S. propinquum* cross between parents which are separated by 1-2 million years. Sometimes, biparental QTL mapping can be more efficient than GWAS to detect alleles that are rare in populations. QTL mapping and GWAS are thus two complementary approaches to determine "saturation" in terms of QTL discovery. Our findings lead to the prediction that the likelihood intervals on chromosome Sb01, Sb03, Sb04 and Sb06 identified by QTL mapping should bear the signature of selection, and may encode important variants with low/rare frequency for plant height and flowering time.

Genetic Correspondence between Sorghum and Rice

Synteny and colinearity have been well conserved between grass species such as rice, maize, and sorghum since their divergence about 50 Mya [21], enabling us to compare causal loci in corresponding regions across taxa. SSR mapping in rice revealed that two intervals affecting awns are located on chromosome Os01, and designated as *An9* and *An10*, respectively (Figure 4b) [63]. *An10* produced shorter and sparser awns than *An9*, and is thought to be more important for awn presence in rice. Recently, a rice gene, *OsETT2*, which colocalizes with *An10*, has been identified using genetic analyses and RNA-silencing experiments, to be involved in awn formation [64]. Chromosome Sb03 harbors the most significant hotspot for awn presence in sorghum, bounded in a region sharing large-scale homoeology with a region on Os01 containing *An10* and *OsETT2* (Figure 4b). The two hotspots do not appear to have direct correspondence across taxa, making it unclear whether there has been some localized rearrangement, functional divergence of different members of a co-regulated cluster of genes, or some other explanation.

Duplication and subsequent rearrangements can result in chromosomes exhibiting a complex pattern of homoeology across taxa. The *Rc* gene is responsible for conditioning red pericarp in rice [52]. Association mapping shows no noteworthy signal around the sorghum ortholog (*Sb02g006380*) of *Rc*, or for another sorghum gene copy (*Sb01g028230*). Rearrangement following polyploidization can lead to a scrambled patchwork in genome organization [65]. In the case of pericarp color, the correspondence of gene arrangement between the most significant hotspot in sorghum and the rice *Rc*region is still discernible through multiple alignment (Figure 4a). The hotspot on sorghum chromosome Sb01 shares homoeology with three small genomic segments on Sb01, one of which has correspondence with the region enclosing*Sb02g006380* on sorghum chromosome Sb02. Thus, despite discernible parallels in genome organization, corresponding genes do not appear to be responsible for phenotypic variation in seed color of sorghum and rice. This may simply reflect that 40-50 My of rice-sorghum divergence, did involve many genetic changes that were different, as well as some that may have been convergent [18].

Both linkage studies and reductions in heterozygosity ratios (Figure 4c and d) implicate two genomic regions in genetic control of plant height and flowering time, which are located in the 58 Mb-64 Mb on chromosome Sb01 [24,26] and in the 62 Mb-64 Mb on chromosome Sb03 [14,26,28] individually. In rice, gene *Osg1* [66], which is located on chromosome Os01 (Figure 4c) orthologus to sorghum chromosome Sb03, has been known to affect plant height via reduction in its expression; genes *Ehd4* [67] and *Hd16* [68] on chromosome Os03 orthologous to sorghum chromosome Sb01, are characterized by their function in flowering time of rice. We observed that sorghum has experienced gene loss for *Ehd4*, and none of three rice genes shows direct correspondence with the two genomic regions in sorghum. These findings indicate that different genes, rather than orthologs of *Osg1*, *Ehd4* and *Hd16*, were identified by QTL mapping on Sb03 and Sb01.

QTL Correspondence Across Homoeologous Regions within Sorghum

QTLs may share homoeology not only across taxa, but also within taxa. Maize, in which many homoeologous chromosome segments have been identified as a result of lineage-specific genome duplication, is a particularly favorable organism in which to show such QTL correspondence [14]. Only occasional cases have been found in sorghum [14], probably due to much greater antiquity of genome duplication [1].

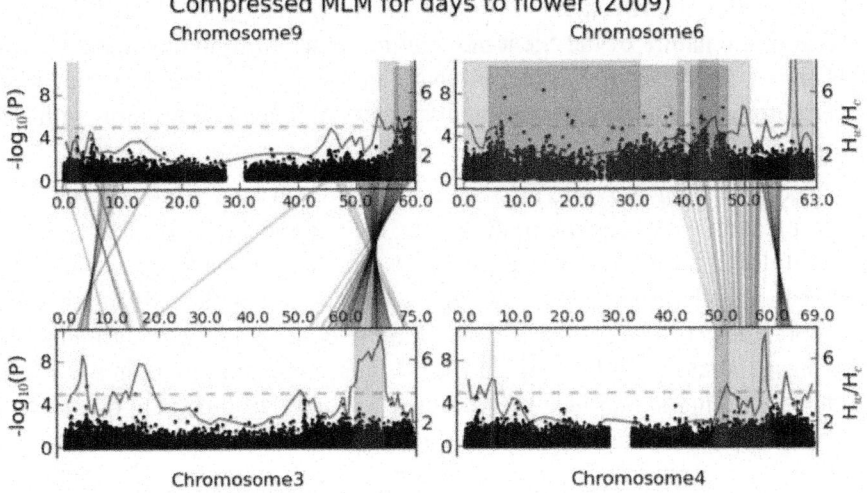

Figure 6: Genetic correspondence within taxa. Intra-genomic genetic correspondence for QTL intervals for flowering time on chromosome pairs of Sb03-Sb09 and Sb04-Sb06.

A total of 4 pairs of homoeologous QTLs for plant height/flowering time are found in our study. For flowering time, we observed two instances of homoeologous QTLs in sorghum (Figure 6). The confidence interval mapped on chromosome Sb03 [28] shares homoeology with the hotspot of 54 Mb-59.6 Mb on chromosome Sb09 which has been identified by QTL mapping [23] and our GWAS. The other correspondence was determined between 48.6 Mb-59.3 Mb on chromosome Sb04 [26,28] and 38 Mb-50.7 Mb on chromosome Sb06 [14,18]. Likewise, we found homoeologous QTLs for plant height on chromosome pairs of Sb03-Sb09 and Sb04-Sb06. The additional cases for flowering time are shown on chromosome pairs of Sb02-Sb07 and

Sb05-Sb08. Such genetic correspondence within sorghum implies that some homoeologous regions created in a genome duplication 70 million years ago [21] may still retain some genes with similar functions. In most instances, we found that one QTL hotspot exhibits much stronger selection signal than its homoeologous counterpart on the basis of genetic diversity reduction in the wild population. Independent episodes of domestication and crop improvement may preferentially select one counterpart, but may or may not select the other.

CONCLUSIONS

Our studies illustrate how GWAS may be used to complement the genetic resolution of causal elements (genes) of quantitative phenotypes that is typically attained from conventional likelihood intervals determined by QTL mapping. The degree of improvement in resolution by GWAS over QTL mapping is related to the nature of the 'genomic environment' surrounding a gene – with substantial improvement in recombinationally-active euchromatin but much less improvement in recombinationally recalcitrant heterochromatin with long LD blocks.

Understanding and utilizing the relative strengths and weaknesses of QTL mapping and GWAS can aid in dissecting the genetic basis of a complex trait. A carefully chosen cross can allow QTL mapping to have better statistical power to detect variants with low/rare frequency in a natural population. For example, the classical maturity locus *Ma3/phyB* was initially identified with a map-based strategy (QTL mapping) [69], but association mapping is unable to detect striking signal near*Ma3* because virtually all members of the panel are wild type (i.e. the mutation is rare). Another case is that a map-based (QTL mapping) strategy was used to determine one major-effect QTL (*sh1*) controlling shattering in sorghum, and revealed three different non-shattering haplotypes widely existing in five major cultivated races that may reflect low/rare frequency of causative polymorphism [2,3]. Hence, association mapping may have reduced power to find phenotype-genotype association for *sh1*. Additionally, complex population structure in unrelated individuals may compromise genetic variation at the true causative loci, resulting in cases of true negative detection in GWAS. More generally, multiple independent domestications may create a scenario under which many QTLs cannot be verified with the current sorghum association panel.

An important fundamental question in gene mapping is why the GWAS approach has revealed so much less variation than anticipated [62]. Even given 'perfect' (i.e. 100% accurate) information about phenotype and genotype, some associations may not be repeatable due to interaction between genotype and environment [57]. We observed that traits related to the inflorescence were more greatly affected by environment than traits related to plant height (Figure 7). The interaction between genotype and environment plays an important role in the variation between individuals in their observable characteristics, even for traits with high heritability.

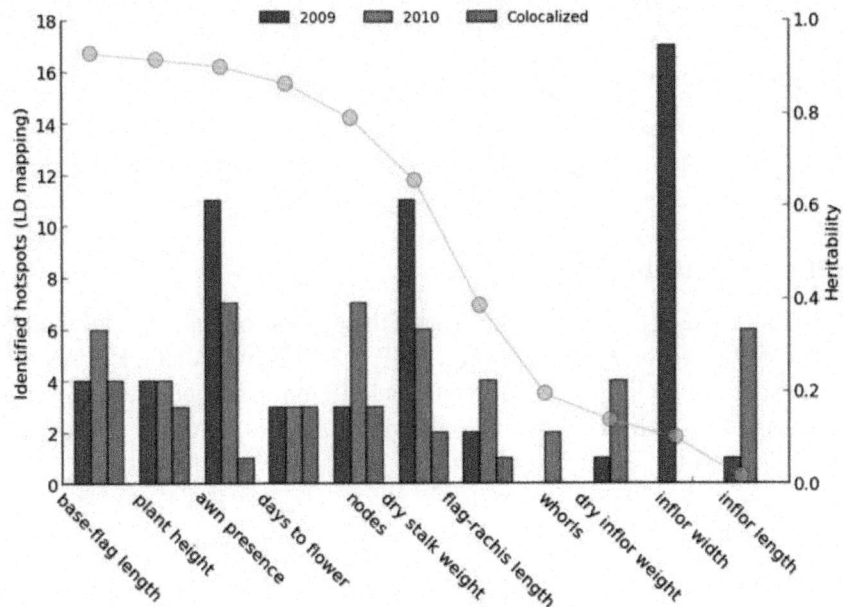

Figure 7: Summary of hotspots identified by GWAS for multi-year trait data. For each trait, numbers of hotspots for 2009, 2010 and both years are indicated with bars. The values of heritability for the traits are indicated with the curve.

Identification of intersections among QTL, GWAS and comparative data advance knowledge of the genetic determinants of variation in sorghum, and provide a finer-scale comparison than was previously possible of the genetics of important traits between sorghum and rice. Non-random correspondence of sorghum GWAS signals to rice QTLs, i.e. between divergent panicoid and oryzoid grasses, adds a new dimension to evidence of the ability to leverage genetic data about the important traits across divergent plants. More variants can be gained through re-sequencing gene candidates. Joint evidence from hotspots and re-sequencing data are likely to be extremely powerful to answer the question of whether such genetic correspondence implicates 'conserved' single genes or co-regulated clusters of genes that exist for the trait of interest after cereal divergence.

Rich SNP data also clarifies the phylogenetic relatedness of sorghum races, improving on several prior studies [38-41]. Based on the high-density genotype in the large population, we showed joint evidence from the striking clustering patterns in population structure and fixation index values, suggesting several hypotheses that might shed light on the unknown ancestry of sorghum races.

DECLARATIONS

Acknowledgements

This work was supported by the United Sorghum Checkoff Program, and the US Department of Energy-US Department of Agriculture Plant Feedstock Program (project grant numbers: 2011-02786, 2012-03304).

Authors' Contributions

AHP conceived of the project and guided the process of analyses. DZ and WK conducted the genome data analyses. AHP, DZ, WK, JR, VHG, EE, AK, GM, JC, YL and CP collected phenotypic data. JR built and curated phenotypic databases. AHP and DZ wrote the paper. All authors read and approved the final manuscript.

REFERENCES

1.   Paterson AH, Bowers JE, Bruggmann R, Dubchak I, Grimwood J, Gundlach H, et al. The Sorghum bicolor genome and the diversification of grasses. Nature. 2009;457:551–6.

2.   Lin Z, Li X, Shannon LM, Yeh C-T, Wang ML, Bai G, et al. Parallel domestication of the Shattering1 genes in cereals. Nat Genet. 2012;44:720–4.

3.   Olsen KM. One gene's shattering effects. Nat Genet. 2012;44:616–8.

4.   Morris GP, Ramu P, Deshpande SP, Hash CT, Shah T, Upadhyaya HD, et al. Population genomic and genome-wide association studies of agroclimatic traits in sorghum. Proc Natl Acad Sci U S A. 2012;110:453–8.

5.   Paterson AH, Lander ES, Hewitt JD, Peterson S, Lincoln SE, Tanksley SD. Resolution of quantitative traits into Mendelian factors by using a complete RFLP linkage map. Nature. 1988;335:721–6.

6.   Kover PX, Valdar W, Trakalo J, Scarcelli N, Ehrenreich IM, Purugganan MD, et al. A multiparent advanced generation inter-cross to fine-map quantitative traits in Arabidopsis thaliana. PLoS Genet. 2009;5.

7. Buckler ES, Holland JB, Bradbury PJ, Acharya CB, Brown PJ, Browne C, et al. The genetic architecture of maize flowering time. Science. 2009;325:714–8.

8. Yu J, Pressoir G, Briggs WH, Vroh Bi I, Yamasaki M, Doebley JF, et al. A unified mixed-model method for association mapping that accounts for multiple levels of relatedness. Nat Genet. 2006;38:203–8.

9. Thornsberry JM, Goodman MM, Doebley J, Kresovich S, Nielsen D, Buckler ES. Dwarf8 polymorphisms associate with variation in flowering time. Nat Genet. 2001;28:286–9.

10. Yu J, Holland JB, McMullen MD, Buckler ES. Genetic design and statistical power of nested association mapping in maize. Genetics. 2008;178:539–51.

11. Myles S, Peiffer J, Brown PJ, Ersoz ES, Zhang Z, Costich DE, et al. Association mapping: critical considerations shift from genotyping to experimental design. Plant Cell. 2009;21:2194–202.

12. Morrell PL, Buckler ES, Ross-Ibarra J. Crop genomics: advances and applications. Nat Rev Genet. 2011;13:85–96.

13. Brown PJ, Klein PE, Bortiri E, Acharya CB, Rooney WL, Kresovich S. Inheritance of inflorescence architecture in sorghum. Theor Appl Genet. 2006;113:931–42.

14. Lin Y, Keith F, Paterson AH. Comparative analysis of QTLs affecting plant height and maturity across the poaceae, in reference to an interspecific sorghum population. Genetics. 1995;141:391–411.

15. Rong J, Feltus FA, Waghmare VN, Pierce GJ, Chee PW, Draye X, et al. Meta-analysis of polyploid cotton QTL shows unequal contributions of subgenomes to a complex network of genes and gene clusters implicated in lint fiber development. Genetics. 2007;176:2577–88.

16. Zhang D, Guo H, Kim C, Lee T-H, Li J, Robertson J, et al. CSGRqtl, a comparative quantitative trait locus database for Saccharinae grasses. Plant Physiol. 2013;161:594–9.

17. Multani DS, Briggs SP, Chamberlin MA, Blakeslee JJ, Murphy AS, Johal GS. Loss of an MDR transporter in compact stalks of maize br2 and sorghum dw3 mutants. Science. 2003;302:81–4.

18. Paterson AH, Lin YR, Li Z, Schertz KF, Doebley JF, Pinson SR, et al. Convergent domestication of cereal crops by independent mutations at corresponding genetic Loci. Science. 1995;269:1714–8.

19. Lee JM, Sonnhammer ELL. Genomic gene clustering analysis of pathways in eukaryotes. Genome Res. 2003;13:875–82.

20. Paterson AH, Wendel JF, Gundlach H, Guo H, Jenkins J, Jin D, et al. Repeated polyploidization of Gossypium genomes and the evolution of spinnable cotton fibres. Nature. 2012;492:423–7.

21. Paterson AH, Bowers JE, Chapman BA. Ancient polyploidization predating divergence of the cereals, and its consequences for comparative genomics. Proc Natl Acad Sci U S A. 2004;101:9903–8.

22. Casa AM, Pressoir G, Brown PJ, Mitchell SE, Rooney WL, Tuinstra MR, et al. Community resources and strategies for association mapping in sorghum. Crop Sci. 2008;48:30–40.

23. Feltus FA, Hart GE, Schertz KF, Casa AM, Casa AM, Kresovich S, et al. Alignment of genetic maps and QTLs between inter- and intra-specific sorghum populations. Theor Appl Genet. 2006;112:1295–305.

24. Hart GE, Schertz KF, Peng Y, Syed NH. Genetic mapping of sorghum bicolor (L.) Moench QTLs that control variation in tillering and other morphological characters. TAG Theor Appl Genet. 2001;103:1232–42.

25. Kebede H, Subudhi PK, Rosenow DT, Nguyen HT. Quantitative trait loci influencing drought tolerance in grain sorghum (Sorghum bicolor L. Moench). TAG Theor Appl Genet. 2001;103:266–76.

26. Ritter KB, Jordan DR, Chapman SC, Godwin ID, Mace ES, Lynne McIntyre C. Identification of QTL for sugar-related traits in a sweet × grain sorghum (Sorghum bicolor L. Moench) recombinant inbred population. Mol Breed. 2008;22:367–84.

27. Srinivas G, Satish K, Madhusudhana R, Reddy RN, Mohan SM, Seetharama N. Identification of quantitative trait loci for agronomically important traits and their association with genic-microsatellite markers in sorghum. Theor Appl Genet. 2009;118:1439–54.

28. Shiringani AL, Frisch M, Friedt W. Genetic mapping of QTLs for sugar-related traits in a RIL population of Sorghum bicolor L. Moench. Theor Appl Genet. 2010;121:323–36.

29. Kong W, Jin H, Franks CD, Kim C, Bandopadhyay R, Rana MK, et al. Genetic analysis of recombinant inbred lines for Sorghum bicolor × Sorghum propinquum. G3 (Bethesda). 2013;3:101–8.

30. Bradbury PJ, Zhang Z, Kroon DE, Casstevens TM, Ramdoss Y, Buckler ES. TASSEL: software for association mapping of complex traits in diverse samples. Bioinformatics. 2007;23:2633–5.

31. Zhang Z, Ersoz E, Lai C-Q, Todhunter RJ, Tiwari HK, Gore MA, et al. Mixed linear model approach adapted for genome-wide association studies. Nat Genet. 2010;42:355–60.

32. Lipka AE, Tian F, Wang Q, Peiffer J, Li M, Bradbury PJ, et al. GAPIT: genome association and prediction integrated tool. Bioinformatics. 2012;28:2397–9.

33. Matthies I, Malosetti M, Röder MS, Eeuwijk FV. Genome-Wide association mapping for kernel and malting quality traits using historical European barley records. PLoS One. 2014;9(11):e110046.

34. Bowers JE, Abbey C, Anderson S, Chang C, Draye X, Hoppe AH, et al. A High-density genetic recombination map of sequence-tagged sites for sorghum, as a framework for comparative structural and evolutionary genomics of tropical grains and grasses. Genetics. 2003;386:367–86.

35. Kawahara Y, de la Bastide M, Hamilton JP, Kanamori H, McCombie WR, Ouyang S, et al. Improvement of the Oryza sativa Nipponbare reference genome using next generation sequence and optical map data. Rice (N Y). 2013;6:4.

36. Doggett H. Sorghum, 2nd edition. UK: Longman; 1988.

37. Wiersema JH, Leon B. World Economic Plants: A standard Reference. New York: CRC press; 1999.

38. Brown PJ, Myles S, Kresovich S. Genetic support for phenotype-based racial classification in sorghum. Crop Sci. 2011;51:224.

39. Bouchet S, Pot D, Deu M, Rami J-F, Billot C, Perrier X, et al. Genetic structure, linkage disequilibrium and signature of selection in Sorghum: lessons from physically anchored DArT markers. PLoS One. 2012;7:e33470.

40. Billot C, Ramu P, Bouchet S, Chantereau J, Deu M, Gardes L, et al. Massive sorghum collection genotyped with SSR markers to enhance use of global genetic resources. PLoS One. 2013;8:e59714.

41. Mace ES, Tai S, Gilding EK, Li Y, Prentis PJ, Bian L, et al. Whole-genome sequencing reveals untapped genetic potential in Africa's indigenous cereal crop sorghum. Nat Commun. 2013;4.

42. Quinby JR. Sorghum Improvement and the Genetics of Growth. College Station: Texas A&M University Press; 1974.

43. Rooney WL, Aydin S. Genetic control of a photoperiod-sensitive response in Sorghum bicolor (L.) Moench. Crop Sci. 1999;39:397–400.

44. The Wellcome Trust Case Control Consortium. Genome-wide association study of 14,000 cases of seven common diseases and 3,000 shared controls. Nature. 2007;447:661–78. **Central**

45. Brady JA. Sorghum Ma5 and Ma6 Maturity Genes. Ph.D. dissertation. Texas A&M University; 2006.

46. Chopra S, Gevens A, Svabek C, Wood KV, Peterson T, Nicholson RL. Excision of the Candystripe1 transposon from a hyper-mutable Y1-cs allele shows that the sorghumY1 gene controls the biosynthesis of both 3-deoxyanthocyanidin phytoalexins and phlobaphene pigments. Physiol Mol Plant Pathol. 2002;60:321–30.

47. Ibraheem F, Gaffoor I, Chopra S. Flavonoid phytoalexin-dependent resistance to anthracnose leaf blight requires a functional yellow seed1 in Sorghum bicolor. Genetics. 2010;184:915–26.

48. Wu Y, Li X, Xiang W, Zhu C, Lin Z, Wu Y, et al. Presence of tannins in sorghum grains is conditioned by different natural alleles of Tannin1. Proc Natl Acad Sci U S A. 2012;109:10281–6.

49. Worzella WW, Khalidy R, Badawi Y, Daghir S. Inheritance of Beta-carotene in grain sorghum hybrids. Crop Sci. 1965;5:591–2.

50. Gorbet DW, Weibel DE. Inheritance and genetic relationships of six endosperm types in sorghum. Crop Sci. 1972;12:378–82.

51. Kinoshita T. Gene analysis and Linkage in Biology of Rice. In: Tsunoda S, Takahashi N, editors. JSSP. Tokyo: Elsevier; 1984. p. 187–274.

52. Sweeney MT, Thomson MJ, Pfeil BE, Mccouch S. Caught red-handed: Rc encodes a basic helix-loop-helix protein conditioning red pericarp in rice. Plant Cell. 2006;18:283–94.

53. Magness JR, Markle GM, Compton CC. Food and feed crops of the United States. Interregional Research Project IR-4, IR Bul. 1 (Bul. 828 New Jersey Agr. Expt. Sta.); 1971.

54. Thurber CS, Ma JM, Higgins RH, Brown PJ. Retrospective genomic analysis of sorghum adaptation to temperate-zone grain production. Genome Biol. 2013;14:R68.

55. Murphy RL, Klein RR, Morishige DT, Brady JA, Rooney WL, Miller FR, et al. Coincident light and clock regulation of pseudoresponse regulator protein 37 (PRR37) controls photoperiodic flowering in sorghum. Proc Natl Acad Sci U S A. 2011;108:16469–74.

56. Mullet JE, Rooney WL, Klein PE, Morishige DT, Murphy R, Brady JA. Discovery and utilization of sorghum genes (MA5/MA6). Office USP ed., vol. US 8,309793 B2. USA: The Texas A&M University System; 2012.

57. Ross-Ibarra J, Morrell PL, Gaut BS. Plant domestication, a unique opportunity to identify the genetic basis of adaptation. Proc Natl Acad Sci U S A. 2007;104 Suppl:8641–8.

58. Doebley JF, Gaut BS, Smith BD. The molecular genetics of crop domestication. Cell. 2006;127:1309–21.

59. Xu X, Liu X, Ge S, Jensen JD, Hu F, Li X, et al. Resequencing 50 accessions of cultivated and wild rice yields markers for identifying agronomically important genes. Nat Biotechnol. 2012;30:105–11.

60. Huang X, Kurata N, Wei X, Wang Z-X, Wang A, Zhao Q, et al. A map of rice genome variation reveals the origin of cultivated rice. Nature. 2012;490:497–501.

61. Sieglinger JB. Seed-color inheritance in certain grain-sorghum crosses. J Agric Res. 1934;XXVII(1).

62. Manolio TA, Collins FS, Cox NJ, Goldstein DB, Hindorff LA, Hunter DJ, et al. Finding the missing heritability of complex diseases. Nature. 2009;461:747–53.

63. Matsushita S, Kurakazu T, Sobrizal Doi K, Yoshimura A. Mapping of genes for awn in rice using Oryza meridionalis introgression lines. Rice Genet Newsl. 2003;20:17.

64. Toriba T, Hirano H-Y. The DROOPING LEAF and OsETTIN2 genes promote awn development in rice. Plant J. 2014;77:616–26.

65. Shoemaker RC, Polzin K, Labate J, Specht J, Brummer EC, Olson T, et al. Genome duplication in soybean (Glycine subgenus soja). Genetics. 1996;144:329–38.

66. Wan L, Zha W, Cheng X, Liu C, Lv L, Liu C, et al. A rice β-1,3-glucanase gene Osg1 is required for callose degradation in pollen development. Planta. 2011;233:309–23.

67. Gao H, Zheng X-M, Fei G, Chen J, Jin M, Ren Y, et al. Ehd4 encodes a novel and Oryza-genus-specific regulator of photoperiodic flowering in rice. PLoS Genet. 2013;9:e1003281.

68. Hori K, Ogiso-Tanaka E, Matsubara K, Yamanouchi U, Ebana K, Yano M. Hd16, a gene for casein kinase I, is involved in the control of rice flowering time by modulating the day-length response. Plant J. 2013;76:36–46.

69. Childs KL, Miller FR, Cordonnier-Pratt MM, Pratt LH, Morgan PW, Mullet JE. The sorghum photoperiod sensitivity gene, Ma3, encodes a phytochrome B. Plant Physiol. 1997;113:611–9.

Chapter 11

GENETIC DIVERSITY IN THE MODERN HORSE ILLUSTRATED FROM GENOME-WIDE SNP DATA

Jessica L. Petersen[1], James R. Mickelson[1] , E. Gus Cothran[2] , Lisa S. Andersson[3] , Jeanette Axelsson[3] , Ernie Bailey[4] , Danika Bannasch[5] , Matthew M. Binns[6] , Alexandre S. Borges[7] , Pieter Brama[8] , Artur da Ca^mara Machado[9] , Ottmar Distl[10], Michela Felicetti[11], Laura Fox-Clipsham[12], Kathryn T. Graves[4] , Ge´ rard Gue´ rin[13], Bianca Haase[14], Telhisa Hasegawa[15], Karin Hemmann[16], Emmeline W. Hill[17], Tosso Leeb[18], Gabriella Lindgren[3] , Hannes Lohi1[6], Maria Susana Lopes[9] , Beatrice A. McGivney[17,] Sofia Mikko[3] , Nicholas Orr[19], M. Cecilia T Penedo[5] , Richard J. Piercy[20], Marja Raekallio[16], Stefan Rieder[21], Knut H. Røed[22], Maurizio Silvestrelli[11], June Swinburne[12,23], Teruaki Tozaki[24], Mark Vaudin1[2], Claire M. Wade[14], Molly E. McCue[1]

[1] University of Minnesota, College of Veterinary Medicine, St Paul, Minnesota, United States of America

[2] Texas A&M University, College of Veterinary Medicine and Biomedical Science, College Station, Texas, United States of America

[3] Swedish University of Agricultural Sciences, Department of Animal Breeding and Genetics, Uppsala, Sweden

[4] University of Kentucky, Department of Veterinary Science, Lexington, Kentucky, United States of America

[5] University of California Davis, School of Veterinary Medicine, Davis, California, United States of America

[6] Equine Analysis, Midway, Kentucky, United States of America

[7] University Estadual Paulista, Department of Veterinary Clinical Science, Botucatu-SP, Brazil

[8] University College Dublin, School of Veterinary Medicine, Dublin, Ireland

[9] University of Azores, Institute for Biotechnology and Bioengineering, Biotechnology Centre of Azores, Angra do Heroı´smo, Portugal

[10] University of Veterinary Medicine Hannover, Institute for Animal Breeding and Genetics, Hannover, Germany

[11] University of Perugia, Faculty of Veterinary Medicine, Perugia, Italy

[12] Animal Health Trust, Lanwades Park, Newmarket, Suffolk, United Kingdom

[13] French National Institute for Agricultural Research-Animal Genetics and Integrative Biology Unit, Jouy en Josas, France

[14] University of Sydney, Veterinary Science, New South Wales, Australia

[15] Nihon Bioresource College, Koga, Ibaraki, Japan

[16] University of Helsinki, Faculty of Veterinary Medicine, Helsinki, Finland

[17] University College Dublin, College of Agriculture, Food Science and Veterinary Medicine, Belfield, Dublin, Ireland

[18] University of Bern, Institute of Genetics, Bern, Switzerland

[19] Institute of Cancer Research, Breakthrough Breast Cancer Research Centre, London, United Kingdom

[20] Royal Veterinary College, Comparative Neuromuscular Diseases Laboratory, London, United Kingdom

[21] Swiss National Stud Farm, Agroscope Liebefeld-Posieux Research Station, Avenches, Switzerland

[22] Norwegian School of Veterinary Science, Department of Basic Sciences and Aquatic Medicine, Oslo, Norway

[23] Animal DNA Diagnostics Ltd, Cambridge, United Kingdom

[24] Laboratory of Racing Chemistry, Department of Molecular Genetics, Utsunomiya, Tochigi, Japan

ABSTRACT

Horses were domesticated from the Eurasian steppes 5,000–6,000 years ago. Since then, the use of horses for transportation, warfare, and agriculture, as well as selection for desired traits and fitness, has resulted in diverse populations distributed across the world, many of which have become or are in the process of becoming formally organized into closed, breeding populations (breeds). This report describes the use of a genome-wide set of autosomal SNPs and 814 horses from 36 breeds to provide the first detailed description of equine breed diversity. F_{ST} calculations, parsimony, and distance analysis demonstrated relationships among the breeds that largely reflect geographic origins and known breed histories. Low levels of population divergence were observed between breeds that are relatively early on in the process of breed development, and between those with high levels of within-breed diversity, whether due to large population size, ongoing outcrossing, or large within-breed phenotypic diversity. Populations with low within-breed diversity included those which have experienced population bottlenecks, have been under intense selective pressure, or are closed populations with long breed histories. These results provide new insights into the relationships among and the diversity within breeds of horses. In addition these results will facilitate future genome-wide association studies and investigations into genomic targets of selection.

INTRODUCTION

With a world-wide population greater than 58 million [1], and as many as 500 different breeds, horses are economically important and popular animals for agriculture, transportation, and recreation. The diversity of the modern horse has its roots in the process of domestication which began 5,000–6,000 years ago in the Eurasian Steppe [2]–[4]. Unlike other agricultural species such as sheep [5] and pigs [6], [7], archaeological and genetic evidence suggests that multiple horse domestication events occurred across Eurasia [2], [8]–[12]. During the domestication process, it is believed that gene flow continued between domesticated and wild horses [13] as is likely to also have been the case during domestication of cattle [14], [15]. Concurrent gene flow between domestic and wild horses would be expected to allow newly domestic stock to maintain a larger extent of genetic diversity than if domestication occurred in one or few events with limited individuals.

Prior genetic work aimed at understanding horse domestication has shown that a significant proportion of the diversity observed in modern maternal lineages was present at the time of domestication [2], [8], [16]. The question of mitochondrial DNA (mtDNA) diversity was further addressed by recent sequencing of the entire mtDNA genome. These studies estimate that, minimally, 17 to 46 maternal lineages were used in the founding of the modern horse [2], [17]; however, those data were unable to support prior studies suggesting geographic structure among maternal lineages [9], [18]. Recent nuclear DNA analyses have utilized "non-breed" horses sampled across Eurasia to attempt to understand the population history of the horse. These microsatellite-based studies suggest a weak pattern of isolation by distance with higher levels of diversity in, and population expansion originating from Eastern Asia [13], [19]. High diversity as observed by both mtDNA and microsatellites and the absence of strong geographical patterns is likely a result of continued gene flow during domestication, the high mobility of the horse, and its prevalent use for transportation during and after the time of domestication. Interestingly, while significant diversity is observed in maternal lineages, paternal input into modern horse breeds appears to have been extremely limited as shown by a lack of variation at the Y-chromosome [20], [21].

Diversity in the founding population of the domestic horse has since been exploited to develop a wealth of specialized populations or breeds.

While some breeds have been experiencing artificial selection for hundreds of years (e.g. Thoroughbred, Arabian), in general, most modern horse breeds have been developed recently (e.g. Quarter Horse, Paint, Tennessee Walking Horse) and continue to evolve based upon selective pressures for performance and phenotype (Table 1). Horse breeds resulting from these evolutionary processes are generally closed populations consisting of individual animals demonstrating specific phenotypes and/or bloodlines. Each breed is governed by an independent set of regulations dictated by the respective breed association. Not all breeds are closed populations. Some breed registries allow admixture from outside breeds (e.g. Swiss Warmblood, Quarter Horse), and others are defined by phenotype (e.g. Miniature). Finally, some populations that are often referred to as breeds are classified simply by their geographic region of origin and may not be actively maintained by a formal registry (e.g. Mongolian, Tuva) (Table 1). Those breeds that may be free ranging and experience lesser degrees of management may more appropriately be termed "landrace populations." Therefore, genetic characteristics within horse breeds are expected to differ based upon differences in the definition of the breed, the diversity of founding stock, the time since breed establishment, and the selective pressures invoked by breeders. The extent of gene flow not only varies within breed, but among horse breeds, the direction and level of gene flow is influenced by breed restrictions/requirements, and potentially by geographic distance.

Table 1: Populations (breeds) included in the study, region of breed origin and sampling location, notes on population history relevant to diversity statistics, and breed classification based upon use and phenotype.

doi:10.1371/journal.pone.0054997.t001

Breed	Geographic Origin	Region Sampled	Population size (approx)	Population Notes	Classification(s)
Akhal Teke	Turkmenistan	US & Russia	3,500	Pedigree records began-1885, Stud book-1941	Riding horse, endurance
Andalusian	Spain	United States	185,000	US registry formed in 1995 including Pura Raza Española & Lusitano bloodlines	Riding horse, sport
Arabian	Middle East	United States	1 million	Arabian type bred for over 3,500 years; US stud book-1908	Riding horse, endurance
Belgian	Belgium	United States	common	US Association began-1887	Draft
Caspian	Persia	United States	rare	Rediscovered in 1965 with N~50, no breeding records prior; Stud book-1966	Riding and driving pony
Clydesdale	Scotland	US & UK	5,000	Registry formed-1877 in Scotland; Stud book-1879	Draft
Exmoor	Great Britain	United Kingdom	2,000	Exmoor Pony Society-1921	Riding and driving pony
Fell Pony	England	United Kingdom	6,000	Fell Pony Society began in 1922; outcrossed with Dale's pony until 1970s	Light draft pony
Finnhorse	Finland	Finland	19,800	Stud book-1907	Light draft; riding horse; trotting
Florida Cracker	United States	United States	rare	Introduced to US in 1500s; association began-1989 with 31 horses	Riding horse, gaited
Franches-Montagnes	Switzerland	Switzerland	21,000	Official stud book-1921; Current breeding association established-1997	Light draft, riding horse
French Trotter	France	France	common	Population closed-1937 although allows some Standardbred influence	Riding horse, trotting
Hanoverian	Germany	Germany	20,000 (Germany)	Outcrossing allowed	Riding horse
Icelandic	Iceland	Sweden	180,000	Isolated >1,000 years; Federation of Icelandic Horse Association began-1969	Riding horse, gaited
Lusitano	Portugal	Portugal	12,000	Stud book-1967 after split from Spanish Andalusian breed	Riding horse, sport
Mangalarga Paulista	Brazil	Brazil	common	Registry began-1934	Riding horse
Maremmano	Italy	Italy	7,000	Breed identification based upon conformation and inspection	Riding horse
Miniature	United States	United States	185,000	Two US registries founded in 1970s; Maximum height restrictions for registration	Driving pony, extreme small size
Mongolian	Mongolia	Mongolia	2 million	Many types based upon purpose and geography	Riding horse, landrace
Morgan	United States	United States	100,000	Founding sire born in 1789; Registry-1894	Riding and driving horse
New Forest Pony	England	United Kingdom	15,000	Stud book-1910 with a variety of sires; No outcrossing since 1930s	Light draft, riding pony, landrace
North Swedish Horse	Sweden	Sweden	10,000	Breed association-1894; Stud book-1915	Draft
Norwegian Fjord	Norway	Norway	common	Stud book-1909	Riding and light draft
Paint	United States	United States	1 million	Registry-1965; One parent can be Quarter Horse or Thoroughbred	Riding horse, stock horse
Percheron	France	United States	20,000	Stud book-1893	Draft
Peruvian Paso	Peru	United States	25,000	Breed type over 400 years old; Closed population	Riding horse, gaited
Puerto Rican Paso Fino	Puerto Rico	Puerto Rico	250,000	Breed type ~500 years old; Association founded-1972	Riding horse, gaited
Quarter Horse	United States	United States	4 million	Association formed-1940; One parent may be Paint or Thoroughbred	Riding horse, stock horse, racing
Saddlebred	United States	United States	75,000	Breed type founded in late 1700s; Association began-1891	Riding and driving horse, some gaited
Shetland	Scotland	Sweden	common	Stud book-1891	Riding pony
Shire	England	United States	7,000	1st Shire organization-1877 (UK); stud book-1880; US assoc-1885	Draft
Standardbred	United States	Norway	common	Stud book-1871; Some outside trotting bloodlines (French Trotter) allowed	Riding horse, harness racing (trot)
Standardbred	United States	United States		Stud book-1871; Harness racing in early 1800s included pacing horses	Riding horse, harness racing (trot or pace)
Swiss Warmblood	Switzerland	Switzerland	15,000	Stud book-1921; Crossed with European Warmbloods, Thoroughbreds, Arabians	Riding horse, sport
Tenn Walking Horse	United States	United States	500,000	Registry-1935; Blood typing and parentage verification mandated in 1993	Riding horse, gaited
Thoroughbred	England	UK & Ireland	common	Stud book-1791; Closed population	Race horse, riding horse, sport
Thoroughbred	England	United States			Race horse, riding horse, sport
Tuva	Siberia	Russia	30,000	Different types depending on region	Light draft, landrace

doi:10.1371/journal.pone.0054997.t001

Considering modern breeds, unlike mtDNA, nuclear markers can discern breed membership[12]. However, studies of nuclear genetic diversity of modern breeds to date have most commonly focused on a single population of interest, sets of historically related breeds, or breeds within a specific geographic region [22]–[36]. Additionally, these analyses of nuclear genetic diversity in horse breeds are largely based upon microsatellite loci, which do not often permit consolidation of data across studies. Thus, large, across-breed investigations of nuclear diversity in the modern, domestic horse are lacking.

The Equine Genetic Diversity Consortium (EGDC), an international collaboration of the equine scientific community, was established in an effort to quantify nuclear diversity and the relationships within and among horse populations on a genome-wide scale. The development of this consortium has facilitated the collection of samples from 36 breeds for genotyping on the Illumina 50K SNP Beadchip. The breeds included in this report represent many of the most popular breeds in the world as well as divergent phenotypic classes, different geographic regions of derivation, and varying histories of breed origin (Table 1). The standardized SNP genotyping platform permits the compilation of data across breeds at a level never before achieved. Results of this collaboration now allow for the detailed description of diversity and assessment of the effects of genetic isolation, inbreeding, and selection within breeds, and the description of relationships among breeds. These data will also facilitate future across breed genome-wide association studies as well as investigations into genomic targets of selection.

RESULTS

Samples

Of the 38 populations sampled, two breeds were represented by geographically distinct populations: the Thoroughbred was sampled in the both the United States (US) and the United Kingdom and Ireland (UK/Ire), and the Standardbred was sampled in the US as well as in Norway. Eight Standardbred horses sampled from the US were noted to be pacing horses as opposed to the Norwegian and remaining US individuals that were classified as trotters. In addition, the International Andalusian and Lusitano Horse Association Registry (IALHA) in the US maintains one stud book but designates whether the individual was derived from Spanish (Pura Raza Española) or Portuguese (Lusitano) bloodlines, or a combination of both. Of the Andalusian horses collected in the US, five were noted to have Portuguese bloodlines.

Phenotypic classifications of the horse breeds include those characterized by small stature (Miniature Horse, pony breeds), breeds characterized by large stature and/or large muscle mass in proportion to size (draft breeds), light horse or riding breeds, gaited breeds, rare breeds, breeds founded in the past 80 years, and populations that are relatively unmanaged ("landrace"). The number of samples, sampling location, region of breed origin, and a list of primary breed characteristics are found in Table 1.

After pruning of individuals for genotyping quality and relationships (see methods), and keeping a similar number of individuals per breed, 814 of the 1,060 horses remained in the analysis. Of the horses removed, 12 had known pedigree relationships at or more recent to the grandsire/dam level, 44 individuals were removed at random from overrepresented breeds to equalize sample size across breeds, 4 failed to genotype at a rate greater than 0.90, and 186 were removed due to pi hat values (pairwise estimates of identity by descent) above the allowed threshold. Of those last 186 horses that were removed, 122 were from disease studies where relationships were common due to sampling bias.

Within Breed Diversity

Diversity indices were calculated using 10,536 autosomal SNPs that remained after pruning for minor allele frequency (MAF), genotyping rate, and linkage disequilibrium (LD) across breeds (referred to as the primary SNP set). Diversity indices were also calculated using three other SNP sets, resulting from different levels of LD-based pruning (see methods). Individuals noted as outliers in parsimony and cluster analyses (see below) were excluded from within-breed diversity calculations.

Using the primary SNP set, diversity, as measured by expected heterozygosity (H_e), ranged from 0.232 in the Clydesdale, to 0.311 in the Tuva (Table 2). Considering the SNP sets pruned less stringently for LD, the diversity within the Thoroughbred increased in relationship to the other breeds, as did that of nine other breeds. Mean and total heterozygosity increased with increased number of loci and less stringent LD pruning (Table 2). Inbreeding coefficients (F_{IS}) calculated on the primary SNP set showed significant excess homozygosity in 17 populations, which was greatest in the Andalusian (0.065). Three of the four lowest F_{IS} values were found in the Thoroughbred samples (UK/Ire, US, and when considered together) (Table 2).

Table 2: Number of samples (N), effective population size (N_e), individual inbreeding estimates (f), inbreeding coefficient (F_{IS}), and expected heterozygosity (H_e) from four SNP sets pruned based upon varying levels of LD.

Breed	N	Ne	FIS	Individual inbreeding (f)			Expected Heterozygosity (He)			
				Min	Max	Mean	r2 0.1 10,536	R2 0.1 6,028	r2 0.2 18,539	r2 0.4 26,171
Akhal Teke	19	302	0.015*	0.015	0.297	0.101	0.287	0.281	0.303	0.311
Andalusian	18a	329	0.065*	0.028	0.274	0.114	0.296	0.293	0.308	0.312
Arabian	24a	346	0.033*	0.060	0.060	0.060	0.287	0.280	0.302	0.310
Belgian	30b	431	−0.002	0.039	0.166	0.111	0.278	0.276	0.284	0.284
Caspian	18	351	−0.022	−0.033	0.136	0.041	0.294	0.292	0.305	0.308
Clydesdale	24	194	0.004	0.128	0.323	0.261	0.232	0.225	0.238	0.236
Exmoor	24	216	0.034*	0.055	0.556	0.239	0.247	0.242	0.253	0.252
Fell Pony	21	289	0.002	0.069	0.178	0.114	0.278	0.272	0.285	0.285
Finnhorse	27	575	−0.004	0.011	0.100	0.052	0.296	0.296	0.302	0.301
Florida Cracker	7	171	0.026*	0.004	0.359	0.159	0.270	0.263	0.284	0.291
Franches-Montagnes	19a	316	0.003	0.018	0.203	0.095	0.284	0.279	0.297	0.301
French Trotter	17a	233	−0.018	0.064	0.173	0.105	0.275	0.262	0.295	0.307
Hanoverian	15a	269	−0.010	0.002	0.087	0.052	0.294	0.280	0.320	0.335
Icelandic	25c	555	0.006*	0.043	0.234	0.083	0.289	0.288	0.290	0.288
Lusitano	24	391	0.039*	0.008	0.220	0.090	0.296	0.292	0.309	0.315
Maremmano	24	341	−0.012	−0.015	0.109	0.038	0.298	0.287	0.318	0.329
Miniature	21	521	0.005	0.043	0.161	0.075	0.291	0.292	0.296	0.295
Mangalarga Paulista	15	155	−0.011	0.176	0.320	0.242	0.235	0.228	0.246	0.250
Mongolian	19a	751	0.001	−0.034	0.055	0.015	0.309	0.308	0.314	0.314
Morgan	40	448	0.040*	0.003	0.307	0.090	0.296	0.287	0.310	0.317
New Forest Pony	15	474	0.000	−0.022	0.066	0.025	0.304	0.300	0.316	0.319
Norwegian Fjord	21a	335	−0.003	0.053	0.168	0.122	0.274	0.274	0.278	0.277
North Swedish Horse	19	369	0.011*	0.069	0.210	0.133	0.275	0.276	0.279	0.278
Percheron	23	451	0.003	0.043	0.143	0.086	0.287	0.284	0.292	0.293
Peruvian Paso	21	433	0.002	0.008	0.134	0.055	0.298	0.293	0.306	0.310
Puerto Rican Paso Fino	20	321	−0.003	0.004	0.298	0.103	0.280	0.278	0.287	0.290
Paint	25	399	0.006*	−0.013	0.101	0.040	0.302	0.289	0.324	0.337
Quarter Horse	40a	426	0.011*	−0.012	0.144	0.047	0.302	0.290	0.323	0.336
Saddlebred	25d	297	−0.008	0.051	0.145	0.103	0.279	0.268	0.297	0.306
Shetland	27	365	0.032*	0.108	0.370	0.182	0.264	0.268	0.268	0.266
Shire	23	357	0.024*	0.130	0.258	0.187	0.261	0.252	0.268	0.267
Standardbred - Norway	25e	232	−0.004	0.063	0.202	0.130	0.272	0.255	0.289	0.298
Standardbred - US	15	179	0.039*	0.097	0.222	0.153	0.276	0.262	0.293	0.303
Standardbred - all	40	290	0.022*	−0.028	0.323	0.130	0.276	0.260	0.293	0.303
Swiss Warmblood	15a	271	0.005	0.023	0.117	0.059	0.296	0.281	0.322	0.337
Thoroughbred - UK/Ire	19a	143	−0.028	0.089	0.171	0.133	0.264	0.245	0.292	0.309
Thoroughbred - US	17a	163	−0.015	0.093	0.182	0.134	0.267	0.250	0.295	0.313
Thoroughbred - all	36	190	−0.019	0.089	0.182	0.134	0.266	0.248	0.294	0.312
Tuva	15	533	0.016*	−0.028	0.116	0.022	0.311	0.309	0.320	0.322
Tennessee Walking Horse	19	230	0.008*	0.065	0.276	0.148	0.269	0.256	0.284	0.291
Mean	22.3	341	0.007	0.039	0.204	0.107	0.282	0.275	0.295	0.300
Total	814						0.313	0.303	0.329	0.336
Min			−0.028	−0.034	0.055	0.015	0.232	0.225	0.238	0.236
Max			0.005	0.176	0.556	0.261	0.311	0.309	0.324	0.337

[a]Individuals from this breed also included in [41];
[b]20 of these individuals were also reported in [41];
[c]17 of these individuals were also reported in [41];
[d]21 of these individuals were also reported in [41];
[e]19 of these individuals were also reported in [41].
F_{IS} and f were calculated based upon the primary SNP set (10,536 loci). Samples also used in [41] are indicated in the footnotes.
*Indicates significance at $\alpha < 0.05$ determined by 10,000 permutations.
doi:10.1371/journal.pone.0054997.t002

Inbreeding coefficients (f) calculated for each individual based upon observed and expected heterozygosity showed several individuals with significant loss of heterozygosity. The highest individual value of f (0.56) was found in an Exmoor pony. Within breeds, average individual estimates of f were greatest in the Clydesdale, Mangalarga Paulista, and Exmoor while the lowest breed means were found in the landrace populations (Table 2).

Effective population size (N_e), as estimated by LD [37] using an autosomal SNP set pruned within each breed for quality, was lowest (143) in the UK/Ire sample of the Thoroughbred (UK/Ire) but also low in the other racing breeds as well as the Clydesdale (Table 2). Highest values of N_e were observed in the Eurasian landrace populations, the Mongolian (743) and Tuva (533), and also in the Icelandic (555), Finnhorse (575), and Miniature (521). Breed-specific decay of LD essentially mirrors the results of the N_e calculation given the relationship between the statistics.

Parsimony and Principal Component Analyses

With a domestic ass designated as the outgroup, parsimony analysis of 10,066 loci pruned for LD of $R^2=0.2$ (see methods) resulted in generally tight clustering and monophyly of samples within breeds, supported by high bootstrap values (Figure 1). Major clades of the tree show grouping of the Iberian breeds (Lusitano and Andalusian), ponies (Icelandic, Shetland, Miniature), Scandinavian breeds (Finnhorse, North Swedish Horse, Norwegian Fjord), heavy draft horses (Clydesdale, Shire, Belgian, Percheron), breeds recently admixed with and/or partly derived from the Thoroughbred (Paint, Quarter Horse, Maremmano, Swiss Warmblood, Hanoverian), modern US breeds (American Saddlebred (hereafter "Saddlebred" and Tennessee Walking Horse), trotting breeds (Standardbred and French Trotter), and Middle Eastern breeds (Akhal Teke and Arabian). Exceptions to monophyly include the Paint and Quarter Horse as well as the Hanoverian and Swiss Warmblood, which are mixed in clades surrounding the Thoroughbred and Maremmano. In addition, the Clydesdale was placed as a clade within the Shire breed and the Shetlands as a clade within the Miniatures. Strong bootstrap support for monophyly is present within a subset each of Lusitanos (83%), and Andalusians (87%); however the remainder of individuals from these breeds were intermixed. No structure was found within the US sample regarding individual Andalusians noted to have Portuguese bloodlines opposed to those with Spanish bloodlines. The Mongolian and most Tuva horses were grouped together while a subset of the Tuvas fell out as a sister clade to the Caspians. Several individuals were not positioned in the clades that represented the majority of the other individuals in the breed (Figure 1). These include three Shires, two Mongolians, a Caspian, and a Norwegian Fjord. In each instance, the outlier status of these individuals was also supported by cluster analysis (see below).

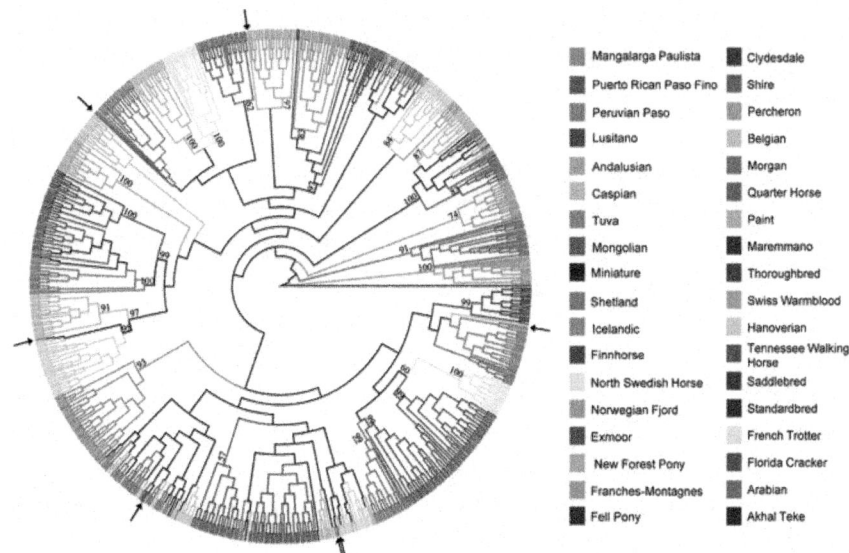

Mangalarga Paulista	Clydesdale
Puerto Rican Paso Fino	Shire
Peruvian Paso	Percheron
Lusitano	Belgian
Andalusian	Morgan
Caspian	Quarter Horse
Tuva	Paint
Mongolian	Maremmano
Miniature	Thoroughbred
Shetland	Swiss Warmblood
Icelandic	Hanoverian
Finnhorse	Tennessee Walking Horse
North Swedish Horse	Saddlebred
Norwegian Fjord	Standardbred
Exmoor	French Trotter
New Forest Pony	Florida Cracker
Franches-Montagnes	Arabian
Fell Pony	Akhal Teke

Figure 1: Individual and breed relationships among 814 horses illustrated by parsimony.

Parsimony tree created from 10,066 SNPs and rooted by the domestic ass. Breeds are listed in the legend in order starting from the root and working counterclockwise. Individual outliers with respect to their breeds are noted with arrows. Bootstrap support calculated from 1,000 replicates is shown for major branches when greater than 50%.

doi:10.1371/journal.pone.0054997.g001

Principal component analysis (PCA) also serves to visualize individual relationships within and among breeds. All Thoroughbred samples, regardless of origin, are separated from the others by PC1 and form a cluster at the top of the figure. Intermediate between the Thoroughbred and central cluster of breeds are the Hanoverian, Swiss Warmblood, Paint, and Quarter Horse. The Shetland, Icelandic, and Miniature split from the remainder of samples in PC2, falling out in the lower left corner, and the British drafts anchor the figure at the lower right. While most breeds cluster tightly, several are dispersed across one or both PCs. The Hanoverian, Swiss Warmblood, Paint, and Quarter Horse, as noted above, are extended along PC1, while the Arabian and Franches-Montagnes show similar spreading, also along PC1. The Tuva, Clydesdale, and Shire individuals also are not as tightly clustered as other populations despite the low within breed diversity of the latter two.

Distance Analysis

An unrooted neighbor joining (NJ) tree of Nei's distance [38] was constructed using SNP frequencies within breeds from the 10,536 SNP data set (Figure 2). The relative placement of breeds reflects that seen in the parsimony tree with several exceptions. The Paint, Quarter Horse, Swiss Warmblood, Hanoverian, Maremmano, and Thoroughbred, are found in one large branch of the tree, although the Maremmano is placed outside of the clade containing the aforementioned breeds. The position of the Morgan with the Saddlebred and Tennessee Walking Horse also deviates from parsimony analysis but reflects historic records of relationships among these breeds. The Scandinavian breeds remain in one branch of the clade, which also includes the Shetland and Miniature. Unlike the parsimony cladogram, the Caspian falls in a clade with the other Middle Eastern breeds, the Arabian and Akhal Teke. Finally, the Exmoor, a British breed, is placed with another British breed, the New Forest Pony, rather than with the Scandinavian breeds as in the parsimony analysis. Each branch shows support of over 50%, with many clades being supported by over 99% of the 1,000 bootstrap replicates.

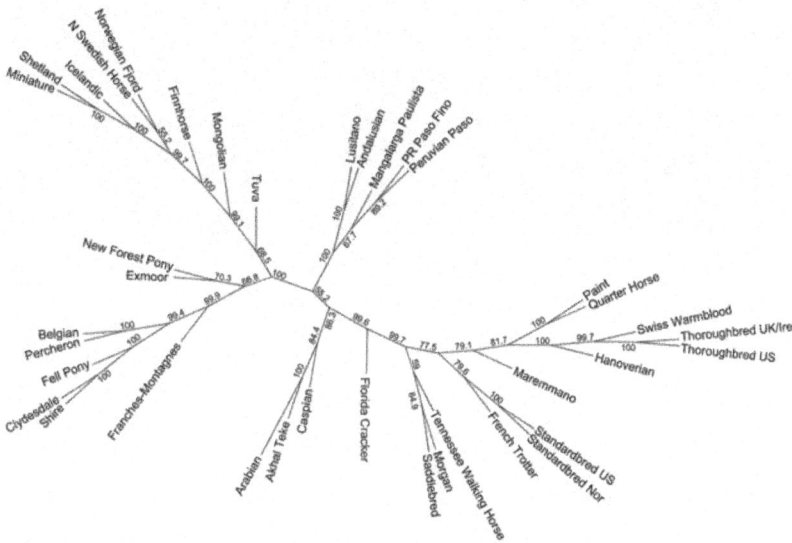

Figure 2: Distance based, neighbor joining tree calculated from SNP frequencies in 38 horse populations.

Majority rule, neighbor joining tree created from 10,536 SNP makers using Nei's genetic distance and allele frequencies within each population. Percent bootstrap support for all branches calculated from 1,000 replicates is shown.

doi:10.1371/journal.pone.0054997.g002

Cluster Analysis

Likelihood scores for runs of various K in Structure showed an increase in overall mean ln P(X|K) until K=35. A clear "true" value of K is not obvious examining the likelihood scores or using the Evanno method [39] (data not shown); however, variance among runs begins to increase with a diminishing increase in likelihood scores after K=29, which is near the peak of the curve. The value of the highest proportion (breed average q-value) of assignment of each breed for each value of K.

The first breeds to have all individuals assign strongly to one cluster are the Thoroughbred and Clydesdale (with Shire) at K=2, followed by the Shetland at K=3; these four breeds do not show signs of admixture at any K value analyzed. Evidence of weak geographic grouping is observed at K =4, which consists of: 1, the Middle Eastern and Iberian breeds (pink); 2, the Thoroughbred and breeds to which it continues to be or was historically crossed (yellow); 3, breeds developed in Scandinavia and Northern Europe (orange); and 4, the British Isles draft breeds (blue) (Figure 3).

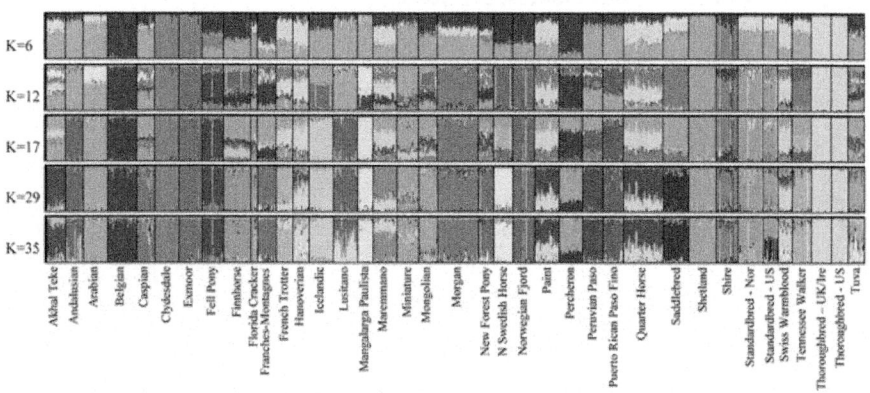

Figure 3: Bayesian clustering output for five values of K in 814 horses of 38 populations.

Structure output for five values of K investigated. Each individual is represented by one vertical line with the proportion of assignment to each cluster shown on the y axis and colored by cluster.

doi:10.1371/journal.pone.0054997.g003

Middle Eastern and Iberian Breeds

As also observed in the NJ tree, clustering of the Iberian and Middle Eastern breeds with the Mangalarga Paulista, Peruvian Paso, and Puerto Rican Paso

Fino (q >0.5) is observed until K =8, after which point the Mangalarga Paulista assigns with q =0.93 to another cluster. The remaining breeds cluster together until K=12, at which time the Middle Eastern breeds (Arabian, Akhal Teke, and Caspian) are assigned to their own cluster, leaving the Iberian breeds clustered with the Peruvian Paso and Puerto Rican Paso Fino. At low values of K (*i.e.* K <6) the Florida Cracker, Saddlebred, Standardbreds, Morgan, and Tennessee Walking Horse fall into the cluster with the Iberian and Middle Eastern breeds with breed mean q >0.5. At K=29, each of these breeds is assigned with q >0.72 to an individual cluster with the exception of the Lusitano and Andalusian, which remaining clustered together.

Thoroughbreds and Thoroughbred Crossed Breeds

Relationships described by the NJ tree among the Thoroughbred, Hanoverian, Swiss Warmblood, Paint, Quarter Horse, and Maremmano are also seen in cluster analysis. Clustering of those breeds with the Thoroughbred is observed throughout the values of K examined although at moderate frequencies (Figure 3). At K =29, the Hanoverian and Swiss Warmblood remain assigned to the cluster defined by the Thoroughbred but with assignment probabilities of 0.51 each. The Quarter Horse and Paint also assign to this cluster with q-values of 0.30 and 0.34, respectively. Neither the Quarter Horse, Paint, Hanoverian, or Swiss Warmblood populations assign to any cluster with q >0.62 at K =29. No evidence of population substructure is observed between the US and UK/Ire Thoroughbreds as also shown by PCA and parsimony analyses.

Scandinavian and Northern European Breeds

As in the NJ and parsimony trees, the Finnhorse, Icelandic, Miniature, North Swedish Horse, Norwegian Fjord, and Shetland are parsed into the same cluster (q-value >0.5) through K =5. However, unlike the NJ tree, at K =4, the highest value of assignment places the Belgian and Percheron into this cluster although with q <0.5 (0.42 and 0.38, respectively). The relationship remains until K =6, at which time the Miniature, Icelandic, and Shetland fall into a different cluster. At K =10, the Icelandic clusters again with the North Swedish Horse and Norwegian Fjord. The Norwegian and United States Standardbred populations, which at K =4 assign with q >0.5 to the cluster containing the Scandinavian breeds, separate from the Scandinavian breeds at K =5. At K =31, substructure appears in the Standardbred samples, which correlates to those individuals identified as pacers and that fall into an individual clade in the parsimony tree. At K =29, the Miniature and Shetland continue to be assigned to the same cluster (q-values =0.55 and 0.95, respectively). The next highest proportions of assignment of the Miniature horse are to the clusters described by the New Forest Pony (q =0.20) and Icelandic (q =0.11). No value of K evaluated eliminated signals of admixture from all populations in the dataset at K=38 (the actual number of populations sampled) or any value of K through 45 (data not shown).

British Isles Draft

The Clydesdale and Shire cluster together, and apart from the other breeds beginning at K =3. In addition, the Fell Pony, which is placed within the same clade in the NJ and parsimony trees, and proximal to the Clydesdale and Shire in PCA, shows moderate assignment to this cluster (0.29< q <0.41) for several values of K from 4 to 14. At K =29, the Shire assigns to the same cluster as the Clydesdale with q =0.69. The individual outliers from the Shire breed also noted in parsimony analysis are evident beginning at K =3. Excluding these outliers, at K =29, the proportion of assignment for the Shires to the cluster with the Clydesdale increases to 0.74.

F_{ST}

All pairwise F_{ST} values calculated between the 37 populations (excluding the Florida Cracker) were significant as tested by 20,000 permutations (Figure 4). The lowest level of differentiation was found between the Paint and Quarter Horse populations (F_{ST} =0.002), while the greatest divergence was observed between the Clydesdale and Mangalarga Paulista (F_{ST}=0.254). The two Thoroughbred populations had an F_{ST} value of 0.004, while the two Standardbred populations had 10-fold greater divergence (F_{ST} =0.020) than the minimum observed value in this dataset; this value is similar to that observed between the Lusitano and Andalusian (0.021). An F_{ST} value of 0.006 was identified between the Tuva and Mongolian populations. The global F_{ST} value was 0.100. AMOVA computed on the set of 37 samples (excluding the outliers identified in Structure and the Florida Cracker) showed that 10.03% of the variance was accounted for among populations (p=<0.001), 0.53% of the variance was among individuals within populations (p=0.19), and 89.44% of the variation was within individuals (p<0.001).

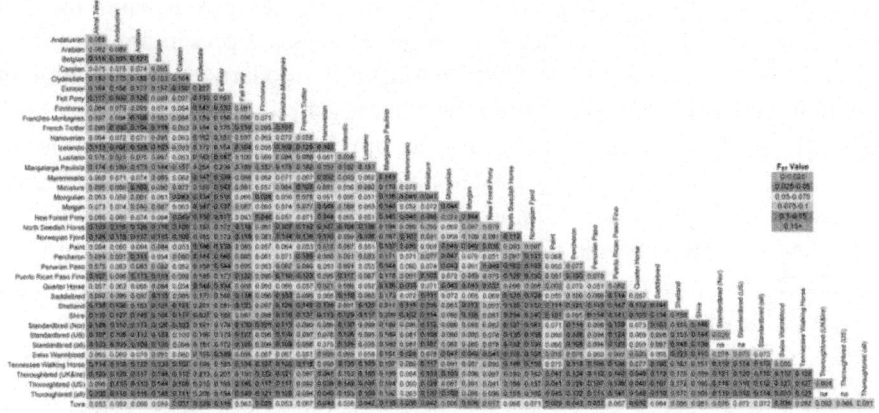

Figure 4: Pairwise F_{ST} **values based upon 10,536 SNPs in 37 horse populations.**

Pairwise F_{ST} values as calculated in Arlequin using 10,536 autosomal SNPs and significance tested using 20,000 permutations. All pairwise values are significantly different from zero. (individual outliers were removed from this analysis).

doi:10.1371/journal.pone.0054997.g004

DISCUSSION

These data are gathered from populations that represent tremendous diversity in phenotype and breed specialization. With breeds sampled across four continents, the resulting relationships observed largely reflect similarities of geographic origin, documented breed histories, and shared phenotypes. In general the highest within breed diversity was observed in breeds that are recently derived, continue to allow introgression of other populations, those that have a large census population size, and landrace populations that experience a lesser degree of controlled breeding. Not surprisingly, low diversity is observed in breeds with small census size, relatively old breeds with closed populations, and those with documented founder effects, whether due to population bottlenecks or selective breeding.

A total of seven individuals were identified by parsimony and cluster analysis as outliers with respect to the breed to which they were assigned. The pedigrees of these individuals were unknown. Because it is possible these horses were unknowingly crossbred or subject to mishandling in the field or laboratory, they were excluded from the within-breed analyses to avoid potential bias in indices of diversity. In addition, the potential impact of SNP ascertainment bias on diversity calculations must be acknowledged. The reference genome is from a Thoroughbred mare [40] and SNP identification was based upon the reference genome and data from seven other horses representing six breeds. Therefore, SNPs are generally derived to identify modern variation within the Thoroughbred as well as between the Thoroughbred and these other breeds. Thus, the SNPs identified may reflect an upward bias in diversity indices in the Thoroughbred and closely related breeds [41]. It seems that ascertainment bias may have particularly influenced the results when considering the data sets that have an increased number of loci resulting from relaxed LD pruning. These results show an increase in the relative diversity of the Thoroughbred, breeds with which the Thoroughbred continues to actively interbreed, and the other SNP discovery breeds, with respect to other breeds in the study. This is opposite of what may be expected given the high levels of genome-wide LD in the Thoroughbred. Without considering SNP ascertainment bias, it is expected that measured diversity would increase in breeds with low LD more quickly than in those with high LD, due to greater independence of markers in the former breeds. These SNPs, derived largely from the Thoroughbred, are apparently detecting a higher proportion of Thoroughbred-specific, rare variants and it appears that as more loci are included, more of these Thoroughbred-based variants are assayed, resulting in the observed increase in variation in the Thoroughbred, Thoroughbred-influenced breeds, and breeds used in SNP ascertainment.

The majority of the analyses were performed using 10,536 SNP markers pruned across breeds for LD of $r^2 > 0.2$ as well as MAF of 0.05 or above. Even though additional markers could have been used for analysis, many population-level statistics assume independence of loci. The stringent pruning for LD was therefore undertaken to eliminate bias in the test statistics that may result from substantial breed-specific differences in LD [40], [41]. A truncated data set also helped to make calculations, especially cluster analysis, computationally feasible. On the other hand, diversity indices were calculated after pruning the full data set to $r^2 = 0.2$ and $r^2 = 0.4$ (using pairwise correlation), and one replicate setting the threshold to $R^2 < 0.1$ (using the variance inflation factor), to examine the effect of allowing for varying levels of LD and therefore varying numbers of loci (see methods).

Within-breed Diversity

Even considering SNP ascertainment, low diversity as measured by H_e was observed in the Thoroughbred as well as the Standardbred, which both experience high selective pressures and are closed populations. Low diversity was also observed in breeds that have undergone a severe population bottleneck, such as the Exmoor and Clydesdale, and breeds that have small census population sizes, such as the Florida Cracker. Although the Thoroughbred is a large population that is widely distributed on a geographic scale, historic records suggest that one sire is responsible for 95% of the paternal lineages in the breed and as few as 30 females make up 94% of maternal lineages [28]. In addition, the population has been largely closed to outside gene flow since the formation of the first stud book in 1791 [42] and individuals within the breed are subject to selective pressure for racing success; therefore low, within-breed diversity is not at all surprising.

Using LD-based calculations, the estimated N_e for the Thoroughbred was similar to that found in a UK sample [43] and among the lowest of the study set despite the large census population size and geographic distribution of this breed. Individual inbreeding values based upon observed vs. expected homozygosity indicate that individual Thoroughbred horses show signs of inbreeding, with a mean loss of heterozygosity of 16.3%. This value is slightly larger than that found in [28] (13.9%). Using the same SNP array, [44] also showed inbreeding in the Thoroughbred, and specifically an increase in inbreeding over time. The only breeds with higher f values were the Exmoor, Clydesdale, Mangalarga Paulista, and Shire. Despite low individual diversity, F_{IS} values do not show significant inbreeding in either of the Thoroughbred populations as a whole, or in the Norwegian Standardbred although F_{IS} is significant in the US Standardbred population (discussed below).

The Clydesdale and Exmoor, in addition to having high individual estimated coefficients of inbreeding, also show the lowest within-breed diversity observed in the dataset. A lack of diversity in the Clydesdale and another British draft breed, the Shire, is likely a result of a severe population bottleneck observed in most draft breeds with the onset of industrialization and after the conclusion of World War II (WWII) as well as selection for size and color [45], [46]. The Exmoor pony, considered to be one of the purest native breeds of Britain, has been naturally selected for survival in harsh winter conditions on the moors in southwest England[45], [47]. Similar to the draft breeds, the Exmoor population decreased significantly after WWII to approximately 50 individuals, undoubtedly influencing the diversity observed in this study. The effect of low population size and selection is also reflected in extremely high individual estimates of f within some individuals. Finally, the Mangalarga Paulista shows low levels of heterozygosity, and as discussed below, the greatest divergence as measured by pairwise F_{ST} of all breeds in the study. While these results could be due to geographic distance between this and other breeds, and/or genetic drift, unfortunately these horses were all sampled from only two farms and likely do not represent the entirety of the diversity present in the breed; therefore we cannot rule out sampling error which would inflate the estimated level of divergence between these individuals and the other breeds and result in a decrease in H_e. However, a lack of diversity in sampling of the breed would not have an effect on estimates of individual inbreeding coefficients, which were among the highest of the entire data set.

Converse to the above examples, high levels of diversity as measured by both H_e and N_e, accompanied by low estimates of inbreeding (f and F_{IS}), are observed in the Mongolian, Tuva, and New Forest Pony. The Mongolian and Tuva are unique in that they represent landrace populations that are less managed than the popular breeds of Western Europe and North America; they occupy a diverse range of habitat, have been selected for meat and milk in addition to use in transportation, and originate in the region where domestication was likely to have occurred. The population of Mongolian horses is large and individuals are phenotypically diverse [48]. In 1985, approximately two million Mongolian horses of four different types were estimated to live within the country [45]. The Tuva is not as numerous as the Mongolian but is similar in its purpose and also has high within-breed phenotypic diversity. In addition, it is suggested that the Tuva has experienced outcrossing in order to increase its size and stamina[45] as may also be the case in the Mongolian [49]. Similarly, the New Forest Pony was historically a free-ranging population in Great Britain, but was crossbred until the 1930's. These traits: old populations, large population size, outcrossing, high phenotypic diversity, and lesser artificial selection/management, result in the high levels of genetic diversity observed. This extent of diversity appears to diminish as populations are restricted by selective pressures into formal breeds.

Other population characteristics are likely the cause of the diversity observed in the Finnhorse, Icelandic, and Miniature. In the case of the Icelandic, the high level of diversity was possibly maintained by a large census population size despite isolation for almost a thousand years and several population bottlenecks due to natural disasters [45]. In the case of the Finnhorse, diversity may be due to within-breed substructure into four sections of the studbook established in 1970: the work horse (draft), trotters, riding horse, and pony [50]. Finally, high diversity in breeds such as the Miniature is likely a result of a diverse founding stock [45], [51], [52]; horses of small size from a variety of geographic regions and bloodlines were utilized in founding the breed, which is defined by phenotype.

All of these three factors, large population size, phenotypic diversity within the breed, and a diversity of founding stock, also lead to the relatively high levels of diversity observed in the Paint and Quarter Horse; in addition, these breeds both allow continued outcrossing between themselves and with the Thoroughbred and have experienced a tremendous population expansion since the formal foundation of the breeds within the past 45–75 years. Due to the relative infancy of these populations, it could be argued that the Paint, Quarter Horse, and other, newly-derived breeds, have not yet had time to undergo the evolutionary processes necessary to be genetically distinct populations as is observed in breeds with longer histories and closed studbooks. However, even with high within-breed diversity and large census population sizes (over 1 million and 4 million worldwide for the Paint and Quarter Horse, respectively), N_e for these breeds account for only a fraction of the census size, demonstrating non-random mating and selection. Outcrossing is also continued in the Swiss Warmblood and Hanoverian breeds, which show similar trends in diversity measures as the Quarter Horse and Paint. The relatively low N_e in these breeds, accompanied by moderate H_e may partially be due to significant crossing with the Thoroughbred, which would contribute long blocks of LD [40],[41], resulting in decreased estimates of N_e.

Of note in breeds such as the Quarter Horse, Lusitano, and Andalusian, is that despite moderate to high relative levels of H_e, and low to moderate estimates of f, F_{IS} values in each breed are significantly positive. Significant F_{IS} was also previously observed in the Iberian breeds using microsatellite markers [53]. While selection and inbreeding may be responsible for significant values of F_{IS} in some of these breeds, another instance in which F_{IS} may be significantly positive is in the presence of subpopulation structure within the sample. Evidence of this in the Lusitano and Andalusian is present in parsimony analysis where individuals of the two breeds fall into one clade, but within that clade are two highly supported branches represented by a subset of each breed. In addition, when forcing high values of K in Structure, such as observed at K =35, Andalusian and Lusitano individuals fall into one of two clusters with q-value >0.5 in a nonbreed-specific manner (data not shown). These results support [54], which showed potential subpopulation structure in the Lusitano via microsatellite analysis. In the Quarter Horse, subpopulation structure is evident through the evaluation of bloodlines and the selection of popular sires for diverse performance classes. This population substructure in the Quarter Horse has also been demonstrated by marked differences in allele frequencies among performance types (cutting, western pleasure, halter, racing, etc.) [55]. A similar instance is found in the US population of the Standardbred, which also has significant excess homozygosity (F_{IS}). Unlike Standardbreds in Europe, which are raced at a trot, those in the US are divergently selected for racing at either the pace or the trot, creating structure within the breed [56].

Finally, several rare populations are included in this dataset. The Caspian is one of the oldest breeds in the Middle East and was thought to be extinct until its recent rediscovery in 1965. The Florida Cracker, a now rare breed, was developed in the United States from feral stock of Iberian descent [57]. The sample size of the Florida Cracker limits the conclusions that can be drawn regarding within-breed diversity. However, the Caspian shows high N_e, H_e, and estimates of f, given its rarity. After rediscovery of the breed, which historically was believed to represent a type of landrace population, [58] describes a three-year survey, which found approximately 50 individuals remaining, noting that many could not be considered "pure." In addition, [33] were unable to show evidence of a recent bottleneck in the Caspian breed. The diversity observed in what are now considered Caspian horses likely stems from high levels of diversity within those individuals that founded the modern population.

Among-breed Diversity

The expectation of homogeneity within breeds due to closed populations and selection is supported by the results of AMOVA, which show significant variation present among populations, but a non-significant proportion of variance within. However, the variation among samples lends information about current and historic relationships. Observed trends include patterning based upon geographic origin and/or phenotypic similarities, and relatively low divergence observed in comparisons that include breeds with high within-breed diversity. In Structure analysis, K =29 was chosen as the most likely value of K; however no single value stood out as the "best" number of clusters to describe these data. Regardless, patterns observed in clustering were also supported by pairwise F_{ST} values, parsimony, PCA, and NJ dendograms.

High Diversity and Low Divergence – Landrace Breeds

The Mongolian and Tuva populations are believed to have been influential in the spread of horses across Asia and Europe [45], [59]; these landrace populations, harboring high levels of within-breed diversity, were found to be similar to one another, with a pairwise F_{ST} value of 0.006. In addition, with the exception of six Tuva individuals that fell into a clade with the Caspians, both parsimony and NJ analyses place the Tuva and Mongolian into the same clade of each tree. Examining all comparisons, low F_{ST} values were observed between the Mongolian and Tuva compared to the other breeds in this study, supporting the potential role of Eurasian horses of similar type in founding modern stocks. This also aligns with high microsatellite diversity observed in Eastern Eurasian "non-breed" (landrace) populations in [19]. On the other hand, breeds with high diversity in general show lower levels of divergence as measured by F_{ST}, while those with low diversity show higher values of F_{ST}. Low divergence in breeds with high diversity is expected as variation within a breed may indicate outcrossing with other populations, and high variation also makes these breeds more likely to share variation with others by chance. In contrast, if a breed has little within-breed variation, it is less likely to share genetic variation with another breed by chance, especially with another breed that is relatively homogeneous itself. As demonstrated in human literature, source populations are expected to contain greater diversity than those populations which they found [60], [61]; this is also suggested in the horse by [11], which showed greater mtDNA diversity in Iberian breeds than the recently founded American breeds. If the argument is made that the low F_{ST} values of Tuvas and Mongolians supports their role in founding modern breeds, the same argument could be made for the Quarter Horse, which also shows low levels of pairwise divergence; however that argument would be unreasonable as the Quarter Horse was developed in only the past century. Therefore, the relative values of F_{ST} are informative, but these F_{ST} values and data, which represent modern breeds generally derived from limited founding stock, and subjected to intense artificial selective pressures, cannot be used independently to elucidate the evolution of the modern horse. Also, as is the case for other analyses, while the relationships observed can shed light on the history of breeds, they cannot distinguish between recent admixture and shared ancestry.

Thoroughbred-influenced Breeds

The Thoroughbred is believed to have founding sires of Arabian, Turk, and Barb ancestry [42], and [28] found that two sires, noted as being Arabian (Godolphin Arabian and Darley Arabian) together contributed to over 20% of the modern population. However, it is likely that the "Arabian" foundation stallions were not Arabians as the breed is known today. It is noted in [62]that the Godolphin Arabian was a Turkoman stallion with partial Arabian blood, while in other work it is suggested he was a Barb [45]. Regardless of the true ancestry of these stallions, restrictions placed upon the export of purebred Arabians during the 16th and 17th centuries, as well as the general use of the term "Arab" for horses of Middle Eastern descent, it is likely other "Arabian" horses with influence on the Thoroughbred breed also had Turkoman, and Barb bloodlines [45], [62]. The pairwise F_{ST} values between the Thoroughbred and the Arabian do not suggest any less divergence than observed between the Thoroughbred and a majority of the other breeds. In addition, at K=29 the Arabian assigns to the Thoroughbred cluster at only 2.3%. If the Arabian did have significant influence on the Thoroughbred breed, there are several possible explanations for why the supposed Arabian influence is not more apparent. The first is related to SNP ascertainment and the bias of SNPs toward modern variation in Thoroughbred. It is possible that the genes derived from the Arabian are at or near fixation in the Thoroughbred, which would reduce the chance that these SNPs, and the variation described within them are present in the dataset. Another possibility is that the current Arabian sample, taken from the United States, may not reflect the Arabian lineage(s) that were influential in the founding of the Thoroughbred. Finally, as noted above and also suggested elsewhere [63], it may simply be that Arabian bloodlines were not as instrumental in the Thoroughbred breed as once thought or that the initial Arabian influence (and genes) have been selected against or lost to drift during the development of the modern Thoroughbred racehorse.

Within the Thoroughbred itself, divergence between the US and European samples had a significant F_{ST} of 0.004, similar to that observed between the Hanoverian and Swiss Warmblood (0.008) and Mongolian and Tuva (0.006), but larger than the minimally observed value seen between the Paint and the Quarter Horse (0.002). Although artificial insemination is prohibited in the Thoroughbred and would be anticipated to limit gene flow to some extent, the founder effect in the original European Thoroughbred by few high-impact sires and dams, accompanied by shared selective pressures, relatively recent importation of the breed to the United States, and ongoing shipment of horses between continents are likely contributing to the lack of geographic population structure identified by parsimony and cluster analyses.

While within-breed diversity of the Thoroughbred was relatively low, and a notable relationship with the Arabian was not observed, among-breed analysis shows a clear influence of the Thoroughbred on many other breeds. Placed with the Thoroughbred in parsimony analysis are the Hanoverian, Maremmano, and Swiss Warmblood. The Maremmano, an Italian breed, shows a q-value of assignment to the Thoroughbred clade of 0.26 at K =29. This is not surprising given reports that Thoroughbreds contributed over 13% of the maternal lineages to the stallion lines within the stud book [64]. Low differentiation of the Maremmano compared to the Hanoverian was also reported in [29], which is logical given similar influence of the Thoroughbred on the Hanoverian. This and the allowed crossing of Thoroughbreds into the Swiss Warmblood population is reflected in minimal measures of divergence between these samples. The continued influence of the Thoroughbred on the Paint and Quarter Horse is also reflected both by low F_{ST} values as well as greater than 30% assignment of the Paint and Quarter Horse to the Thoroughbred cluster. As each of these breeds experience continued gene flow from the Thoroughbred, and had Thoroughbred founding stock, these results are not unexpected. While outcrossing of these breeds is allowed (with restrictions), it is likely that even in cases of breeds with closed stud books, some outcrossing, intentional, unintentional, and/or undocumented, has occurred; it has been demonstrated that historical pedigrees, while helpful, are not always accurate [65]. Issues of outcrossing and individual identification can now be more easily addressed using genetic testing and have the potential to assist managers in decisions regarding breeding and registration.

Finally, the Standardbred samples, which also represent two continents, had a significant pairwise F_{ST} value of 0.020, five-fold greater than that observed between the two Thoroughbred samples. This comparison may reflect geographic structure, the influence of French Trotter bloodlines in the European sample, and selection for pacing horses in the US population. Within the horses included in this study, eight were noted to be pacers. These pacers all fall within one clade of the parsimony tree, supported by a bootstrap value of 98%. The limited sample size does not allow a thorough comparison of the pacing vs. trotting Standardbreds, however significant genetic differentiation between horses of the two racing groups has been reported [56].

Middle Eastern and Iberian Breeds

The Middle Eastern breeds, Arabian, Akhal Teke, and Caspian, were placed into a single clade of the NJ tree, with the Arabian and Akhal Teke in their own, highly supported clade. This relationship was supported by low values of K in Structure, which placed the Iberian and Middle Eastern breeds into the same cluster until K =12. However, parsimony analysis did not support this relationship between the Middle Eastern and Iberian breeds as the Arabian and Akhal Teke individuals were placed into one clade, apart from the Iberian samples and from the Caspians.

In Europe, the Iberian breeds (Lusitano and Andalusian) have only recently been distinguished from one another depending upon the region in which they are bred and divergent selective pressures. In the US, horses of each breed are occasionally interbred [51]. The close relationship between the Andalusian and Lusitano samples in this study is reflected in the minimum F_{ST} value observed in either breed, 0.021. The parsimony tree and PCA also shows that individuals cannot necessarily be distinguished from one another regardless of whether they were sampled in the US or Portugal. Of the two clades that appear to suggest population substructure, one includes only Portuguese Lusitanos while the other includes only US samples, although two are of Portuguese ancestry.

Iberian and Gaited Breeds

A horse that is considered to be "gaited" naturally moves in a means other than the traditional walk, trot, canter, and gallop. Alternative gaits in horses are distinguished from traditional gaits by their unique footfall pattern and/or rhythm. The genetic basis of gait has recently been investigated and suggests that all modern gaited breeds share a common ancestor as supported by a shared, extended haplotype spanning a variant significantly associated with the ability to pace [66]; [67]. There is a great deal of historical evidence that the shared ancestry of gaited breeds traces back to Iberian bloodlines, in particular to the Spanish Jenette [45], [51]. Influence of the Iberian breeds on modern gaited breeds is seen in early clustering in Structure analysis as well as in the NJ tree where the Puerto Rican Paso Fino and Peruvian Paso are placed on the same branch as the Andalusian and Lusitanos. In addition to Iberian lines, the Narragansett Pacer is often named as instrumental in the founding of American breeds that may gait including the Saddlebred, Standardbred, and Tennessee Walking Horse [45]. Within those breeds, the Tennessee Walking Horse was documented to be greatly influenced by the Saddlebred, Standardbred, and Morgan [68], [69]; and the Standardbred itself had influence from the Thoroughbred and Morgan (among others) [70]. While we do not have samples of the now extinct Narragansett Pacer, our data set does support a close relationship between the Tennessee Walking Horse and Saddlebred, as observed in the NJ and parsimony analysis as well as with the Morgan and Standardbred. At low values of K in cluster analysis, the Tennessee Walking Horse and Saddlebred cluster strongly. In NJ analysis, the Florida Cracker, which has many individuals that demonstrate the ability to gait, is found intermediate to the Iberian and the modern US gaited breeds. Interestingly, the Icelandic, a four- or five-gaited breed, does not show any significant affinity to the other gaited breeds although they share the recently identified major locus that appears to be essential to the ability to perform alternate gaits [66]; [67]. It thus seems that the genetic variant associated with the gait phenotype arose well before the separation of breeds. Instead of clustering with the other gaited breeds, the Icelandic clusters with the Shetland through K=16 and also is within a highly supported branch of the NJ tree with the Shetland and Miniature. Finally, the influence of the Shetland on the development of the Miniature is observed at all values of K as well as in the parsimony tree where both breeds occupy the same clade.

Drafts

The Shire and Clydesdale populations share assignment to the same cluster throughout Structure runs. The similarity between these breeds is also seen in a lack of monophyly in the parsimony tree, the sharing of a branch of the NJ tree, positioning in PCA, and a pairwise F_{ST} value of 0.037. The Fell Pony, a British breed, falls out as sister taxa to the British draft horses, the Shire and Clydesdales. However F_{ST} values show that divergence between the Fell Pony and either the Clydesdale or Shire is not significantly less than seen with most other populations. The other branch of the "draft" clade of the NJ tree contains the breeds from the European mainland, the Belgian, Percheron, and Franches-Montagnes; each of these breeds shows monophyly in parsimony analysis. In addition, similarities among draft and light draft breeds are reflected in cluster analyses at K =6, which show the populations from the European mainland and Scandinavian Peninsula (Belgian, Finnhorse, Franches-Montagnes, North Swedish Horse, Norwegian Fjord, and Percheron) assign to one cluster with q >0.5. The grouping of the Scandinavian breeds is similar to that previously reported [31]. Geographic relationships are also suggested by the two British breeds, the New Forest Pony and Exmoor that fall just basal to the draft clade in the NJ tree.

Summary

This data set resulting from a large international collaboration represents the first study in the horse to provide an extensive overview of nuclear genetic diversity within, and relationships among a diverse sample of breeds and landrace populations. These data are now available for use in subsequent studies of population-level relationships and provide a baseline for monitoring changes in breed diversity. With high mtDNA diversity but limited paternal input during domestication, this increased understanding of nuclear diversity within the horse will allow for the identification of genomic regions of importance to breed derivation and will be instrumental in guiding across-breed gene discovery projects.

METHODS

Ethics Statement

DNA sampling was limited to the collection of blood by jugular venipuncture performed by a licensed veterinarian or from hairs pulled from the mane or tail by the horse owner or researcher. All animal work was conducted in accordance with and approval from the international and national governing bodies at the institutions in which samples were collected (the University of Minnesota Institutional Animal Care and Use Committee (IACUC); the University of Kentucky IACUC; the University College Dublin, Animal Research Ethics Committee; Swiss Law on Animal Protection and Welfare; the Ethical Board of the University of Helsinki; the Animal Health Trust Clinical Research Ethics Committee; Norwegian Animal Research Authority; UK Home Office License; and the Lower Saxon state veterinary office).

Samples and Genotyping

Tissue samples and previously collected genotypes from 1,060 horses were obtained from members of the EGDC or were obtained by our laboratory. 814 samples representing the 38 populations included in this study were selected from the EDGC sample collection with the goal of obtaining as random of a sample as possible and to minimize close relationships among individuals. In some cases, genotypes were available from breeds collected for genome-wide association studies (GWAS). In all cases, when pedigree information was available, no relationships were allowed at or more recent to the grandsire/ dam level. If no pedigrees were available, once genotyping was performed, individuals were removed from the analyses to reduce genome sharing as measured by autosomal estimates of identity by descent (pi hat) values in PLINK [71] greater than 0.3 (after pruning for MAF>0.05). In samples that were obtained as a result of GWAS, "control" individuals were preferentially chosen for inclusion in these analyses. When necessary, DNA isolation from hair roots took place using a modification of the Puregene (Qiagen) protocol for DNA purification from tissue. Modifications include the addition of 750 µl of isopropanol rather than 300, increasing the precipitation spin time to 15 m at 4°C, and washing the pellet twice. Approximately 1 µg of DNA was used for SNP genotyping using the Illumina SNP50 Beadchip according to the manufacturer's protocol. All genotype calls were extracted from the raw intensity data using GenomeStudio (Illumina) with the minimum gencall score threshold of 0.15. The raw intensity scores were available for all populations with the exception of the Lusitano and Maremmano.

Data Pruning

SNP discovery was conducted using horses from seven breeds (Akhal Teke, Andalusian, Arabian, Icelandic, Quarter Horse, Standardbred, Thoroughbred) as well as the reference genome of a Thoroughbred mare [41]. To eliminate ascertainment bias as much as possible, horses from the discovery breeds were removed from the dataset, which was then pruned to exclude SNPs with MAF less than 0.05. All horses were then replaced and those SNPs removed from all analyses. In this new, complete data set, SNP markers that failed to genotype in at least 99% of the individuals and SNPs that had a MAF of 0.05 or less across all samples were removed as well as SNPs on ECAX. SNPs that were in LD across breeds were also removed; files used for basic diversity indices were pruned for $r^2<0.1$ in PLINK [71] considering 100 SNP windows and moving 25 SNPs per set (–indep-pairwise 100 25 1.11). Allowing for additional LD, data sets were also created for r2<0.2 and 0.4. An additional data set, used for Structure analysis was pruned for $R^2<0.1$ in Plink (–indep;

R=multiple correlation coefficient), which is similar to the above method but instead of analyzing pairwise relationships of SNPs as in the former method, uses a multiple regression approach upon the SNPs in the analysis window. Files were converted for usage between analyses programs using PLINK, perl script, CONVERT [72] and/or PGDSpider 2.0.1.4 [73].

Within-breed Diversity

Expected heterozygosity (H_e) and AMOVA were calculated in Arlequin3.5 [74] on all four data sets. AMOVA was conducted on the primary data set with breeds designated as populations and excluding the Florida Cracker due to small sample size. Analyses were also conducted grouping the two Thoroughbred and Standardbred samples together by breed. F_{IS} was calculated and significance tested on the primary data set, with 10,000 permutations of the data in Genetix [75]. Individual inbreeding coefficients (f) were calculated in PLINK based upon loss of heterozygosity (–het).

Among Breed Relationships

Pairwise F_{ST} values were calculated on the primary, 10,536 SNP dataset in Arlequin3.5 [74]using Reynolds' distance [76] with significance tested using 20,000 permutations.

A neighbor joining (NJ) cladogram was built using breed allele frequencies calculated from the primary SNP set using the packages seqboot, gendist, neighbor, and consense and Nei's genetic distance [38] in PHYLIP ver3.69 [77]. Bootstrap support from 1,000 iterations of the data was used to assess support for the resulting majority rule consensus cladogram.

A parsimony cladogram was constructed using 10,066 SNP markers pruned from the original data set using the MAF and genotyping rate criteria as above and allowing for $R^2<0.2$ (–indep-pairwise). The domestic ass was included as an outgroup for traditional and new-technology searches in TNT [78].

Principal component analysis (PCA) was conducted in snpStats in R (http://cran.r-project.org) on the full SNP set consisting of all 814 individuals and 38,755 autosomal SNPs (pruned only for MAF and genotyping rate).

Cluster Analysis

Clustering of breeds into genetic groups was examined using the program Structure 2.3.3 [79],[80] assuming K =1 to 45. The Structure algorithm included the admixture model and correlated allele frequencies. Three iterations of each K value were conducted with 35,000 MCMC repetitions (15,000 burn-in). The convergence of Structure runs was evaluated by equilibrium of alpha and likelihood scores. The value of K most suitable to explain the diversity in these data was predicted by the highest mean estimated ln P(X|K) while minimizing variance and also making biological sense [80], [81]. The replicates from each run of K were input into CLUMPP [82] and the average cluster membership calculated using the LargeK Greedy algorithm. Output from CLUMPP was visualized in Distruct [83].

Effective Population Size

To estimate effective population size (N_e), the full set of 54,602 SNP markers was pruned within each population to remove those with MAF <0.01 and genotyping rate of <0.05. Pairwise r^2 values between remaining SNPs were calculated in Haploview [84], for each population considering intermarker distances from 0 to 4 Mb in 50 kb increments. Values of N_e were calculated using the method of [37], which includes a correction for small sample size and the assumption that 1 Mb =1 cM.

Data Access

All SNP genotype data are available at the NAGPR Community Data Repository (animalgenome.org) for the purpose of reconstructing the analyses. The only exception is the data collected from the Tennessee Walking Horse, which, under agreement from the granting agency (to the University of Minnesota from the Foundation for the Advancement of the Tennessee Walking Show Horse (FAST) and the Tennessee Walking Horse Foundation (TWHF)), is only available under a Material Transfer Agreement (MTA) between interested individuals and the University of Minnesota.

ACKNOWLEDGMENTS

The computational support provided by Rob Schaefer and Aaron Rendahl is gratefully acknowledged. Finally, the EGDC thanks the Havemeyer Foundation for their continued support of equine genomics research.

AUTHOR CONTRIBUTIONS

Conceived and designed the experiments: JLP JRM MEM. Performed the experiments: JLP MEM. Analyzed the data: JLP. Contributed reagents/materials/analysis tools: JLP JRM EGC LSA JA EB DB MMB ASB PB ACM OD MF LFC KTG GG BH TH KH EWH TL GL HL MSL BAM SM NO MCTP RJP MR SR KHR MS JS TT MV CMW MEM. Wrote the paper: JLP JRM EGC MEM.

REFERENCES

1.   .FAOSTAT (2010). Available: http://www.faostat.fao.org.Accessed 2012 Jun 5.

2.   .Lippold S, Matzke NJ, Reissmann M, Hofreiter M (2011) Whole mitochondrial genome sequencing of domestic horses reveals incorporation of extensive wild horse diversity during domestication. BMC Evol Biol 11: 328. doi: 10.1186/1471-2148-11-328

3.   .Ludwig A, Pruvost M, Reissmann M, Benecke N, Brockmann GA, et al. (2009) Coat color variation at the beginning of horse domestication. Science 324: 485. doi: 10.1126/science.1172750

4.   .Outram AK, Stear NA, Bendrey R, Olsen S, Kasparov A, et al. (2009) The earliest horse harnessing and milking. Science 323: 1332–1335. doi: 10.1126/science.1168594

5.   .Pedrosa S, Uzun M, Arranz JJ, Gutierrez-Gill B, Primitivo FS, et al. (2005) Evidence of three maternal lineages in near eastern sheep supporting multiple domestication events. P R Soc B 272: 2211–2217. doi: 10.1098/rspb.2005.3204

6.   .Larson G, Dobney K, Albarella U, Fang M, Matisoo-Smith E, et al. (2005) Worldwide phylogeography of wild boar reveals multiple centers of pig domestication. Science 307: 1618–1621. doi: 10.1126/science.1106927

7.   Wu GS, Yao YG, Qu KX, Ding ZL, Li H, et al. (2007) Population phylogenomic analysis of mitochondrial DNA in wild boars and domestic pigs revealed multiple domestication events in East Asia. Genome Biol 8: R245. doi: 10.1186/gb-2007-8-11-r245

8.   .Cieslak M, Pruvost M, Benecke N, Hofreiter M, Morales A, et al. (2010) Origin and history of mitochondrial DNA lineages in domestic horses. PLOS One 5: e15311. doi: 10.1371/journal.pone.0015311

9.   .Jansen T, Forster P, Levine MA, Oelke H, Hurles M, et al. (2002) Mitochondrial DNA and the origins of the domestic horse. PNAS 99: 10905–10910. doi: 10.1073/pnas.152330099

10. .Lei CZ, Su R, Bower MA, Edwards CJ, Wang XB, et al. (2009) Multiple maternal origins of native modern and ancient horse populations in China. Anim Genet 40: 933–944. doi: 10.1111/j.1365-2052.2009.01950.x

11. .Lira J, Linderholm A, Olaria C, Brandstrom Durling M, Gilbert MT, et al. (2010) Ancient DNA reveals traces of Iberian Neolithic and Bronze Age lineages in modern Iberian horses. Mol Ecol 19: 64–78. doi: 10.1111/j.1365-294x.2009.04430.x

12. .Vila C, Leonard JA, Gotherstrom A, Marklund S, Sandberg K, et al. (2001) Widespread origins of domestic horse lineages. Science 291: 474–477. doi: 10.1126/science.291.5503.474

13. .Warmuth V, Eriksson A, Bower MA, Barker G, Barrett E, et al. (2012) Reconstructing the origin and spread of horse domestication in the Eurasian steppe. PNAS 109: 8202–8206. doi: 10.1073/pnas.1111122109

14. .Beja-Pereira A, Caramelli D, Lalueza-Fox C, Vernesi C, Ferrand N, et al. (2006) The origin of European cattle: evidence from modern and ancient DNA. PNAS 103: 8113–8118. doi: 10.1073/pnas.0509210103

15. .Gotherstrom A, Anderung C, Hellborg L, Elburg R, Smith C, et al. (2005) Cattle domestication in the Near East was followed by hybridization with aurochs bulls in Europe. P R Soc B 272: 2345–2350. doi: 10.1098/rspb.2005.3243

16. .Keyser-Tracqui C, Blandin-Frappin P, Francfort HP, Ricaut FX, Lepetz S, et al. (2005) Mitochondrial DNA analysis of horses recovered from a frozen tomb (Berel site, Kazakhstan, 3rd Century BC). Anim Genet 36: 203–209. doi: 10.1111/j.1365-2052.2005.01316.x

17. .Achilli A, Olivieri A, Soares P, Lancioni H, Kashani BH, et al. (2012) Mitochondrial genomes from modern horses reveal the major haplogroups that underwent domestication. PNAS 109: 2449–2454. doi: 10.1073/pnas.1111637109

18. .McGahern A, Bower MA, Edwards CJ, Brophy PO, Sulimova G, et al. (2006) Evidence for biogeographic patterning of mitochondrial DNA sequences in Eastern horse populations. Anim Genet 37: 494–497. doi: 10.1111/j.1365-2052.2006.01495.x

19. .Warmuth V, Manica A, Eriksson A, Barker G, Bower M (2012) Autosomal genetic diversity in non-breed horses from eastern Eurasia provides insights into historical population movements. Anim Genet. In press. doi:10.1111/j.1365–2052.2012.02371.x.

20. .Lindgren G, Backstrom N, Swinburne J, Hellborg L, Einarsson A, et al. (2004) Limited number of patrilines in horse domestication. Nat Genet 36: 335–336. doi: 10.1038/ng1326

21. .Ling Y, Ma Y, Guan W, Cheng Y, Wang Y, et al. (2010) Identification of y chromosome genetic variations in Chinese indigenous horse breeds. J Hered 101: 639–643. doi: 10.1093/jhered/esq047

22. .Aberle KS, Hamann H, Drogemuller C, Distl O (2004) Genetic diversity in German draught horse breeds compared with a group of primitive, riding and wild horses by means of microsatellite DNA markers. Anim Genet 35: 270–277. doi: 10.1111/j.1365-2052.2004.01166.x

23. .Achmann R, Curik I, Dovc P, Kavar T, Bodo I, et al. (2004) Microsatellite diversity, population subdivision and gene flow in the Lipizzan horse. Anim Genet 35: 285–292. doi: 10.1111/j.1365-2052.2004.01157.x

24. .Bjornstad G, Gunby E, Roed KH (2000) Genetic structure of Norwegian horse breeds. J Anim Breed Genet 117: 307–317. doi: 10.1046/j.1439-0388.2000.00264.x

25. .Bömcke E, Gengler N, Cothran EG (2010) Genetic variability in the Skyros pony and its relationship with other Greek and foreign horse breeds. Genet and Mol Biol 34: 68–76. doi: 10.1590/s1415-47572010005000113

26. .Canon J, Checa ML, Carleos C, Vega-Pla JL, Vallejo M, et al. (2000) The genetic structure of Spanish Celtic horse breeds inferred from microsatellite data. Anim Genet 31: 39–48. doi: 10.1046/j.1365-2052.2000.00591.x

27. .Cothran EG, Canelon JL, Luis C, Conant E, Juras R (2011) Genetic analysis of the Venezuelan Criollo horse. Genet Mol Res 10: 2394–2403. doi: 10.4238/2011.october.7.1

28. .Cunningham EP, Dooley JJ, Splan RK, Bradley DG (2001) Microsatellite diversity, pedigree relatedness and the contributions of founder lineages to thoroughbred horses. Anim Genet 32: 360. doi: 10.1046/j.1365-2052.2001.00785.x

29. .Felicetti M, Lopes MS, Verini-Supplizi A, Machado Ada C, Silvestrelli M, et al. (2010) Genetic diversity in the Maremmano horse and its relationship with other European horse breeds. Anim Genet 41 Suppl 253–55. doi: 10.1111/j.1365-2052.2010.02102.x

30. .Glowatzki-Mullis ML, Muntwyler J, Pfister W, Marti E, Rieder S, et al. (2005) Genetic diversity among horse populations with a special focus on the Franches-Montagnes breed. Anim Genet 37: 33–39. doi: 10.1111/j.1365-2052.2005.01376.x

31. .Leroy G, Callede L, Verrier E, Meriaux JC, Ricard A, et al. (2009) Genetic diversity of a large set of horse breeds raised in France assessed by microsatellite polymorphism. Genet Sel Evol 41: 5. doi: 10.1186/1297-9686-41-5

32. .Marletta D, Tupac-Yupanqui I, Bordonaro S, Garcia D, Guastella AM, et al. (2006) Analysis of genetic diversity and the determination of relationships among western Mediterranean horse breeds using microsatellite markers. J Anim Breed Genet 123: 315–325. doi: 10.1111/j.1439-0388.2006.00603.x

33. .Shasavarani H, Rahimi-Mianji G (2010) Analysis of genetic diversity and estimation of inbreeding coefficient within Caspian horse population using microsatellite markers. Afr J Biotechnol 9: 293–299.

34. .Solis A, Jugo BM, Meriaux J-C, Iriondo M, Mazon LI, et al. (2005) Genetic diversity within and among four South European native horse breeds based on microsatellite DNA analysis: Implications for conservation. J Hered 96: 670–678. doi: 10.1093/jhered/esi123

35. .Tozaki T, Takezaki N, Hasegawa T, Ishida N, Kurosawa M, et al. (2003) Microsatellite variation in Japanese and Asian horses and their phylogenetic relationships using a European horse outgroup. J Hered 94: 374–380. doi: 10.1093/jhered/esg079

36. .Luis C, Juras R, Oom MM, Cothran EG (2007) Genetic diversity and relationships of Protuguese and other horse breeds based on protein and microsatellite loci variation. Anim Genet 38: 20–27. doi: 10.1111/j.1365-2052.2006.01545.x

37. .Weir BS, Hill WG (1980) Effect of mating structure on variation in linkage disequilibrium. Genetics 95: 477–488.

38. Nei M (1972) Genetic distance between populations. Amer Nat 106: 283–292. doi: 10.1086/282771

39. Evanno G, Regnaut S, Goudet J (2005) Detecting the number of clusters of individuals using the software STRUCTURE: a simulation study. Mol Ecol 14: 2611–2620. doi: 10.1111/j.1365-294x.2005.02553.x

40. .Wade CM, Giulotto E, Sigurdsson S, Zoli M, Gnerre S, et al. (2009) Genome sequence, comparative analysis, and population genetics of the domestic horse. Science 326: 865–867. doi: 10.1126/science.1178158

41. .McCue ME, Bannasch DL, Petersen JL, Gurr J, Bailey E, et al. (2012) A high density SNP array for the domestic horse and extant perissodactyla: utility for association mapping, genetic diversity, and phylogeny studies. PLOS Genet 8: e1002451. doi: 10.1371/journal.pgen.1002451

42. .Weatherby and Sons (1791) An Introduction to a General Stud Book. London, UK.

43. .Corbin LJ, Blott SC, Swinburne JE, Vaudin M, Bishop SC, et al. (2010) Linkage disequilibrium and historical effective population size in the Thoroughbred horse. Anim Genet 41 Suppl 28–15. doi: 10.1111/j.1365-2052.2010.02092.x

44. .Binns MM, Boehler DA, Bailey E, Lear TL, Cardwell JM, et al. (2012) Inbreeding in the Thoroughbred horse. Anim Genet 43: 340–342. doi: 10.1111/j.1365-2052.2011.02259.x

45. .Hendricks BL (2007) International Encyclopedia of Horse Breeds. Norman, OK: University of Oklahoma Pres. 486 p.

46. .Weatherley L (1978) Great horses of Britain. Hindhead: Spur Publications. viii, 269 p.

47. .Gates S (1979) A study of the home ranges of free-ranging Exmoor ponies. Mammal Rev 9: 3–18. doi: 10.1111/j.1365-2907.1979.tb00228.x

48. .Peilieu C (1984) Livestock Breeds of China. FAO: Animal Production and Health paper 46: 1–217.

49. .Hund A (2008) The Stallion's Mane: The Next Generation of Horses in Mongolia. Available: http://digitalcollections.sit.edu/cgi/viewcontent. cgi?article=1543&context=isp_collection. Accessed 2012 Dec 31.

50. .Ticklen M, editor (2006) Get to know the Finnhorse. Available:http:// www.suomenhevonen.info/hippos/sh2007/pdf/SHjulkaisu_englanti_ nettiin.pdf. Accessed 2012 Dec 31.

51. .Bowling AT, Ruvinsky A (2000) The genetics of the horse. Wallingford: CABI Pub. viii, 527 p.

52. .Lynghaug F (2009) The Official Horse Breeds Standards Guide: The Complete Guide to the Standards of All North American Equine Breed Associations. Stillwater, MN: Voyageur Press.

53. .Conant EK, Juras R, Cothran EG (2012) A microsatellite analysis of five Colonial Spanish horse populations of the southeastern United States. Anim Genet 43: 53–62. doi: 10.1111/j.1365-2052.2011.02210.x

54. .Lopes MS (2011) Molecular tools for the characterisation of the Lusitano horse. Angra do Heroísmo: University of Azores. 219 p.

55. .Tryon RC, Penedo MC, McCue ME, Valberg SJ, Mickelson JR, et al. (2009) Evaluation of allele frequencies of inherited disease genes in subgroups of American Quarter Horses. J Am Vet Med Assoc 234: 120–125. doi: 10.2460/javma.234.1.120

56. .Cothran EG, MacCluer JW, Weitkamp LR, Bailey E (1987) Genetic differentiation associated with gait within American standardbred horses. Anim Genet 18: 285–296. doi: 10.1111/j.1365-2052.1987.tb00772.x

57. .Florida Cracker Horse Association. Available:http://www. floridacrackerhorses.com/index.htm. Accessed 2012 Jun 5.

58. .Firouz L (1969) Conservation of a domestic breed. Biol Conserv 1: 1–2. doi: 10.1016/0006-3207(69)90117-7

59. .Bjornstad G, Nilsen NO, Roed KH (2003) Genetic relationship between Mongolian and Norwegian horses? Anim Genet 34: 55–58. doi: 10.1046/j.1365-2052.2003.00922.x

60.   .Conrad DF, Jakobsson M, Coop G, Wen X, Wall JD, et al. (2006) A worldwide survey of haplotype variation and linkage disequilibrium in the human genome. Nat Genet 38: 1251–1260. doi: 10.1038/ng1911

61.   .Li JZ, Absher DM, Tang H, Southwick AM, Casto AM, et al. (2008) Worldwide human relationships inferred from genome-wide patterns of variation. Science 319: 1100–1104. doi: 10.1126/science.1153717

62.   .Mackay-Smith A (2000) Speed and the thoroughbred : the complete history. Lanham, MD: Derrydale Press. xxvii, 193 p.

63.   .Bower MA, Campana MG, Whitten M, Edwards CJ, Jones H, et al. (2011) The cosmopolitan maternal heritage of the Thoroughbred racehorse breed shows a significant contribution from British and Irish native mares. Biol Letters 7: 316–320. doi: 10.1098/rsbl.2010.0800

64.   .Silvestrelli M (1991) The Maremmano horse. Animal Genetic Resources Information (FAO/UNEP) 8: 74–83. doi: 10.1017/s1014233900003126

65.   .Bower MA, Campana MG, Nisbet RER, Weller R, Whitten M, et al. (2012) Truth in the bones: Resolving the identify of the founding elite Thoroughbred racehorses. Archaeometry 54: 916–925. doi: 10.1111/j.1475-4754.2012.00666.x

66.   .Andersson LS, Larhammar M, Memic F, Wootz H, Schwochow D, et al. (2012) Mutations in DMRT3 affect locomotion in horses and spinal circuit function in mice. Nature 488: 642–646. doi: 10.1038/nature11399

67.   .Petersen JL, Mickelson JR, Rendahl AK, Valberg SJ, Andersson LS, et al. (2012) Genome-wide analysis reveals selection for important traits in domestic horse breeds. PLOS Genet. In press.

68.   .Fletcher JL (1946) A study of the first fifty years of Tennessee walking horse breeding. J Hered 37: 369–373.

69.   .Tennessee Walking Horse Breeders and Exhibitors Association. Available:http://www.twhbea.com/breed/history.php. Accessed 2012 Jun 5.

70.   .MacCluer JW, Boyce AJ, Dyke B, Weitkamp LR, Pfennig DW, et al. (1983) Inbreeding and pedigree structure in Standardbred horses. J Hered 74: 394–399.

71.   .Purcell S, Neale B, Tood-Brown K, Thomas L, Ferreira MAR, et al. (2007) PLINK: a toolset for whole-genome association and population-based linkage analysis. Amer J Hum Genet 81: 559–575. doi: 10.1086/519795

72.   .Glaubitz JC (2004) CONVERT: A user-friendly program to reformat diploid genotypic data for commonly used population genetic

software packages. Mol Ecol Notes 4: 309–310. doi: 10.1111/j.1471-8286.2004.00597.x

73. .Lischer H, Excoffier L (2012) An automated data conversion tool for connecting population genetics and genomics programs. Bioinformatics 28: 298–299. doi: 10.1093/bioinformatics/btr642

74. .Excoffier L, Laval G, Schneider S (2005) Arlequin (version 3.0): An integrated software package for population genetics data analysis. Evol Bioinform 1: 47–50.

75. .Belkhir KP, Borsa P, Chikhi L, Raufaste N, Bonhomme F (2001) GENTETIX, logiciel sous Windows TM pour la genetique des population. Universite de Montpellier II, Montpellier, France.

76. .Reynolds J, Weir BS, Cockerham CC (1983) Estimation of the coancestry coefficient: basis for a short-term genetic distance. Genetics 105: 767–779.

77. .Felsenstein J (1989) PHYLIP-Phylogeny Inference Package (version 3.2) Cladistics. 5: 164–166.

78. .Goloboff PA, Farris JS, Nixon K (2003) TNT: tree analysis using new technology. Syst Biol 54: 176–178.

79. .Falush D, Stephens M, Pritchard JK (2003) Inference of population structure using multilocus genotype data: linked loci and correlated allele frequencies. Genetics 164: 1567–1587.

80. .Pritchard JK, Stephens M, Donnelly P (2000) Inference of population structure using multilocus genotype data. Genetics 155: 945–959.

81. .Pritchard JK, Wen X, Falush D (2010) Documentation for Structure software: version 2.3.38.

82. .Jakobsson M, Rosenberg NA (2007) CLUMPP: a cluster matching and permutation program for dealing with label switching and multimodality in analysis of population structure. Bioinformatics 23: 1801–1806. doi: 10.1093/bioinformatics/btm233

83. .Rosenberg NA (2004) DISTRUCT: a program for the graphical display of population structure. Mol Ecol Notes 4: 137–138. doi: 10.1046/j.1471-8286.2003.00566.x

84. .Barrett JC, Fry B, Maller J, Daly MJ (2005) Haploview: analysis and visualization of LD and haplotype maps. Bioinformatics 21: 263–265. doi: 10.1093/bioinformatics/bth457

Chapter 12

Y-CHROMOSOME AND MTDNA GENETICS REVEAL SIGNIFICANT CONTRASTS IN AFFINITIES OF MODERN MIDDLE EASTERN POPULATIONS WITH EUROPEAN AND AFRICAN POPULATIONS

Danielle A. Badro[1], Bouchra Douaihy[1]., Marc Haber[1,2], Sonia C. Youhanna[1], Ange´lique Salloum[1], Michella Ghassibe-Sabbagh[1], Brian Johnsrud[3], Georges Khazen[1], Elizabeth Matisoo-Smith[4], David F. Soria-Hernanz[2,5], R. Spencer Wells[5], Chris Tyler-Smith[6], Daniel E. Platt[7], Pierre A. Zalloua[1,8]

[1]The Lebanese American University, Chouran, Beirut, Lebanon

[2]Institut de Biologia Evolutiva (CSIC-UPF), Departament de Cie`ncies de la Salut i de la Vida, Universitat Pompeu Fabra, Barcelona, Spain

[3]Modern Thought and Literature, Stanford University, Stanford, California, United States of America

[4]Allan Wilson Centre for Molecular Ecology and Evolution, University of Otago, Dunedin, New Zealand, [5]The Genographic Project, National Geographic Society, Washington, DC, United States of America

[6]The Wellcome Trust Sanger Institute, Wellcome Trust Genome Campus, Hinxton, United Kingdom

[7]Computational Biology Centre, IBM TJ Watson Research Centre, Yorktown Heights, New York, United States of America

[8]Harvard School of Public Health, Boston, Massachusetts, United States of America

ABSTRACT

The Middle East was a funnel of human expansion out of Africa, a staging area for the Neolithic Agricultural Revolution, and the home to some of the earliest world empires. Post LGM expansions into the region and subsequent population movements created a striking genetic mosaic with distinct sex-based genetic differentiation. While prior studies have examined the mtDNA and Y-chromosome contrast in focal populations in the Middle East, none have undertaken a broad-spectrum survey including North and sub-Saharan Africa, Europe, and Middle Eastern populations. In this study 5,174 mtDNA and 4,658 Y-chromosome samples were investigated using PCA, MDS, mean-linkage clustering, AMOVA, and Fisher exact tests of F_{ST}'s, R_{ST}'s, and haplogroup frequencies. Geographic differentiation in affinities of Middle Eastern populations with Africa and Europe showed distinct contrasts between mtDNA and Y-chromosome data. Specifically, Lebanon's mtDNA shows a very strong association to Europe, while Yemen shows very strong affinity with Egypt and North and East Africa. Previous Y-chromosome results showed a Levantine coastal-inland contrast marked by J1 and J2, and a very strong North African component was evident throughout the Middle East. Neither of these patterns were observed in the mtDNA. While J2 has penetrated into Europe, the pattern of Y-chromosome diversity in Lebanon does not show the widespread affinities with Europe indicated by the mtDNA data. Lastly, while each population shows evidence of connections with expansions that now define the Middle East, Africa, and Europe, many of the populations in the Middle East show distinctive mtDNA and Y-haplogroup characteristics that indicate long standing settlement with relatively little impact from and movement into other populations.

INTRODUCTION

As a crossroad between Africa, Arabia, Asia, and Europe, the Levant has been a primary historical stepping stone in the first modern human expansions out of Africa and for later migrations into and out of Europe, Asia, and Africa [1]–[7]. As such, it has also become a land of remarkable human diversity. The earliest fossil and archaeological evidence of modern humans outside of the African continent are from the Levant, presumably indicating a migration via the northern route, and date to 125–95 kya [8], [9]. Additionally, genetic studies suggest that the initial peopling of Eurasia occurred through the northern Levantine (modern day Lebanon and Syria) route [10]–[12]. Two proposed routes chart the dispersal of anatomically modern humans out of the African continent: (1) a northern route, reaching west and central Asia through the Sinai Peninsula and the Levant, and (2) a southern route via the Bab el-Mandeb Strait and along the south Asian coast, ultimately reaching Australia [13]–[15].

While the out-of-Africa migrations have been major determining factors, other migratory events have strongly influenced genetic marker distributions throughout the Levant and the surrounding geographical areas. During the last glacial maximum (LGM, 26.5–19 kya), most of the Levant was an uninhabitable desert, with forested hills in Levantine Mediterranean coastal areas [7]. The genetics of the modern Levant were largely determined by subsequent repopulation (especially during the Neolithic agricultural revolution) and mass movements associated with empire building. Neolithic expansions in particular, beginning around 10 kya, induced gene flow between the Fertile Crescent and Europe, which shaped the genetic structure of both regions [16]–[22].

Most genetic studies of the Levant as a geographical area have focused exclusively on either Y-chromosome [23]–[27] or mitochondrial markers [28]. Further, contrasts between Y-chromosome and mtDNA data provide distinct insights into human expansions unavailable to somatic genome analyses [29]. While comparative analyses among the two marker types have been undertaken in the Middle East and Africa [30]–[34], none of these studies have explored the contrasting relationships of expansions throughout Europe, North Africa, the Levant and Arabian Peninsula after the LGM. Building on a previous study that reported phylogeographic characteristics of Y-chromosome markers in the Levantine region [25], we now compare and contrast Y and mtDNA phylogeographic distributions in the Levant and investigate the affinities of Middle Eastern populations with European and African populations.

MATERIALS AND METHODS

Ethics Statement

The samples were collected from donors after they had given their written informed consent to the project and to the data analysis, which was approved by the IRB of the Lebanese American University.

mtDNA Data

3,663 mtDNA records collected from the literature represented populations from Burkina-Faso[35], Cyprus [36], Egypt [37], Ethiopia [38], France [39], [40], Greece [36], Iraq [31], Jordan[30], Kenya [41], Libyan Sahara [42], Mali [35], Morocco [43], [44], Niger [35], Saudi Arabia[45], Slovakia [46], Tunisia [44], [47], and Yemen [38], [48]. In addition to this data, we added 1,511 new samples from Lebanon, Libya, Jordan, Palestine, and Syria. Samples were collected from unrelated blood donors from five countries. Surname repetitions were avoided and used as a criterion for absence of relatedness among volunteers, appropriate for Y-chromosome analysis. All demographic data were provided by self-assignment.

Given the broad cultural and genetic diversity in the region, terms such as "Middle East" may be problematical. Historically, the term evolved during the era of European Imperialism, and included all lands between Arabia and India, but came to include Turkey through Saudi Arabia, extending east through Afghanistan and Pakistan. In this report, "Middle Easterners" refers to Iraqis,

Jordanians, Lebanese, Palestinians, Saudis, Syrians, and Yemenis. The Greek data represent the Southeastern Europe region for mtDNA analyses, and is labelled "Southeastern Europe" in the rest of this report. France represents Western Europe mtDNA, and is labelled "Western Europe" throughout the rest of this report. All haplogroups were reduced to the most informative sets for the purpose of homogeneous representation and comparative analyses. mtDNA haplogroup frequencies are displayed inTable 1 and shown as pie charts in Figure 1.

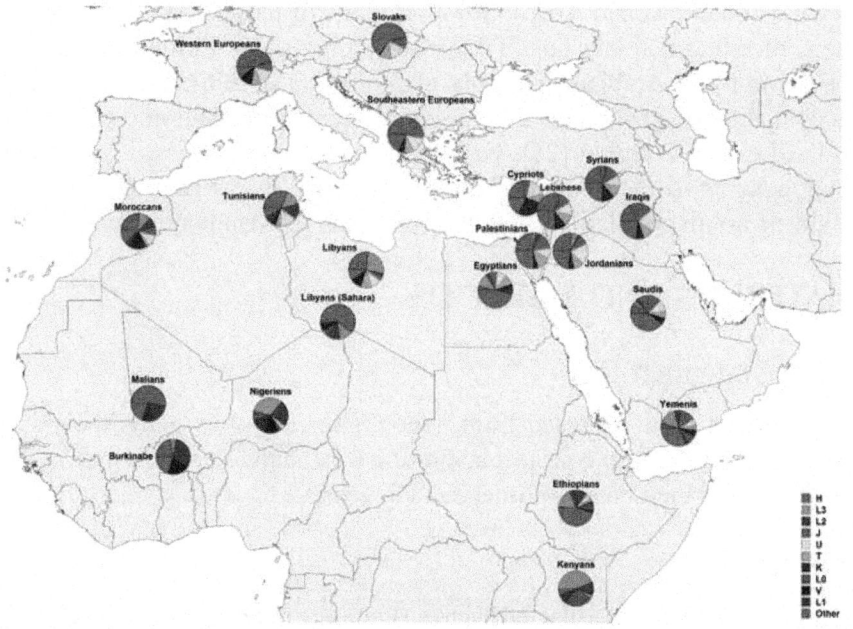

Figure 1: Geographic distribution of mtDNA haplogroups.

Frequencies distribution from the current study and from the published data [30], [31],[35]–[48] as reported in Table 1.

doi:10.1371/journal.pone.0054616.g001

Table 1: mtDNA Haplogroup frequencies of 1509 newly sequenced Levantine samples and 3665 samples collected from the literature

Country	n	L*	L3*	M*	D	N*	R*	R0	HV	H	U	References
Levant												
Lebanon	979	0.6	1.4	1.4	0.2	10.5	21.2	5.0	8.5	29.9	21.1	This study
Syria	234	3.4	2.6	0.9	0.0	11.1	25.6	2.6	9.0	24.4	20.5	This study
Palestine	120	5.0	2.5	3.3	0.0	8.3	27.5	5.8	8.3	25.0	14.2	This study
Cyprus	79	3.8	0.0	2.5	0.0	12.7	17.7	0.0	0.0	22.8	40.5	Irwin et al. [36]
Europe												
France	871	0.6	0.3	0.1	0.0	5.4	18.0	0.5	6.9	45.4	22.8	Dubut et al. [39] Richard et al. [40]
Greece	372	0.0	0.0	1.1	0.0	9.1	19.4	1.9	5.9	42.2	20.4	Irwin et al. [36]
Slovakia	200	1.0	0.0	1.0	0.0	10.0	20.0	0.0	4.0	47.0	17.0	Malyarchuk et al. [46]
Arabian Peninsula												
Jordan	290	4.1	5.9	1.4	0.3	10.7	17.9	3.4	6.9	26.2	23.1	This study Gonzales et al. [30]
Iraq	52	0.0	0.0	1.9	0.0	0.0	34.6	5.8	11.5	17.3	28.8	Al-Zahery et al. [31]
Saudi Arabia	539	6.5	3.5	7.1	0.0	12.2	28.0	18.2	0.7	8.7	15.0	Abu-Amero et al. [45]
Yemen	300	22.0	16.3	5.3	0.0	6.7	21.7	9.3	2.3	4.7	11.7	Kivisild et al. [38] Cerny et al. [48]
North Africa												
Egypt	278	8.3	12.2	6.8	0.0	10.4	21.6	17.27	6.1	0.0	13.7	Saunier et al. [37]
Libya	31	6.5	25.8	0.0	0.0	3.2	12.9	6.5	9.7	25.8	9.7	This study
Libyan Sahara	129	21.7	11.6	1.6	0.0	0.0	0.0	0.0	3.9	61.2	0.0	Ottoni et al. [42]
Morocco	137	18.2	10.9	7.3	0.0	2.2	10.2	3.6	7.3	23.4	16.8	Turchi et al. [44] Harich et al. [43]
Tunisia	160	20.6	14.4	1.9	0.0	0.6	7.5	0.0	7.5	31.3	16.3	Cherni et al. [47] Turchi et al. [44]
East sub Saharan												
Ethiopia	232	36.6	18.1	14.7	0.0	3.9	5.2	11.6	2.6	0.9	6.5	Kivisild et al. [38]
Kenya	84	47.6	44.0	4.8	0.0	0.0	1.2	1.2	0.0	0.0	1.2	Brandstätter et al. [41]
West sub Saharan												
Burkina-Faso	40	35.0	5.0	17.5	0.0	0.0	0.0	0.0	20.0	22.5	0.0	Pereira et al. [35]
Mali	21	28.6	0.0	19.0	0.0	0.0	0.0	0.0	0.0	52.4	0.0	Pereira et al. [35]
Niger	25	44.0	32.0	4.0	0.0	0.0	0.0	0.0	12.0	4.0	4.0	Pereira et al. [35]

doi:10.1371/journal.pone.0054616.t001

doi:10.1371/journal.pone.0054616.t001

Y-Chromosome Data

1,774 previously published Y-chromosome records were obtained from the literature representing populations from the Balkans [49], Burkina-Faso [50], Ethiopia [50], Italy [51], Kenya [50], Saudi Arabia [52], Slovakia [53], and Yemen [52]. In addition 2,884 previously published data from our laboratory representing populations from Cyprus, Egypt, Lebanon, Libya, Jordan, Morocco, Palestine, Syria and Tunisia were added to this study. The Italian Y-chromosome samples represent Western Europe in this study, and are labelled "Western European" through the rest of this report. The Balkan samples represent the Southeastern Europe region, and are labelled "Southeastern European" in the rest of this report. Haplogroups of the Saudi, Yemeni, and Slovak populations were not available, thus we have predicted those haplogroups using the populations haplotypes and the online haplogroup prediction tool [54], [55]:www.hprg.com/hapest5/hapest5a/hapest5.htm. We have also computed STR-predicted Y haplogroups across populations that had been SNP defined to ascertain STR-based haplogroup assignment accuracy, and identify geographically correlated trends in assignment error rates that may impact our conclusions.

mtDNA Sequencing and in Silico Prediction of Haplogroups

Total DNA was extracted from the peripheral leukocyte fraction of whole blood drawn in EDTA anticoagulant or cheek swab samples using a standard phenol/chloroform extraction procedure. The hypervariable region I (HVS-I) was amplified using primers designed by Maca-Meyer et al. [12]. Amplified HVS-I products were sequenced using a forward primer at position 15876 and a reverse primer at position 639 with ABI Big Dye Terminator v3.1 Cycle Sequencing kit (Applied Biosystems) and analysed on an Applied Biosystems 3130 xl Genetic Analyser.

Mutations in the HVS-I region were defined by aligning and comparing the sequences to the revised Cambridge Reference Sequence (rCRS) using the SeqScape software.

mtDNA haplogroups were predicted using the Genographic Project's online haplogroup prediction tool: nnhgtool.nationalgeographic.com.

mtDNA Genotyping of Samples

Haplogroup affiliations were confirmed using the Taqman approach with customized primers and probe sets to identify the SNPs listed (Applied Biosystems). Samples with incompatible prediction and Taqman results were excluded from the study. Mitochondrial nomenclature was assigned according to prior studies since established as standards [38],[56]–[59]. Data archiving was manually organized and edited.

Reduction to Most Informative Derived Set of Haplogroups

MtDNA Haplogroups reported in the literature were updated and reconciled to the 2009 phylogeny reported by van Oven et al. [59]. Construction of the most informative derived set was achieved by identifying the maximum level of resolution shared across all included studies. If some subhaplogroup markers were not typed in any given study, but no samples in that study resolved to a less-derived paragroup, then the most derived resolution was retained for the constructed most informative derived set.

Further, HVS-I regions reported by different sources varied. The range representing the largest common subset of HVS-I SNPs reported included 16090 through 16365.

Statistical Analyses

Fisher Exact Tests

Fisher exact tests were performed for haplogroup frequencies within populations. These tests were performed against a background of all populations, as well as among Middle Eastern populations (from Iraq, Jordan, Lebanon, Palestine, Saudi Arabia, Syria, and Yemen) only, with very low-power tests excluded.

PCA

Numbers of samples bearing mtDNA and Y-Chromosome reduced haplogroups within each population, and relative haplogroup frequencies within populations, were computed using R[60]. Principal Component Analysis was computed using prcomp in R [61]. Results were displayed with principal component contributions from each haplogroup using biplot. Agglomerative clustering with mean linkage (UPGMA) was applied to Euclidean distances computed between relative frequency vectors for each population using agnes and displayed in Figure 2 for mtDNA Haplogroups for Y-Haplogroups. These dendrograms should not be taken as population histories, but rather provide a repeatable description of population similarities also visible in the PCA.

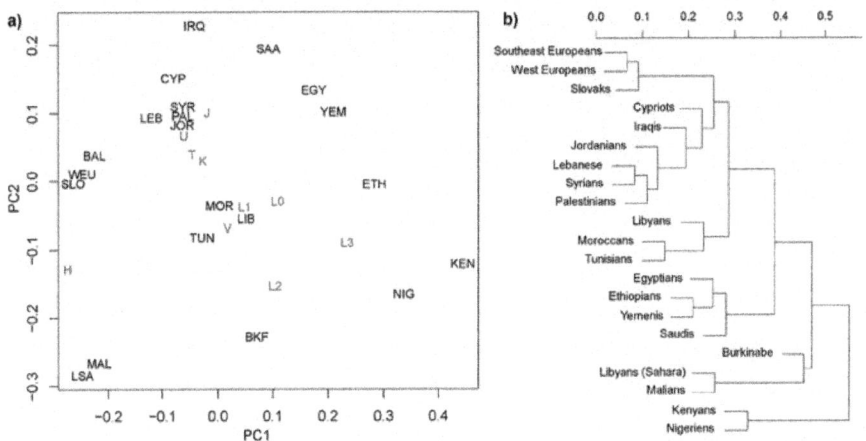

Figure 2: Populations comparison based on mtDNA haplogroups.

a) Principal Component Analysis of relative frequencies of haplogroups within populations, b) with mean-linkage (UPGMA) dendrogram determined from Euclidean distances.

doi:10.1371/journal.pone.0054616.g002

MDS

HVS-I SNPs were constructed against CRS [62] as revised rCRS [63], and the subrange common to all publications was selected. ARLEQUIN [64] was employed to compute F_{ST}'s[65], which were used as distances for non-metric MDS analysis [66], as implemented in isoMDS [61] in R. Agglomerative clustering with mean linkage was applied to the F_{ST}distances in the same way that they were applied to the Euclidean distances as described in the PCA section. An identical MDS and clustering were applied to Slatkin's R_{ST} distances [67]obtained from Y-chromosome samples. These results were displayed in Figures 3.

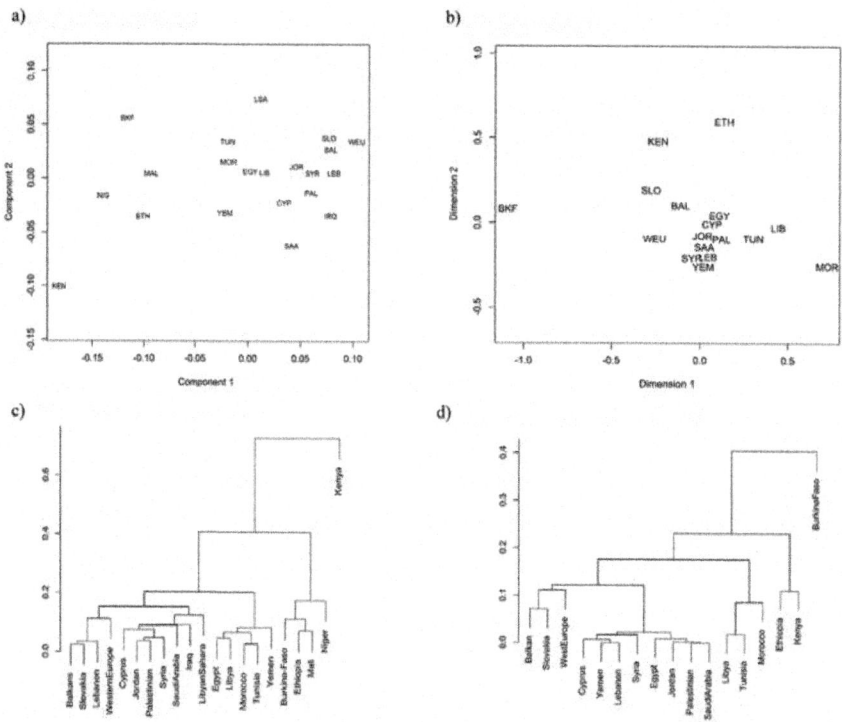

Figure 3: Nonmetric Multidimensional Scaling.

a) mtDNA F_{ST} and b) Y-STR R_{ST} distances with c) mtDNA F_{ST} and d) Y-STR R_{ST}mean-linkage dendrogram.

doi:10.1371/journal.pone.0054616.g003

Heatmap

Relative comparisons between mtDNA F_{ST} and Y-chromosome R_{ST} distances were constructed using a heatmap based on the normalized ratio of the Y-chromosome R_{ST} distance with respect to the total distance $(R_{ST}/(R_{ST}+F_{ST}))$ for mtDNA HVS-1 F_{ST} and Y-chromosome R_{ST} distances (Figure 4). The dendrograms are obtained using complete linkage hierarchical clustering with Euclidean distances between Y/(mtDNA HVS-I+Y) scores.

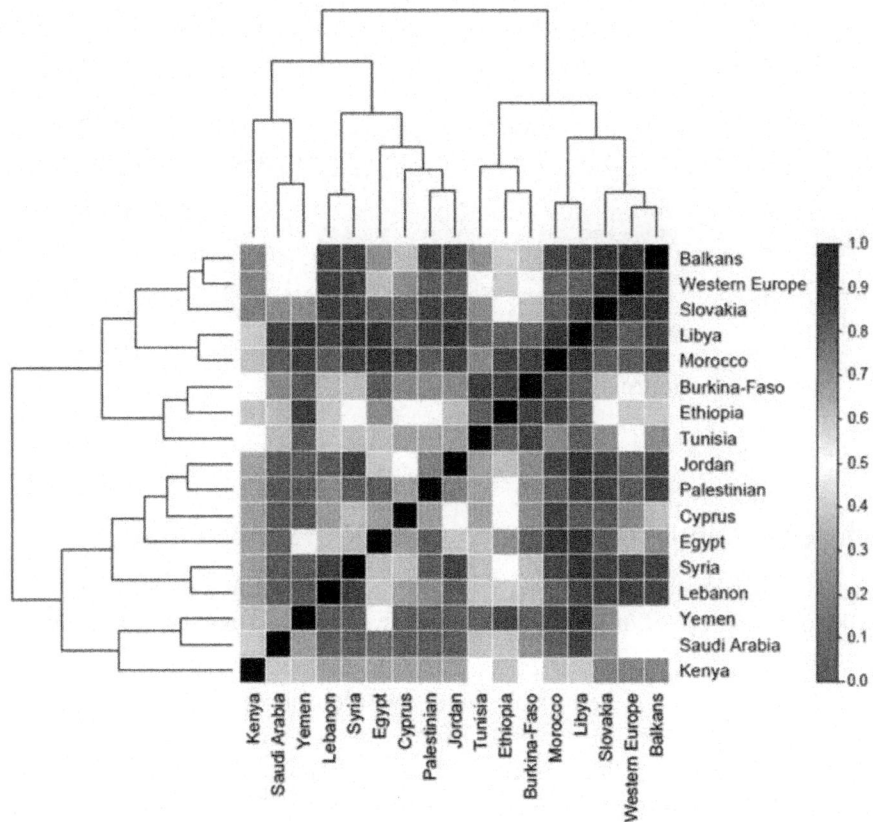

Figure 4: Heatmap of the Y(Y+mtDNA) distances.

The heatmap shows the normalized ratio of the Y F_{ST} distance with respect to the total distance (Y R_{ST}+mtDNA F_{ST} distances). The dendrograms are obtained using complete linkage hierarchical clustering with the Euclidean distance measure.

doi:10.1371/journal.pone.0054616.g004

AMOVA

The agglomerative clusters reflecting the results from the MDS analysis were used to identify groups of populations representing affinity of Middle Eastern populations with European and African populations observed in Y-chromosome and mtDNA genetics. AMOVA [68] was applied to the mtDNA and Y sets for each of the mtDNA and Y affinity sets, yielding a 2 by 2 measure of the differences between mtDNA and Y affinities, reported in Table 2.

Table 2: mtDNA vs. Y-chromosome AMOVA results contrasting mtDNA and Y dendrogram-based classifications

		mtDNA dendrogram based affinities				Y-Chromosome dendrogram based affinities			
		Variation	% Variation	F	p-value	Variation	% Variation	F	p-value
mtDNA AMOVA	Within Populations	2.43	97.15	0.029	<10⁻⁴	2.43	97.59	0.024	<10⁻⁴
	Within Groups	0.028	1.11	0.014	<10⁻⁴	0.045	1.84	0.018	<10⁻⁴
	Between Groups	0.043	1.74	0.017	3.6×10⁻³	0.014	0.57	0.006	1.1×10⁻³
Y-Chromosome AMOVA	Within Populations	6.56	87.05	0.129	<10⁻⁴	6.03	89.04	0.11	<10⁻⁴
	Within Groups	0.456	6.05	0.065	<10⁻⁴	0.329	4.86	0.052	<10⁻⁴
	Between Groups	0.519	6.89	0.068	6.3×10⁻³	0.414	6.11	0.061	<10⁻⁴

doi:10.1371/journal.pone.0054616.t002

doi:10.1371/journal.pone.0054616.t002

RESULTS

Phylogeographic Distribution of mtDNA Haplogroups

A total of 185 distinct HVS-I SNPs were identified across all populations. The distribution of mtDNA haplogroups shows systematic variation with geography.

The haplogroups' geographical distribution shows affinity between the Northern Levant (modern day Lebanon and Syria) and Europe with clear distinctions between the Levant and the Arabian Peninsula with regards to Africa (Fig. 1, Table 1). The main mtDNA haplogroups for both Europe and the Northern Levant are H and R*. The subhaplogroup H is more frequent in Europe (45%) than in the Levant (25%). Among the Levantine populations, only the Lebanese share Western Europe's overrepresentation of H.

Fisher exact tests were applied to determine when haplogroup frequency differences among populations over both Pan-Mediterranean tests, and regional Middle Eastern tests, were significant. They reveal patterns of significant over- and under- representation of haplogroups marking regional affinities.

In Lebanese, haplogroups H, HV, T, and K are over-represented, while Syrians are overrepresented in haplogroups T and K. Western Europeans show overrepresentation of haplogroups H, K.

By contrast, haplogroups J, R0, and M are significantly overrepresented in Saudis, and underrepresented in Western Europeans. Haplogroup J was also significantly overrepresented in Iraqis, among Palestinians, and Yemenis.

The African haplogroups L* and L3* are very rare (frequencies less than 1%) and underrepresented in Europe, noting that rarity reduces the power of these tests. In the Levant, Lebanese have the lowest frequency for these haplogroups with generally highly significant underrepresentation, The L haplogroups show rather broad penetration into Yemen, with most being significantly overrepresented, with Yemenis being the only population with an overrepresentation of L6. We have not found haplogroup L6 in our Lebanese (N=980), Syrian (N=234), and Jordanian samples (N=290). Further, they are absent from Abu-Amero et al's samples from Saudi Arabia, as well as other published results included in his study. A lower bound on the relative frequency of L6 in non-Yemeni Middle East of $f{\geq}8.29{\cdot}10^{-4}$ would guarantee that at least one or more would be observed at least 95% of the time out of the 3614 samples collected across the data on which we are reporting. Assuming independent sampling following a binomial test, there is 5% or less chance of seeing zero L6 by chance with a relative frequency of f or higher, establishing $8.29{\cdot}10^{-4}$ as an upper bound to the relative frequency of L6 with a 95% confidence.

We note that subtypes HV0, HV1 and HV2 are generally too weakly represented among our populations to yield tests with adequate power. HV0 and HV1 show sufficient power when pooled into regions. We note that HV0 appears primarily in Europe (34 out of 42 HV0 samples are European, with p<0.0001), HV1 is primarily non-European (4 out of 60 were European, with 20 among the African sample, and 36 among Middle Eastern samples, p<0.0001). HV2 were very rare, with no significant p-values.

Some U subhaplogroups show regional localization, but none of them rose to sufficient frequency to make any significant contribution to the PCA. U3 appears most frequently in Jordan (Fisher)s test: p=2.39e-8), with representation throughout the Middle East. U4 (p=1.49e-7) and U5 (p=2.2e-16) appear to be more heavily European.

The two leading principal components displayed in Figure 2 capture 47.9% and 26.9% of the variance showing a well-defined separation between Mediterranean African populations and sub-Saharan populations (Fig 2a). There is a clear cluster of North African populations comprised of Libyans, Moroccans, and Tunisians. The Nile River marks another boundary of mtDNA differentiation within Africa, linking Egypt, Ethiopia and Kenya but also extending through to Yemen. Yemenis and Saudis both associate strongly with Egyptians, whereas the Jordanian, Lebanese, Palestinian, and Syrian populations clustered together. Thus, the Arabian Peninsula population clusters were relatively differentiated from the more northern Levantine populations.

Mitochondrial DNA Haplogroups showing significant contributions to the principal components include H, L3, L2, L0, V, L1, M, J, U, T, K, HV, and R0. The principal vectors for HV, T, K, J, and U point almost directly at the Levantine cluster (Fig 2a). H marks Western Europe and is a significant contributor to Libyan Sahara and Mali mtDNA diversity. L2 and L3 frequencies distinguish the populations of Kenya, Niger, Burkina Faso, Mali, Tunisia, and Libyan Sahara, with a decrease in frequencies of L haplotypes from Kenya through Saudi Arabia.

The dendrogram based on mtDNA haplogroup frequencies (Fig 2b) reveals the strongest differentiation across the Sahara, showing the northern populations differentiated from the southern ones (with Nigeria, Kenya, Mali, Libyan South Sahara, and Burkina-Faso). Egyptian, Yemeni, Saudi Arabian, and Ethiopian populations form a cluster that is distinct from the rest of North Africa, the remaining parts of the Middle East, and Europe. Among these, Libyans, Moroccans, and Tunisians, form a cluster.

UPGMA and PCA showed Yemenis and Saudis (two of the STR predicted Hg populations) closely associated, forming a clear outlier to clusters identifying more northerly Middle Eastern populations and Europe. Slovaks (the third predicted population) also formed a distinct outlier to all of these. Africans were partitioned into northern African populations and Sahel populations, and distinct from the other populations. Burkinabe formed a very distinct outlier to every other population.

MDS analyses were performed using mitochondrial HVS-I based F_{ST} and Y-chromosome STR R_{ST} data (Figure 3a & 3b). The F_{ST}'s computed with mtDNA HVS-I data and R_{ST}'s computed from Y-STRs of Ethiopian and Levantine populations tended to be less than 1/3, with $Nm>1$ [69], roughly corresponding to gene flow between populations over the course of time [5], [20], [38], [70], [71].

For the mtDNA HVS-I F_{ST} MDS analysis, the European populations formed a clear cluster very close to the Cypriots, Jordanians, Lebanese, Palestinians, and Syrians. Egyptian, Libyan, Moroccan, and Tunisian populations form a clear cluster. Significantly, Yemenis are on the far side of North Africans, distinct from the Levantine populations and the Libyan Sahara population stands significantly separated from the North African group. The sub-Saharan populations are clearly distinguished from the Mediterranean populations and show significant distances between them in comparison to the Mediterranean populations. The mtDNA HVS-I MDS and dendrogram show most of the

Levantine and Arabian Peninsula populations clustering together. Significantly, Yemenis do not seem to cluster with proximal African populations or with Saudis. The entire Levant population seems to cluster with Western Europeans, Southeastern Europeans, and Slovaks.

In contrast to mtDNA, the Y-STR-based MDS shows a tight cluster of Cypriots, Egyptians, Jordanians, Lebanese, Palestinians, Saudis, Syrians, and Yemenis, though Libyans, Tunisians, and Moroccans extend away from this cluster. The Southeastern Europeans, Slovaks, and Western Europeans lie in the opposite direction. The dendrogram shows a European cluster closer to the Levant/Arabian Peninsula cluster and the North African cluster acting as out-group to those.

In general, the MDS plots for mtDNA and Y-STRs show general agreement of European populations extending from the Levant in one direction and North Africans tending to extend in another direction. This places the Levant as a middle ground, either by averaging of in-migration, as a source feeding both North African populations and European populations, or both. The Y and mtDNA MDS plots differ in identifying affinities of Lebanese with Europeans and Yemenis with Egyptians.

Comparative Analyses of Paternal and Maternal Lineages in the Levant

The relative distance heatmap plot (Figure 4) shows proportion of genetic distances of mtDNA vs. Y. Red colors indicate greater distance of mtDNA vs. Y, while blue colors indicate greater distance of Y vs. mtDNA. Hierarchical clustering organizes the plot relating populations showing similar profiles of Y vs. mtDNA isolation. Most striking is that Saudis, Kenyans, and Yemenis cluster together away from Lebanese, Syrians, Palestinians, Cypriots and Jordanians in terms of showing relatively high differentiation of mtDNA vs. Y-chromosome genetics. Dendrograms provide a consistent description of the organization of data that may be easily compared with PCA or MDS plots. The application of mean-linkage dendrograms to Y STR data, mtDNA HVS-I data, and mtDNA haplogroup frequency data provides a consistent basis of comparison. Application of AMOVA to clustering results provides an independent test characterized by p-values and percent variances between vs. within groups. We are not inferring relationships of heritage among populations by application of mean-linkage clustering.

In order to preserve normalization, common subsets comprised of the 11 populations in common in both dendrograms were included. Each of the candidate partitions marking mtDNA affinities and Y affinities formed three groups. The groups representing mtDNA affinities were: (1) Southeastern Europeans, Lebanese, Slovaks, and Western Europeans vs. (2) Cypriots, Jordanians, Palestinians, Saudis, and Syrians, vs. (3) Egyptians and Yemenis. The groups representing Y affinities were: (1) Southeast Europeans, Slovaks, and Western Europeans vs. (2) Cypriots, Lebanese, Syrians, and Yemenis, vs. (3) Egyptians, Jordanians, Palestinians, and Saudis. These two affinity groupings were applied to both the Y and the mtDNA data, yielding results presented in Table 2. Both Y and mtDNA tend to cluster African, European, and Middle Eastern populations separately, and all combinations showed highly significant between-group vs. within-group variations. This reflects the dominating clustering distinguishing Africa, Europe, and the Middle East populations that mean-linkage clustering is picking up. Affinities of Lebanese and the Levantine populations with Europeans vs. Africans depend on comparisons of AMOVA variations within and between groups. Notably, the mtDNA affinity grouping increased AMOVA between-group variation of mtDNA HVS-I data by a factor of 3.05 compared to the result obtained applying the Y affinity grouping to the mtDNA HVS-I data, and decreased AMOVA within-group variation by a factor of 1.66. However, application of the Y affinity grouping reduced AMOVA between-group variation in Y STR data by a factor of 1.13 while reduced AMOVA within-group variations in the Y STR data by a factor of nearly 1.2 compared to the mtDNA affinity grouping. These factors are relatively neutral in contrasting Lebanese Y-chromosome affinity with Europe vs. North Africa, and actually place Lebanese Y-chromosome organization closer to Europeans than Africans. It is expected that mean-linkage clustering would minimize AMOVA within-groups variation, leading to larger AMOVA between-groups variation. Observation did not meet expectation. Instead, the AMOVA within groups' variations for the Y-chromosomes were reduced using the mtDNA clustering compared to Y clustering, suggesting reduced discrimination using the Y clustering for Y-STR genetics.

Limitations

Y Chromosome haplogroup frequency analyses are limited by a relatively high misclassification rate, with more than half of the populations showing more than 10% misclassification, and Ethiopia showing nearly 50%. Since PCA is a non-linear computation which folds in all populations, the apparent locations of any two populations may shift relative to each other when a third population is added or distorted.

Discussion

Here we present mitochondrial characteristics of a large group of newly typed samples from five populations (Lebanese, Libyans, Jordanians, Palestinians, and Syrians) and compare their geographical affinity, distribution, and frequency with those of Y-chromosome markers from populations across the broader region of Africa, Europe and the Arabian Peninsula.

The Y-chromosome results of the current study are in agreement with previous studies, suggesting a Middle Eastern gene pool with greater affinity to Africa. Maternal lineages of the Levantine populations studied here, however, reveal stronger European genetic affinities, while not showing Arabian peninsular influences.

The Contrast between the Two Lineages

Our results show a contrast of mtDNA affinities with previous Y-DNA results [25]. While our Y-DNA MDS and mean-linkage clustering showed a much greater proportion of East African and Near East Y-chromosomes in the Levant, evidence of much less mtDNA affinity, however, was found between the Levant and its southern neighbours.

European mtDNA affinity with the Levant was established in haplogroup frequency data through Fisher exact tests, PCA, and mean-linkage clustering based on Euclidean distances, and in HVS-I derived F_{ST} distances via MDS and mean-linkage cluster analysis. The mtDNA results are distinct from the Y-STR R_{ST}-based mean-linkage cluster analysis that showed closer affinity of the Levant populations with Cypriots, North Africans, and Yemenis, than to Europeans.

This cluster analysis suggests that the position of Lebanese relative to European Y-chromosome genetics represented in STR haplotype data is also much more ambiguous than suggested entirely by frequency analysis, revealing otherwise cryptic relationships between Lebanese's Y-STR structure and that of Europeans. Cluster analysis of Y-chromosome frequency based data shows similar partitioning of Europe, Africa, and Middle East, with the Levant much more strongly associated with the Middle East than Europe. As with mtDNA, African Y-chromosome haplogroup data also shows a clear partition between Northern populations and Sahel populations. Due to uncertainties in haplogroup inference from STRs, affinities of Yemenis with Ethiopians vs. Egyptians are uncertain, as are the relationships of Saudi Arabian haplogroups both similar to Yemenis or differentiated from Yemenis in affinity with African populations.

The Levant and Europe

Beyond the associations noted above, Lebanese show affinity with Europeans for mtDNA haplogroups H, HV, T, K, J, and U, all of which have been identified as markers of agricultural expansions from the Fertile Crescent into Europe [17].

Colonization of West Eurasia by modern humans is believed to have been a consequence of the Out-of-Africa dispersal and to have occurred via the Levant [6]. Indeed, migrating modern humans are believed to have settled near the Arabian Sea until climate changes allowed them to reach the Levant and then Europe [17], [72]–[74]. The LGM, followed by re-expansions from smaller LGM communities relatively isolated by widespread arid conditions, further impacted the coastal-inland contrast of Y-chromosome genetics [25], [75]. The significant overrepresentation of mtDNA haplogroup HV among Levantine populations compared to their southern neighbours has suggested these lineages were most likely derived from a single maternal Levantine source population [17].

Arabian genetic expansions: Arabia East Africa, and North Africa

From the 7th millennium B.C.E., empire expansions and trade, including the slave trade, heavily influenced genetic migration between Yemen and East Africa. Alternatively, known trade networks linking Egypt with Yemen included those for obsidian, and later through Aksum, spices, incense and other precious materials, as well as slaves [76]–[78]. It is particularly clear from prior mtDNA studies of this region that East African migration into Arab populations involved females to an extensive degree [79]. While Ethiopian and other East African populations may appear to be better candidates for the origins of modern Yemeni populations, our PCA and MDS analyses, and their associated mean-linkage clustering of Yemen›s mtDNA, show greater affinity between Yemenis, Egyptians and North Africans. They share in common haplogroups J, L0, L2, and N1. Comparison of mtDNA HVS-I F_{ST} distances also suggest that Yemen appears more similar to Egypt than Ethiopia.

Two haplogroups in this region show significant evidence of relative isolation. First, mtDNA patterns for haplogroup J reflect relatively moderate genetic outflow from Saudi Arabia, and haplogroup L6 is strongly localized within Yemen. Haplogroup J is evenly distributed throughout the Middle East, except in Saudi Arabia where it is significantly overrepresented.

It is likely the pattern of Hg J›s significant penetration, and the shared underrepresentation of Hg H, tips the balance for Yemenis› mtDNA affinity with Egyptians. Given the significant underrepresentation of Hg J in East

Africa, while not being significantly uncommon in Egypt, it is therefore plausible that Arabian female gene flow followed well established trade routes on the Red Sea with Egypt and North Africa while avoiding assimilation of Yemeni L6›s on the way.

The most striking feature of the heat map (Fig 4) is the relative isolation of mtDNA genetics of Yemenis and Saudis from the other populations in the Middle East in comparison to Y-chromosome variation. While Yemenis appear to share overrepresented haplogroups that characterize each of its neighbouring populations, none of the African populations have become dominated by Saudi Arabian J›s, nor have Middle Eastern populations been differentially dominated by the in-migration of African L›s the way Yemenis have.

The expansion of trade through the Red Sea and into the Indian Ocean basin starting in Classical times has provided the largest opportunities for genetic transfers from Africa into Yemen, being dominated by the Red Sea superpower: Egypt. The distribution of mtDNA haplogroup L6 provides a measure of the limited impact of genetic outflow from Yemen, and this flow seems to have been primarily unidirectional. This establishes the upper limits for Yemeni female-mediated gene flow during the Muslim Expansions, as well as identifying possible routes for the expansions.

Whether considering haplogroup composition revealed in Fisher tests, PCA, or F_{ST} based MDS analysis of HVS-I data, mtDNA shows a much stronger affinity between Levantine populations and Europeans compared with the rest of the Middle Eastern populations, or with North Africans. While Lebanese and Yemeni mtDNA epitomize very distinct affinities to different populations and regions well outside of the Middle East, Saudi Arabia seems to display strong local over-representation haplogroup J, while Yemen is even more localized in its L6. Further, these large-scale differences in affinity between mtDNA genetics appear in sharp contrast to regional affinities seen in their Y-chromosomal counterparts. While the mtDNA signal is sharp and clear in its affinities, the Y-chromosome results show somewhat more ambiguous associations in R_{ST} based analyses, with Lebanese showing less within-group variation when organized consistently with mtDNA and demonstrating associations closer to Europeans than Africans. This would suggest that while male migrants accompanied female migrants, especially to Europe, females did not always accompany male migrants, especially into North Africa. This leaves a more ambiguous signal for male compared to female migrations.

The historical and archaeological record reveals how trade and labour, colonization and settlement events, and military expansions all contributed to the immigration and displacement of individuals throughout these regions. As a distinct crossroad between geographic regions and civilizations, the Levant and the Near East harbour unique genetic affinities which are revealed most clearly through the comparison of Y-chromosome and mtDNA data.

Due to uncertainties in haplogroup inference from STRs, specific questions regarding affinities of Yemen with Ethiopia vs. Egypt are inaccessible, as are questions regarding the relationship of Saudi Arabian haplogroups both similar to Yemenis or differentiated from Yemenis in affinity with African populations.

ACKNOWLEDGMENTS

We thank the sample donors for taking part in this study. We would like to thank Professor Colin Renfrew for his insights, critical comments, and suggestions which helped us improve this manuscript significantly.

The Genographic Consortium includes: Janet S. Ziegle (Applied Biosystems, Foster City, California, United States); Li Jin & Shilin Li (Fudan University, Shanghai, China); Pandikumar Swamikrishnan (IBM, Somers, New York, United States); Asif Javed, Laxmi Parida & Ajay K. Royyuru (IBM, Yorktown Heights, New York, United States); Lluis Quintana-Murci (Institut Pasteur, Paris, France); R. John Mitchell (La Trobe University, Melbourne, Victoria, Australia); Syama Adhikarla, ArunKumar GaneshPrasad, Ramasamy Pitchappan & Arun Varatharajan Santhakumari (Madurai Kamaraj University, Madurai, Tamil Nadu, India); Angela Hobbs & Himla Soodyall (National Health Laboratory Service, Johannesburg, South Africa); Elena Balanovska & Oleg Balanovsky (Research Centre for Medical Genetics, Russian Academy of Medical Sciences, Moscow, Russia); Daniela R. Lacerda & Fabrício R. Santos (Universidade Federal de Minas Gerais, Belo Horizonte, Minas Gerais, Brazil); Pedro Paulo Vieira (Universidade Federal do Rio de Janeiro, Rio de Janeiro, Brazil); Jaume Bertranpetit, David Comas, Begoña Martínez-Cruz & Marta Melé (Universitat Pompeu Fabra, Barcelona, Spain); Christina J. Adler, Alan Cooper, Clio S. I. Der Sarkissian & Wolfgang Haak (University of Adelaide, South Australia, Australia); Matthew E. Kaplan & Nirav C. Merchant (University of Arizona, Tucson, Arizona, United States); Colin Renfrew (University of Cambridge, Cambridge, United Kingdom); Andrew C. Clarke & Elizabeth A. Matisoo-Smith (University of Otago, Dunedin, New Zealand); Matthew C. Dulik, Jill B. Gaieski, Amanda C. Owings, Theodore G. Schurr & Miguel G. Vilar (University of Pennsylvania, Philadelphia, Pennsylvania, United States).

AUTHOR CONTRIBUTIONS

Conceived and designed the experiments: PZ CTS RSW EMS. Performed the experiments: DB BD SY AS. Analyzed the data: DP MH DSH GK. Contributed reagents/materials/analysis tools: BJ MGS. Wrote the paper: PZ DP DB.

REFERENCES

1. Stringer CB, Grun R, Schwarcz HP, Goldberg P (1989) ESR dates for the hominid burial site of Es Skhul in Israel. Nature 338: 756–758. doi: 10.1038/338756a0

2. Bar-Yosef O (1992) The role of western Asia in modern human origins. Philosophical transactions of the Royal Society of London Series B, Biological sciences 337: 193–200. doi: 10.1098/rstb.1992.0097

3. Lahr MM, Foley RA (1998) Towards a theory of modern human origins: geography, demography, and diversity in recent human evolution. American journal of physical anthropology Suppl 27: 137–176. doi: 10.1002/(sici)1096-8644(1998)107:27+<137::aid-ajpa6>3.3.co;2-h

4. Tchernov E (1994) New comments on the biostratigraphy of the Middle and Upper Pleistocene of the southern Levant. In: Ben-Yosef O, Kra RS, editors. Late Quaternary Chronology and Paleoclimates of the Eastern Mediterranean: Radiocarbon. pp. 333–350.

5. Luis JR, Rowold DJ, Regueiro M, Caeiro B, Cinnioglu C, et al. (2004) The Levant versus the Horn of Africa: evidence for bidirectional corridors of human migrations. American journal of human genetics 74: 532–544. doi: 10.1086/382286

6. Olivieri A, Achilli A, Pala M, Battaglia V, Fornarino S, et al. (2006) The mtDNA legacy of the Levantine early Upper Palaeolithic in Africa. Science 314: 1767–1770. doi: 10.1126/science.1135566

7. Bar-Yosef O (1998) The Natufian Culture in the Levant, Threshold to the Origins of Agriculture. Evolutionary Anthropology 6: 159–177. doi: 10.1002/(sici)1520-6505(1998)6:5<159::aid-evan4>3.0.co;2-7

8. Valladas H, Reyss JL, Joron JL, Valladas G, Bar-Yosef O, et al. (1988) Thermoluminescence data of Mousterian Troto-Cro-Magnon remains from Israel and the origin of modern man. Nature 331: 614–616. doi: 10.1038/331614a0

9. Mercier N, Valladas H, Bar-Yosef O, Stringer CB, Joron JL (1993) Thermoluminescence dates for the Mousterian Burial Site of Es-Skhul, Mt. Carmel. Journal of Archaeological Science 20: 169–174. doi: 10.1006/jasc.1993.1012

10. Cann RL, Stoneking M, Wilson AC (1987) Mitochondrial DNA and human evolution. Nature 325: 31–36. doi: 10.1038/325031a0

11. Vigilant L, Stoneking M, Harpending H, Hawkes K, Wilson AC (1991) African populations and the evolution of human mitochondrial DNA. Science 253: 1503–1507. doi: 10.1126/science.1840702

12. Maca-Meyer N, Gonzalez AM, Larruga JM, Flores C, Cabrera VM (2001) Major genomic mitochondrial lineages delineate early human expansions. BMC genetics 2: 13. doi: 10.1186/1471-2156-2-13

13. Nei M, Roychoudhury AK (1993) Evolutionary Relationships of Human Populations on a Global Scale. Molecular biology and evolution 10: 927–943.

14. Cavalli-Sforza LL, Menozzi P, Piazza A (1994) The History and Geography of Human Genes. Princeton, NJ: Princeton University Press.

15. Foley RA, Lahr MM (1997) Mode 3 technologies and the evolution of modern humans. Cambridge Archaeological Journal 7: 3–36. doi: 10.1017/s0959774300001451

16. Cavalli-Sforza LL (1997) Genes, peoples, and languages. Proceedings of the National Academy of Sciences of the United States of America 94: 7719–7724. doi: 10.1073/pnas.94.15.7719

17. Richards M, Macaulay V, Hickey E, Vega E, Sykes B, et al. (2000) Tracing European founder lineages in the Near Eastern mtDNA pool. American journal of human genetics 67: 1251–1276. doi: 10.1016/s0002-9297(07)62954-1

18. Richards M, Macaulay V, Torroni A, Bandelt HJ (2002) In search of geographical patterns in European mitochondrial DNA. American journal of human genetics 71: 1168–1174. doi: 10.1086/342930

19. Simoni L, Calafell F, Pettener D, Bertranpetit J, Barbujani G (2000) Geographic patterns of mtDNA diversity in Europe. American journal of human genetics 66: 262–278. doi: 10.1086/302706

20. Semino O, Magri C, Benuzzi G, Lin AA, Al-Zahery N, et al. (2004) Origin, diffusion, and differentiation of Y-chromosome haplogroups E and J: inferences on the neolithization of Europe and later migratory events in the Mediterranean area. American journal of human genetics 74: 1023–1034. doi: 10.1086/386295

21. Plaza S, Calafell F, Helal A, Bouzerna N, Lefranc G, et al. (2003) Joining the pillars of Hercules: mtDNA sequences show multidirectional gene flow in the western Mediterranean. Annals of human genetics 67: 312–328. doi: 10.1046/j.1469-1809.2003.00039.x

22. Scozzari R, Cruciani F, Pangrazio A, Santolamazza P, Vona G, et al. (2001) Human Y-chromosome variation in the western Mediterranean

area: implications for the peopling of the region. Human immunology 62: 871–884. doi: 10.1016/s0198-8859(01)00286-5

23. Di Giacomo F, Luca F, Popa LO, Akar N, Anagnou N, et al. (2004) Y chromosomal haplogroup J as a signature of the post-neolithic colonization of Europe. Human genetics 115: 357–371. doi: 10.1007/s00439-004-1168-9

24. Flores C, Maca-Meyer N, Gonzalez AM, Oefner PJ, Shen P, et al. (2004) Reduced genetic structure of the Iberian peninsula revealed by Y-chromosome analysis: implications for population demography. European journal of human genetics : EJHG 12: 855–863. doi: 10.1038/sj.ejhg.5201225

25. El-Sibai M, Platt DE, Haber M, Xue Y, Youhanna SC, et al. (2009) Geographical structure of the Y-chromosomal genetic landscape of the Levant: a coastal-inland contrast. Annals of human genetics 73: 568–581. doi: 10.1111/j.1469-1809.2009.00538.x

26. Zalloua PA, Platt DE, El Sibai M, Khalife J, Makhoul N, et al. (2008) Identifying genetic traces of historical expansions: Phoenician footprints in the Mediterranean. American journal of human genetics 83: 633–642. doi: 10.1016/j.ajhg.2008.10.012

27. Cadenas AM, Zhivotovsky LA, Cavalli-Sforza LL, Underhill PA, Herrera RJ (2008) Y-chromosome diversity characterizes the Gulf of Oman. European journal of human genetics : EJHG 16: 374–386. doi: 10.1038/sj.ejhg.5201934

28. Rowold DJ, Luis JR, Terreros MC, Herrera RJ (2007) Mitochondrial DNA geneflow indicates preferred usage of the Levant Corridor over the Horn of Africa passageway. Journal of human genetics 52: 436–447. doi: 10.1007/s10038-007-0132-7

29. Underhill PA, Kivisild T (2007) Use of y chromosome and mitochondrial DNA population structure in tracing human migrations. Annual review of genetics 41: 539–564. doi: 10.1146/annurev.genet.41.110306.130407

30. Gonzalez AM, Karadsheh N, Maca-Meyer N, Flores C, Cabrera VM, et al. (2008) Mitochondrial DNA variation in Jordanians and their genetic relationship to other Middle East populations. Annals of human biology 35: 212–231. doi: 10.1080/03014460801946538

31. Al-Zahery N, Semino O, Benuzzi G, Magri C, Passarino G, et al. (2003) Y-chromosome and mtDNA polymorphisms in Iraq, a crossroad of the early human dispersal and of post-Neolithic migrations. Molecular phylogenetics and evolution 28: 458–472. doi: 10.1016/s1055-7903(03)00039-3

32. Pilkington MM, Wilder JA, Mendez FL, Cox MP, Woerner A, et al. (2008) Contrasting signatures of population growth for mitochondrial DNA and Y chromosomes among human populations in Africa. Molecular biology and evolution 25: 517–525. doi: 10.1093/molbev/msm279

33. Coelho M, Sequeira F, Luiselli D, Beleza S, Rocha J (2009) On the edge of Bantu expansions: mtDNA, Y chromosome and lactase persistence genetic variation in southwestern Angola. BMC evolutionary biology 9: 80. doi: 10.1186/1471-2148-9-80

34. Wood ET, Stover DA, Ehret C, Destro-Bisol G, Spedini G, et al. (2005) Contrasting patterns of Y chromosome and mtDNA variation in Africa: evidence for sex-biased demographic processes. European journal of human genetics : EJHG 13: 867–876. doi: 10.1038/sj.ejhg.5201408

35. Pereira L, Cerny V, Cerezo M, Silva NM, Hajek M, et al. (2010) Linking the sub-Saharan and West Eurasian gene pools: maternal and paternal heritage of the Tuareg nomads from the African Sahel. European journal of human genetics : EJHG 18: 915–923. doi: 10.1038/ejhg.2010.21

36. Irwin J, Saunier J, Strouss K, Paintner C, Diegoli T, et al. (2008) Mitochondrial control region sequences from northern Greece and Greek Cypriots. International journal of legal medicine 122: 87–89. doi: 10.1007/s00414-007-0173-7

37. Saunier JL, Irwin JA, Strouss KM, Ragab H, Sturk KA, et al. (2009) Mitochondrial control region sequences from an Egyptian population sample. Forensic science international Genetics 3: e97–103. doi: 10.1016/j.fsigen.2008.09.004

38. Kivisild T, Reidla M, Metspalu E, Rosa A, Brehm A, et al. (2004) Ethiopian mitochondrial DNA heritage: tracking gene flow across and around the gate of tears. American journal of human genetics 75: 752–770. doi: 10.1086/425161

39. Dubut V, Chollet L, Murail P, Cartault F, Beraud-Colomb E, et al. (2004) mtDNA polymorphisms in five French groups: importance of regional sampling. European journal of human genetics : EJHG 12: 293–300. doi: 10.1038/sj.ejhg.5201145

40. Richard C, Pennarun E, Kivisild T, Tambets K, Tolk HV, et al. (2007) An mtDNA perspective of French genetic variation. Annals of human biology 34: 68–79. doi: 10.1080/03014460601076098

41. Brandstätter A, Peterson CT, Irwin JA, Mpoke S, Koech DK, et al. (2004) Mitochondrial DNA control region sequences from Nairobi (Kenya): inferring phylogenetic parameters for the establishment of a forensic

database. International journal of legal medicine 118: 294–306. doi: 10.1007/s00414-004-0466-z

42. Ottoni C, Martinez-Labarga C, Loogvali EL, Pennarun E, Achilli A, et al. (2009) First genetic insight into Libyan Tuaregs: a maternal perspective. Annals of human genetics 73: 438–448. doi: 10.1111/j.1469-1809.2009.00526.x

43. Harich N, Costa M, Fernandes V, Kandil M, Pereira J, et al. (2010) The trans-Saharan slave trade - clues from interpolation analyses and high-resolution characterization of mitochondrial DNA lineages. BMC evolutionary biology 10: 138. doi: 10.1186/1471-2148-10-138

44. Turchi C, Buscemi L, Giacchino E, Onofri V, Fendt L, et al. (2009) Polymorphisms of mtDNA control region in Tunisian and Moroccan populations: an enrichment of forensic mtDNA databases with Northern Africa data. Forensic science international Genetics 3: 166–172. doi: 10.1016/j.fsigen.2009.01.014

45. Abu-Amero KK, Larruga JM, Cabrera VM, Gonzalez AM (2008) Mitochondrial DNA structure in the Arabian Peninsula. BMC evolutionary biology 8: 45. doi: 10.1186/1471-2148-8-45

46. Malyarchuk B, Grzybowski T, Derenko M, Perkova M, Vanecek T, et al. (2008) Mitochondrial DNA phylogeny in Eastern and Western Slavs. Molecular biology and evolution 25: 1651–1658. doi: 10.1093/molbev/msn114

47. Cherni L, Loueslati BY, Pereira L, Ennafaa H, Amorim A, et al. (2005) Female gene pools of Berber and Arab neighboring communities in central Tunisia: microstructure of mtDNA variation in North Africa. Human biology 77: 61–70. doi: 10.1353/hub.2005.0028

48. Cerny V, Mulligan CJ, Ridl J, Zaloudkova M, Edens CM, et al. (2008) Regional differences in the distribution of the sub-Saharan, West Eurasian, and South Asian mtDNA lineages in Yemen. American journal of physical anthropology 136: 128–137. doi: 10.1002/ajpa.20784

49. Bosch E, Calafell F, Gonzalez-Neira A, Flaiz C, Mateu E, et al. (2006) Paternal and maternal lineages in the Balkans show a homogeneous landscape over linguistic barriers, except for the isolated Aromuns. Annals of human genetics 70: 459–487. doi: 10.1111/j.1469-1809.2005.00251.x

50. de Filippo C, Barbieri C, Whitten M, Mpoloka SW, Gunnarsdottir ED, et al. (2011) Y-chromosomal variation in sub-Saharan Africa: insights into the history of Niger-Congo groups. Molecular biology and evolution 28: 1255–1269. doi: 10.1093/molbev/msq312

51. Ferri G, Ceccardi S, Lugaresi F, Bini C, Ingravallo F, et al. (2008) Male haplotypes and haplogroups differences between urban (Rimini) and rural area (Valmarecchia) in Romagna region (North Italy). Forensic science international 175: 250–255. doi: 10.1016/j.forsciint.2007.06.007

52. Alshamali F, Pereira L, Budowle B, Poloni ES, Currat M (2009) Local population structure in Arabian Peninsula revealed by Y-STR diversity. Hum Hered 68: 45–54. doi: 10.1159/000210448

53. Petrejcikova E, Sotak M, Bernasovska J, Bernasovsky I, Rebala K, et al. (2011) Allele frequencies and population data for 11 Y-chromosome STRs in samples from Eastern Slovakia. Forensic science international Genetics 5: e53–62. doi: 10.1016/j.fsigen.2010.08.003

54. Athey T (2005) Haplogroup Prediction from Y-STR Values Using an Allele-Frequency Approach. Journal of Genetic Genealogy 1: 1–7.

55. Athey T (2006) Haplogroup Prediction from Y-STR Values Using a Bayesian-Allele-Frequency Approach. Journal of Genetic Genealogy 2: 34–39.

56. Salas A, Richards M, De la Fe T, Lareu MV, Sobrino B, et al. (2002) The making of the African mtDNA landscape. American journal of human genetics 71: 1082–1111. doi: 10.1086/344348

57. Salas A, Richards M, Lareu MV, Scozzari R, Coppa A, et al. (2004) The African diaspora: mitochondrial DNA and the Atlantic slave trade. American journal of human genetics 74: 454–465. doi: 10.1086/382194

58. Trejaut JA, Kivisild T, Loo JH, Lee CL, He CL, et al. (2005) Traces of archaic mitochondrial lineages persist in Austronesian-speaking Formosan populations. PLoS biology 3: e247. doi: 10.1371/journal.pbio.0030247

59. van Oven M, Kayser M (2009) Updated comprehensive phylogenetic tree of global human mitochondrial DNA variation. Human mutation 30: E386–394. doi: 10.1002/humu.20921

60. R Development Core Team (2011) R: A Language and Environment for Statistical Computing. Vienna, Austria: R Foundation for Statistical Computing.

61. Venables WN, Ripley BD (2002) Modern Applied Statistics with S. New York, NY: Springer-Verlag.

62. Anderson S, Bankier AT, Barrell BG, de Bruijn MH, Coulson AR, et al. (1981) Sequence and organization of the human mitochondrial genome. Nature 290: 457–465. doi: 10.1038/290457a0

63. Andrews RM, Kubacka I, Chinnery PF, Lightowlers RN, Turnbull DM, et al. (1999) Reanalysis and revision of the Cambridge reference sequence for human mitochondrial DNA. Nature genetics 23: 147. doi: 10.1038/13779

64. Excoffier L, Lischer HE (2010) Arlequin suite ver 3.5: a new series of programs to perform population genetics analyses under Linux and Windows. Molecular ecology resources 10: 564–567. doi: 10.1111/j.1755-0998.2010.02847.x

65. Reynolds J, Weir BS, Cockerham CC (1983) Estimation of the coancestry coefficient: basis for a short-term genetic distance. Genetics 105: 767–779.

66. Cox TF, Cox MAA (2001) Multidimensional Scaling, Second Edition. New York, NY: Chapman and Hall.

67. Slatkin M (1995) A measure of population subdivision based on microsatellite allele frequencies. Genetics 139: 457–462.

68. Excoffier L, Smouse PE, Quattro JM (1992) Analysis of molecular variance inferred from metric distances among DNA haplotypes: application to human mitochondrial DNA restriction data. Genetics 131: 479–491.

69. Wright S (1951) The genetical structure of populations. Ann Eugenics 15: 323–354. doi: 10.1111/j.1469-1809.1949.tb02451.x

70. Semino O, Santachiara-Benerecetti AS, Falaschi F, Cavalli-Sforza LL, Underhill PA (2002) Ethiopians and Khoisan share the deepest clades of the human Y-chromosome phylogeny. American journal of human genetics 70: 265–268. doi: 10.1086/338306

71. Underhill PA, Shen P, Lin AA, Jin L, Passarino G, et al. (2000) Y chromosome sequence variation and the history of human populations. Nature genetics 26: 358–361.

72. Macaulay V, Hill C, Achilli A, Rengo C, Clarke D, et al. (2005) Single, rapid coastal settlement of Asia revealed by analysis of complete mitochondrial genomes. Science 308: 1034–1036. doi: 10.1126/science.1109792

73. Mellars P (2006) Why did modern human populations disperse from Africa ca. 60,000 years ago? A new model. Proceedings of the National Academy of Sciences of the United States of America 103: 9381–9386. doi: 10.1073/pnas.0510792103

74. Van Andel TH, Tzedakis PC (1996) Paleolithic landscapes of Europe and environs: 150,000–25,000 years ago: an overview. Quaternary Science Reviews 15: 481–500. doi: 10.1016/0277-3791(96)00028-5

75. Chiaroni J, King R, Underhill PA (2008) Correlation of annual precipitation with human Y-chromosome diversity and the emergence of Neolithic agricultural and pastoral economies in the Fertile Crescent. Antiquity 82: 281–289.

76. Fattovich R (1997) The Near East and Eastern Africa: Their Interaction. In: Vogel JO, editor. Encyclopedia of precolonial Africa. Walnut Creek: AltaMira Press. pp. 479–484.

77. Segal R (2001) Islam's Black Slaves: The Other Black Diaspora. New York: Farrar, Straus, and Giroux.

78. Lewis B (1990) Race and slavery in the Middle East: an historical enquiry. New York: Oxford University Press.

79. Richards M, Rengo C, Cruciani F, Gratrix F, Wilson JF, et al. (2003) Extensive female-mediated gene flow from sub-Saharan Africa into near eastern Arab populations. American journal of human genetics 72: 1058–1064. doi: 10.1086/374384

Chapter 13

IDENTIFICATION OF GENETIC VARIATION ON THE HORSE Y CHROMOSOME AND THE TRACING OF MALE FOUNDER LINEAGES IN MODERN BREEDS

Barbara Wallner, Claus Vogl, Priyank Shukla, Joerg P. Burgstaller, Thomas Druml, Gottfried Brem

Institute of Animal Breeding and Genetics, Department of Biomedical Sciences, University of Veterinary Medicine Vienna, Vienna, Austria

ABSTRACT

The paternally inherited Y chromosome displays the population genetic history of males. While modern domestic horses (*Equus caballus*) exhibit abundant diversity within maternally inherited mitochondrial DNA, no significant Y-chromosomal sequence diversity has been detected. We used high throughput sequencing technology to identify the first polymorphic Y-chromosomal markers useful for tracing paternal lines. The nucleotide variability of the modern horse Y chromosome is extremely low, resulting in six haplotypes (HT), all clearly distinct from the Przewalski horse (*E. przewalskii*). The most widespread HT1 is ancestral and the other five haplotypes apparently arose on the background of HT1 by mutation or gene conversion after domestication. Two haplotypes (HT2 and HT3) are widely distributed at high frequencies among modern European horse breeds. Using pedigree information, we trace the distribution of Y-haplotype diversity to particular founders. The mutation leading to HT3 occurred in the germline of the famous English Thoroughbred stallion "Eclipse" or his son or grandson and its prevalence demonstrates the influence of this popular paternal line on modern sport horse breeds. The pervasive introgression of Thoroughbred stallions during the last 200 years to refine autochthonous breeds has strongly affected the distribution of Y-chromosomal variation in modern horse breeds and has led to the replacement of autochthonous Y chromosomes. Only a few northern European breeds bear unique variants at high frequencies or fixed within but not shared among breeds. Our Y-chromosomal data complement the well established mtDNA lineages and document the male side of the genetic history of modern horse breeds and breeding practices.

INTRODUCTION

Mitochondrial DNA (mtDNA) and the paternally transmitted portion of the Y chromosome (NRY) are inherited uniparentally and do not recombine. They accumulate mutations, such as single nucleotide polymorphisms (SNPs), insertions and deletions (Indels) and structural rearrangements [1]–[3]. Genealogies inferred from the distribution of mutations on mtDNA and NRY haplotypes reflect gender-specific population genetic forces. Furthermore, gene conversion i.e. the transfer of a short sequence from the homologous, but non-recombining region on the X- to the Y-chromosome can also contribute to Y-chromosomal variation [4], [5]. mtDNA- and NRY-variant distributions have been widely analysed in humans and in a broad range of wild and domesticated animals to provide indications of the origin of species, domestication processes, the characterization of genetic diversity within and between populations and sex specific demographic behaviors [6]. Depending on the genetic regions analysed, uniparentally inherited markers can be used to trace individual founder lines or families [7]. In livestock, Y-chromosomal and mtDNA variation has been used to study the domestication and population structure of the domestic dog [8]–[10], pig [11], [12], sheep [13]and cattle [14].

Populations of the domestic horse (*E. caballus*) show high levels of mtDNA diversity with limited geographic structure [15]–[19]. Estimates of the coalescence times of mtDNA variants of extant horses far predate the date of domestication [18], [19]. Hence, multiple female lineages contributed to the domestic horse gene pool. Nuclear microsatellite data are also consistent with a scenario of high variation that predates domestication [20].

Nevertheless, genetic variability in the domestic horse represents a paradox: although horses have the largest diversity of maternal mtDNA among domestic species, no noteworthy sequence diversity can be detected on the NRY [21]–[23]. Whereas Y chromosomes of modern horses are clearly distinct from that of the Przewalski horse (*E. przewalskii*) [22] and ancestral Y-chromosomal diversity is found in prehistoric horses [24], the only polymorphism observed in modern horses is a microsatellite mutation found in autochthonous Chinese horse breeds [25]. Screening approaches for the identification of polymorphisms on the domestic horse NRY have proven unsuccessful [21]–[23]. The low NRY diversity in horses is assumed to be the result of the extremely low effective population size of males due to the specific mating behaviour and to several bottlenecks that occurred early in the domestication process[20], [21]. Moreover, Y-chromosomal variation might be further diminished in modern horse breeds by regulated breeding programmes and intensive horse-trading. Today's European horse breeds (e.g. the Lipizzan horse and the English Thoroughbred) are largely the result of centralized and organized breeding over the past 200 years. The breeding effort has mainly focused on stallions. Breeding programmes have been based on so-called "multiplier studs", which were founded at the end of the 18th century and were responsible for the production of stallions needed in rural regions. As a result of these early breeding programmes, modern European horse breeds show the consequences of several waves of introgression of imported stallions into local breeds. These include (a) the "Neapolitan" wave from the 15th to the 18[th]century, when the now extinct "Neapolitan horse" was introgressed; (b) the "Oriental" wave from the late 18th to the late 19th century, when "Original Arabian" stallions, imported from Syria to Egypt, were introgressed and (c) the "English" wave from the early 19th century to the present, when Thoroughbred stallions were introgressed [26]–[28]. In the early 19th century new breeding practices were introduced and rapidly supplanted existing practices. Inbreeding and line breeding concepts became popular and with the integration of private breeding into state breeding programmes the entire population of male horses became highly selected. The result of the modern breeding practices may have been the complete replacement of autochthonous Y-chromosomal variants by imported bloodlines.

In the Lipizzan breed, 89 different sire lines existed in the late 18th century, while only eight are present today [29]. Among these eight male lines, one can be attributed to the "Oriental" wave from the Original Arabian stallion "Siglavy" (born 1810, imported to Lipizza 1814). The other Lipizzan stallion lines derive from the earlier, less documented phase (the "Neapolitan" wave). The male gene pool is even more restricted in the English Thoroughbred, with only three paternal lines remaining [30]. All three lines can be attributed to the "Oriental" wave through the import of the three stallions Byerley Turk, Darley Arabian and Godolphin Arabian. Pedigrees in other European breeds are less well documented but reductions in male lines are similar and an influence of Neapolitan and Original Arabian horses can be observed or is presumed to have occurred. Pedigree information on northern European horse breeds also indicates reduced male diversity. Nevertheless, these breeds, notably the Icelandic horse, might be the only European breeds not to have been subjected to the recent introgression waves [30].

Due to the lack of polymorphic markers on the NRY, it is not possible to trace male-mediated population-genetic dynamics in modern horses. We now describe a systematic screen for horse Y-chromosomal variants in modern domestic horse breeds. As there is only scarce sequence information relating to the horse Y chromosome, we sequenced Y-chromosomal BAC clones to obtain reference sequences for our screen. Based on de novo assembled contigs, we amplified long-range PCR (LRP) products covering about 186 kb for targeted resequencing [31]. To identify variants, we selected (a) nine male horses from phenotypically, genetically and geographically highly distinct breeds, (b) eight Lipizzan stallions, each representing a classical founder line [26], and (c) one Przewalski horse (*E. przewalskii*). LRP products were pooled ("seq-pools") and sequenced with lllumínas Solexa technology. Based on the mutations identified, we describe the first Y-chromosomal haplotypes (HT) in domestic horses and their phylogenetic relationships (a detailed workflow is given in Fig. S1). Furthermore, we present the results of a screen for the distribution of the HTs among purebred modern horse breeds. Using pedigree data, we are able to trace the paternal roots of the extant males. Our data show the strong influence of influential founders, mainly from the Near East and a certain Thoroughbred line, on extant horse breeds.

MATERIALS AND METHODS

Ethics Statement

1) Blood Samples

Genomic DNA samples from the breed pool and the Przewalski horse isolated from blood were collected as part of routine diagnostics at the Institute of Animal Breeding and Genetics, University of Veterinary Medicine, Vienna, during the 1970's, 80's, and 90's. For the breed pool, information on breeds was available but the dates of collection, owners and horse identification were not recorded. The Przewalski horse was provided by Dr. Meltzer (Zoo Munich) and the Icelandic horse by B. Wallner.

The Lipizzan horse blood samples were collected before 1999 during an INCO Copernicus project (1996–2001). Permission for the scientific use of the samples was granted by all involved stud farms, which were partners in the project. A summary of the findings was published in "Der Lipizzaner im Spiegel der Wissenschaft" [26], edited by G. Brem, Publisher: Österreichische Akademie der Wissenschaften. We are not aware whether the blood samples were taken during routine diagnostics.

Samples were collected before the establishment in 2004 of the ethics commission of the University of Veterinary Medicine, Vienna.

2) Hair Root Samples

165 hair samples were recently taken and permission was granted by the private owners or breeding associations. The following breeding associations gave their consent: Associacion française du Lipizzan (France), Escola Portuguesa de Arte Equestre (Portugal), Fundación Real Escuela Andaluza del Arte Ecuestre (Spain), Lipizzanergestüt Piber and the Spanish Riding School (Austria).

450 genomic DNA samples isolated from hair roots derive from routine parentage testing from the archive of the company AGROBIOGEN GmbH, Hilgertshausen. DNA samples kindly provided by Agrobiogen are several years old.

All samples were anonymized.

454 Sequencing of BAC Clones

BAC DNA from 5 Y-chromosomal BAC clones (Fig. 1) [32], [33] was isolated as described at http://dga.jouy.inra.fr/grafra/BAC_DNA_midiprep. htm and treated with RNAseA. BAC DNA was purified with the Mammalian Genomic DNA Miniprep Kit (Sigma-Aldrich, cat. no G1N350-1KT). Parallel sequencing on the 454 Roche system was performed according to standard procedures at the Core Facility Molekularbiologie (Meduni Graz, AT). 50 ng of total BAC DNA were prepared with the Nextera™ DNA Sample Prep Kit (Epicentre, cat.no. NT09115) according to the manufacturer's instructions. DNA was incubated in the enzyme buffer for 5 minutes at 55°C and purified with the Qiagen MinElute PCR Purification Kit (Qiagen cat.no. 28004) and titanium adaptors were ligated to the fragments. After emulsion PCR (GS titanium emPCR reagents Lib-L Roche cat. no. 05618444001), bead recovery and enrichment sequencing was performed with the GS Titanium Sequencing Reagent XLR70 (Roche, cat. no. 05233526001) according to Roche standard 454 protocols.

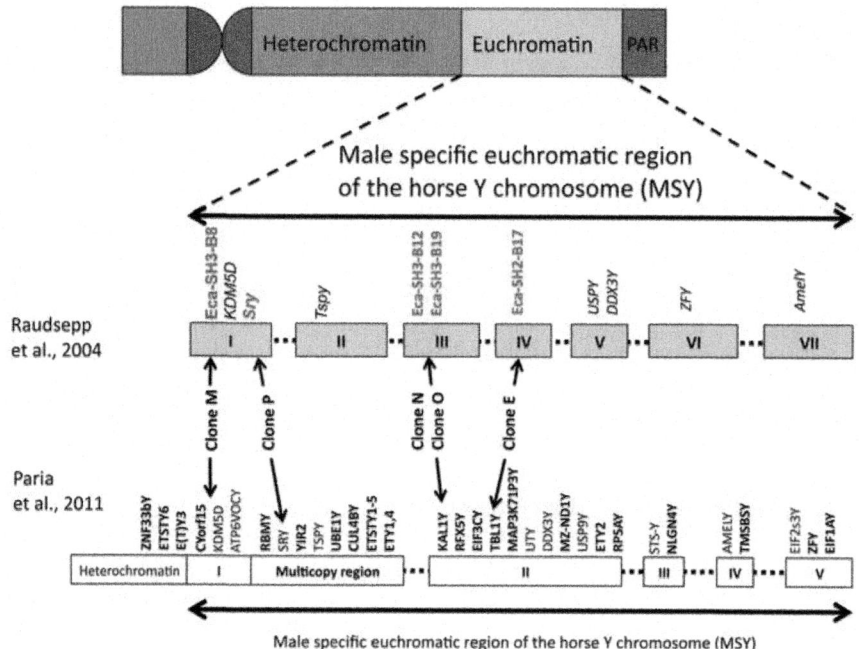

Figure 1: The localisation of the BAC-clones on the horse Y chromosome (ECAY).

The heterochromatic, male specific euchromatic (MSY) and the pseudoautosomal region (PAR) of the horse Y chromosome are shown at the top. The seven MSY contigs (I-VII) described in the study of Raudsepp et al., 2004 [33] are illustrated in the middle and indicated as boxes; gaps between contigs are spanned with dotted lines. The locations of some MSY genes are given (italics). The chromosomal position of each BAC clone used in this study is marked by an arrow. On the bottom the approximate positions of the BAC clones on the horse Y-chromosomal gene map after Paria et al. 2011 [40] are indicated.

doi:10.1371/journal.pone.0060015.g001

The raw sequence reads were filtered and trimmed to reject too short or bad quality sequences. No ambiguous base was allowed and the minimum sequence length exceeded 50 bases. Separate assemblies were performed for BAC clones M, P and E using Roche GS De Novo Assembler v2.3 to generate the contigs. Overlapping clones N and O were assembled together. Contig sequences were trimmed for BAC vector sequence and *E.coli* genome sequences. Assembled sequences longer than 5000 bp were selected for further analysis.

Analysis of Contig Sequences

Contig consensus sequences generated in each library were screened for homologies by manual blast search (blastN) against the complete nucleotide collection in Genbank and by BLAT (http://genome.ucsc.edu/cgi-bin/hgBlat) against the horse genome. Repetitive elements were identified with Repeat Masker (http://www.repeatmasker.org/).

Long–Range PCRs

Primers were designed with Primer3. Amplicon length ranged from 5.5 to 11.8 kilobase pairs (kb) and some contigs overlapped. PCR products were amplified with the Expand Long Template PCR System (Roche, cat. no. 11681842001) as described in the Kit and checked for male specificity by comparative amplification of blood genomic DNA from male and female horses (Fig. S2). Sequence specificity was controlled by restriction enzyme digests.

DNA Amplification for Sequence Analysis

To maximize diversity in the sample set, nine male purebred domestic horses from phenotypically different breeds and geographically distinct regions were selected for the first seq-pool ("breed"). The second pool ("lipp") contained 8 Lipizzan stallions, each representing a particular paternal founder line [26]. In the third pool ("prz"), only one Przewalski horse was sequenced. To detect all variants in a pooled sample [31], single LRP products were generated from each individual using high molecular-weight genomic DNA isolated from whole blood. The LRP products were visualized on a 0.8% agarose gel and the concentration of each product was measured with the Qubit ds DNA HS Assay (Invitrogen cat. no. Q32851). In case of multiple PCR products the Y chromosome specific amplicon was isolated from the agarosegel prior NGS library preparation. Y-specific LRP products were pooled equimolarily and cleaned with the High Pure PCR Product Purification Kit (Roche, cat. no. 11732668001), resulting in 2 µg clean LRP product per pool.

Illumina Library Preparation and Data Generation

The pools were fractionated to a size range of 300–700 bp using a Covaris sonicator and fragments were purified using the High Pure PCR Product Purification Kit. For high-throughput sequencing, libraries were prepared with NEBNext DNA Library Prep (NEB, cat. no. E6000S) using NEBNext Multiplex Oligos (NEB cat. no. E7335L) to index the three pools for multiplex sequencing according to the manufacturer's instructions. Indexed pools were submitted together on one lane of 76 bp paired-end sequencing at the CSF NGS

Unit (http://csf.ac.at/) using the Illumina GA II system [34]. The quality of the data was checked with FASTQC [35], sequences were aligned to the reference BAC sequence with BWA [36] and quality control (filtering) was performed with SAMtools [37]. After quality control and removal of duplicate reads, ~4500×, ~5000× and ~2500× mapped sequence coverage of the available Y reference sequence was obtained from Lipizzan, domestic and Przewalski samples, respectively. SNP calling (identifying positions that differed from the reference sequence) was performed with SAMtools and in-house python scripts.

Data Filtering to Identify Candidate Mutations

As there were nine individuals in the pool "breed", theoretically 11.1% of genomic DNA was contributed by each horse in the sequencing sample. Considering random errors and experimental bias at the various stages of library preparation, 8% was decided as a safe threshold for SNP calling. As only one individual was sequenced in the prz-pool, sites that differed from the reference in the same proportion in all three pools were assumed to be base-calling or alignment errors. For the second-class SNPs, we lowered the threshold for SNP calling to 6% minor allele frequency and/or allowed 3% in the prz-pool horse.

Verification of Candidate Polymorphic Sites by Sanger Sequencing and Sequence Analysis

For the filtered candidate mutations, we designed PCR primers by using Primer3 to amplify 240–760 bp fragments. PCR products were amplified from genomic DNA from each horse from the breed-, lipp-, and prz-pools and from a second Shetland pony (n=18). Amplicons were resequenced by conventional Sanger sequencing for forward and reverse strands at LGC genomics. See Figs. S3, S4, S5, S6 and S7 for the sequence alignment with primer binding sites. PCR products amplified from male and female DNA and the results of the Sanger sequencing for each polymorphic region. *E. przewalskii*-determining sites (positions 4086 and 4161 on contig YM23) and Eca-Y2B17 [22] and intron2 from AMELY [24], representing regions harbouring ancestral diversity, were sequenced for all horses in the pools. Sequences were aligned with the CLC workbench. Nucleotide diversity was calculated in R according to the formula of Nei [38].

Microsatellite Analysis

Individuals were characterized using five equine Y-chromosomal microsatellites described previously [23] . Genotypes were determined with MegaBACE 500 at the University of Veterinary Medicine, Vienna. Electropherograms were evaluated using MegaBACE Genetic Profiler v2.2 (GE Healthcare). Microsatellite variation was investigated for 100 domestic horses from various breeds, including the horses that were used for the seq-pools and three Przewalski horses.

Screening of Domestic Horses for Y Haplotypes

615 male horses, representing 58 mainly European horse breeds, were screened for their Y-chromosomal haplotypes. Genomic DNA samples isolated from hair roots (purified with nexttec™) derive from routine parentage testing. Genotyping was performed using the Sequenom MassARRAY iPLEX system (Sequenom, Germany) at the Department for Agrobiotechnology, IFA Tulln. A section of DNA containing the variant position was amplified from each individual by PCR, before a high-fidelity single-base primer extension reaction over the SNP being assayed was undertaken, using nucleotides of modified mass. The different alleles therefore produce oligonucleotides with mass differences that can be detected using highly accurate Matrix-Assisted Laser Desorption/Ionization Time-Of-Flight mass spectrometry [39]. The five SNP multiplex assay was designed using the Sequenom Assay Design 3.1 software, SNP genotyping was performed using the iPLEX® GOLD Complete Genotyping kit with SpectroCHIPs® II in the 384 format (Sequenom, Germany) in duplicate and female genomic DNA was included as a negative control. We followed the manufacturer's protocol with a single modification: to reduce unspecific primer extension, 5 ng sheared salmon sperm DNA (Invitrogen, Austria) per reaction was added to the PCR mastermix. Results were analysed with the Sequenom Typer 4.0 software (Sequenom, Germany) and the results validated by SANGER resequencing of ten (when available) alleles from each position.

The screening criteria for Shetland pony HT6 were (a) no results in Y-E17.1mut_SNP1 & SNP2, and (b) amplification of the 966 bp deletion and visualization on an Agarose gel.

Pedigree Analysis

For pedigree analysis, web-based databases were used: for Thoroughbreds the Pedigree Online Thoroughbred Database (availiable at: www.pedigreequery. com/) and the Galoppsieger database (www.galopp-sieger.de/); for Shagya

Arabians the Shagya Database (www.shagya-database.ch/hengste.php); for Icelandic horses the Stormhestar Database (www.stormhestar.de/german/default.asp) and for many other breeds the Pedigree Online All Breed Database (www.allbreedpedigree.com) or the Sporthorse Horse Show and Breed Database (www.sporthorse-data.com/breed.htm). All database informations were last accessed in October 2012.

Thirteen breeds (Appaloosa, Barb, Konik, Mangalarga Paulista, Mangalarga Marchador, New Forest, Paint, Russian Arabian, Saddlebred, Shire, Tinker horse, Wuerttemberg, Camargue) are not included in the pedigree analysis, as we had no males with a proven ancestry. For founder classification we call Arabian, so-called Oriental and Turkoman horses (i.e. horses from the middle East ranging from Turkmenistan to Egypt) that were imported to Europe in the 18th and 19th century "Original Arabians".

RESULTS

Identification of Y Chromosome Polymorphisms

To detect polymorphisms, we generated reference sequences from the horse Y chromosome by 454 sequencing of 5 Y-chromosomal BACs from certain regions on the chromosome (Fig. 1) [23], [32], [33], [40]. Sequence reads were de novo assembled and 11 contigs and an *Sry* containing BAC sequence (AC215855.2) were selected for designing 21 LRPs, covering a total of 186,122 bp. Y-chromosomal specificity was checked by comparative amplification of the LRPs on male and female DNA (Fig. S2).

We produced Y-chromosomal LRP products from 17 domestic and one Przwalski horse separately and pooled the products. LRPs of nine males from different breeds (pool-breed), of eight Lipizzan stallions (pool-lipp) and the Przewalski horse (pool-prz) were sequenced with the Illumina platform to high depth. Whereas the Przewalski horse showed 37 base substitutions and one 3051 bp deletion compared to the domestic horse samples, only 4 positions/regions promised to be polymorphic within the domestic horse pools. With relaxed criteria (see Materials and methods) another 19 second-class SNPs were added. For validation, we amplified the region spanning each polymorphic candidate from each domestic horse in the seq-pools and sequenced them by conventional Sanger sequencing. The four top candidate regions were confirmed but none of the 19 second-class SNPs were. The resequencing results and the male specificity of the confirmed candidates are shown in Figs. S3, S4, S5, S6 and S7. The variants detected in domestic horses were two SNPs, one single base deletion and a complex variant consisting of 20 SNPs and four indels. In addition we identified a 966 bp deletion in a Shetland pony and a transition in a second Przewalski horse during the screening procedures.

The two SNPs were found on discrete contigs (YE17 and YXX_24I23) but the indel polymorphism, the complex mutation and the 966 bp deletion are all located in close proximity or even overlapping in a restricted region on contig YE3. Screening for homologies to this region in the horse genome using BLAT revealed high similarities (>95%) with the homologous region on the horse X chromosome and the occurrence of a LINE element approximately 700 bp upstream of the region that is mutated in the domestic horse (Fig. S9). The alignment of X - and Y-sequences shows the need for careful selection of PCR amplification primers to amplify Y-chromosomal sequences. Furthermore, one can see from the alignment (Fig. S5) that the multiple variants on locus YE3 - Pos 1007–12040 conform to the X- rather than to the Y chromosome for a length of 300 bases. We thus assume that the complex variant on YE3 - Pos 1007–12040 comprising 25 SNPs and four indels is the result of a gene conversion event between the X and Y chromosome [5]. The deletion of a single T on locus YE3 - Pos 10594 is also found on the homologous X-chromosomal sequence (Fig. S4) and thus another putative gene conversion event.

Y-Chromosomal Diversity in Modern Horses - Haplotype Network and Estimation of Divergence Time

The polymorphisms result in six haplotypes in the domestic and two in the Przewalski horse (Tables S9, S10 and S11). The ancestral and the derived status in modern horses were determined by comparison to the Przewalski horse sequence. The haplotype network in Fig. 2gives the relationship between the HTs. *E. caballus* haplotypes are separated by only one mutational or gene conversion step, i.e. there is no deep bifurcation with many segregating mutations. Three haplotypes (HT1-3) occur at relatively high frequencies. HT1 is the ancestral haplotype when rooted with the Przewalski Y haplotype. HT2 is differentiated from HT1 by a single nucleotide mutation. The deletion that defines HT3 arose on the background of HT2. Clearly, HT1 is the most prominent haplotype in the evolution of modern horse breeds: of the five mutations observed in the panel, four arose on the background of HT1 and only one on the background of HT2. Thus, although the proportion of HT1 and HT2 in our sample is about equal, all except one variant arose on the background of the older and thus evolutionarily more important HT1. Lippold et al. found a Y-chromosomal HT in an ancient domestic horse that differed at three positions from the common domestic horse HT chromosome [24]. We observed none of these ancient bases when screening a set of extant horse breeds that represent all six haplotypes. The Y chromosome of the Przewalski horse forms a separate clade with a sequence divergence of 0.021% from *E. caballus*.

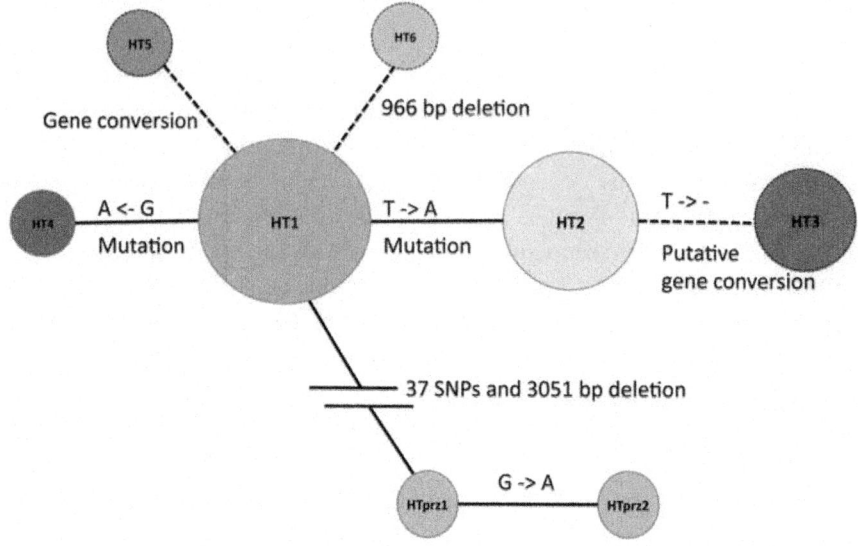

Figure 2: Haplotype network of the six modern and two Przewalski horse HTs.

Circles represent the haplotypes with the area proportional to the observed frequency in 20 male horses in the initial Y-chromosomal sequence analysis. HT1, n=7 (three Lipizzan, two Arabian, one Shetland pony, one Shire horse); HT2, n=5 (five Lipizzan); HT3, n=3 (one Thoroughbred, one Trakehner, one Quarter horse); HT 4 (one Icelandic horse), HT5 (one Norwegian Fjord horse), HT6 (one Shetland pony), HTPrz1 (one Przewalski horse), HTPrz2 (one Przewalski horse). A dashed line between the haplotypes indicates, that the polymorphism is located on the highly variable contig YE3, which was omitted when estimating divergence time and nucleotide diversity.

doi:10.1371/journal.pone.0060015.g002

For inference of population genetic diversity and divergence time estimates, we rely only on putative single nucleotide polymorphisms, i.e., we exclude the region affected by gene conversion (YE3). The rate of gene conversion is known to depend strongly on the genomic location, and is particularly high at translocation hotspots [41], [42]. The prediction of these hotspots seems complex [4], but the occurrence of a LINE element and the X/Y-sequence similarity indicate that the YE3 region is a candidate. Furthermore, we infer from pedigree data, that the deletion on YE3 leading to HT3 arose, presumably by a gene conversion event, very recently (see below). In addition, the occurrence of an independent 966 bp deletion in HT6 points to a general instability of the YE3 region. We thus base our estimates of diversity and dating solely on the observation of two SNPs (YXX_24I23 - Pos 25345 and YE17.1 - Pos 1277) per 170 731 bp of total length. We assume a rate of single nucleotide mutations per generation of about $1.1–5×10^{-8}$ as found for humans [1], [43], [44] and expect the rate of mutations per generation in our dataset to be: $\mu×$ (sequence length) $≈1,8$ to $8×10^{-3}$. Thus only 234 to 1064 meioses are theoretically necessary to give rise to the two SNPs over the region under investigation. We conclude that the present Y-chromosomal diversity in modern breeds most likely arose by mutations from HT1 after domestication, about 6000 years or 1000 generations ago [45]. We note, however, that the NRY is inherited as a single locus, so inferences based on the NRY are subject to large stochastic errors for parameters such as the overall tree depth.

Considering the frequencies of haplotypes with SNPs, pairwise nucleotide diversity [38] is $\pi=3.71×10^{-6}$, an extremely low value compared to those observed in domestic pigs ($\pi=1.38×10^{-3}$) [11], [12] and dogs ($\pi=2.91×10^{-4}$) [46].

Phylogeography of Horse Y-chromosomal Haplotypes

We examined the distribution of the Y-chromosomal haplotypes in 615 males from 56, mainly European, domestic horse breeds. To avoid father-son pairs or paternal brothers in the dataset, only purebred horses with confirmed paternity were investigated. 91% of the male horses were found to carry one of the three major haplotypes. The ancestral HT1 is distributed across almost all breeds and the entire geographical region under investigation, underlining its importance (Fig. 3A). HT2 is also found at high frequencies across a broad range of breeds, although not in the northern European breeds and not in horses from the Iberian Peninsula. HT3 is almost fixed in the English Thoroughbred and is further distributed across many warm-blooded breeds. HT4-6 are only found in three local northern European breeds but in high frequencies in these breeds: HT4 in half the Icelandic horses and HT6 in 74% of the Shetland ponies while HT5 is fixed in Norwegian Fjord horses. We undertook genotyping of five Y-specific microsatellite markers [23] from a subset of 100 horses distributed over all HTs. Microsatellite analysis showed a uniform pattern of amplification over all domestic horses, distinct from the Przewalski horse at one locus, as in previous studies [23], [25].

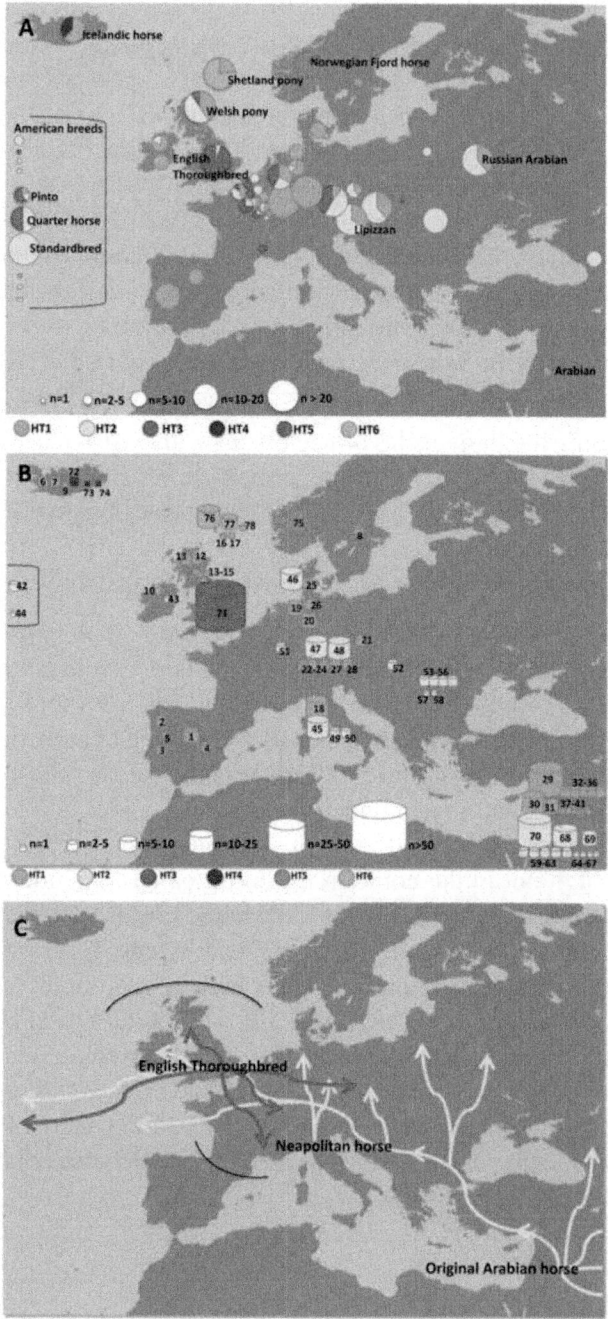

Figure 3: Geographic distribution and history of Y-haplotypes in modern horse breeds.

(a) Geographic distribution of Y-chromosomal haplotypes in a set of modern horse breeds. Only a few important breeds are specified, (b) Origin of modern domestic horse founders deduced from pedigree data. Each founder is represented by a drum with its size proportionally to the number of offspring in the dataset. The number in the drums serve as founder identifiers. Detailed information on founders (name, year of birth, breed, origin, information on import) is listed in Table 1. (c) Male introgression routes deduced from the pedigree and the distribution of HT2 and HT3 in our dataset. HT2 (yellow arrows) arrived from South-East at early times and has been spread during the Neapolitan and Oriental introgression waves, but did not reach Northern Europe and the Iberian peninsula. The English wave in red is well documented through pedigree data and the spread of HT3 (red arrows). Due to the ubiquitous occurrence of HT1, this haplotype is not considered. The black solid lines reflect the limits of the observation of HT2 and HT3.

doi:10.1371/journal.pone.0060015.g003

Incorporation of Pedigree Data to Uncover Founder Traces

We used the documentation of horse ancestry in studbooks and pedigree data from databases to trace the origin of the purebred horses. Sufficient pedigree information was available for 418 (67.97%) males from our dataset. Whereas the paternal ancestry is accurately documented back to the 18th century for old historic breeds such as the English Thoroughbred and the Lipizzan horse, data on most other modern breeds only extend to about 1900. Pedigree analysis revealed that the 418 males from 43 breeds originate from 78 paternal founders whose geographical/breed origin is shown in Fig. 3B. Founder specifications and haplotype information on renowned stallions are available in Table 1. No founder from the Iberian Peninsula, the British Isles (with the exception of a Connemara pony born in 1921, whose origin and possible Arabian influence is unclear) or northern Europe shows HT2. Instead, HT1 predominates, although autochthonous haplotypes (HT4, 5 and 6) are present at high frequencies in local breeds or are even fixed (e.g. in the Norwegian Fjord horse). In contrast, central, eastern and southern European as well as Original Arabian founders from the near East exclusively show HT1 and HT2 in about equal proportions.

Table 1: Founders contributing to modern horses

Founder identifier (Fig. 3B)	Founder name	Born	Inferred haplotype	Breed and/or geographical origin	Documented import	Descendants in dataset (n)	Modern breeds distributed
1	Favorita	1889	HT1	Spanish purebred/Spain		6	Pura Raza Espanola
2	Perola	1917	HT1	Lusitano/Portugal		4	Lusitano
3	Nice	1915	HT1	Lusitano/Portugal		3	Lusitano
4	Descindida	1840	HT1	Spanish purebred/Spain		2	Pura Raza Espanola
5	Maroto	1905	HT1	Alter Real/Portugal		1	Lusitano
6	Brúnn frá Svaðastöðum	1900	HT1	Icelandic horse		5	Icelandic horse
7	Brúnn frá Ármanesi	1910	HT1	Icelandic horse		4	Icelandic horse
8	Hingst frá Gubrandsdalen	1646	HT1	Dole Gudbrandsdal/Norway		2	Swedish coldblood trotter
9	Kjerul frá Sauðarkesli	1981	HT1	Icelandic horse		1	Icelandic horse
10	Mountain Lad	1928	HT1	Connemara/Ireland		3	Connemara
11	Defence	1906	HT1	Welsh/UK		3	Connemara pony
12	Charlie	1880	HT1	Welsh mountain/UK		3	Riding Pony, Dartmoor Pony
13	Trotting Comet	1836	HT1	Dale or Exmoor/UK		1	Fell Pony
14	Broccoli	1855	HT1	Welsh/UK		1	Welsh-D
15	Prince Llewelly	1904	HT1	Welsh/UK		1	Connemara pony
16	Fairy Prince	1940	HT1	Shetlandpony/UK		1	Tigerpony
17	Pallieter de Bevelbonkhof	1902	HT1	Shetlandpony/UK		1	Shetland pony
18	Conversano	1767	HT1	Neapolitan Horse/Italy		12	Lipizzan horse
19	Nemo	1805	HT1	Friesian/Netherlands		6	Baroque Pinto, Friesian, Kladruby
20	Old Flyer	1830	HT1	Brandenburg/Germany		6	Welsh-B, Welsh-D
21	Sacromonte Chontoux	1830	HF1	Kladruber/Czech Republic		4	Kladruby
22	Amor	1888	HT1	Pinzgauer horse/Austria		4	Noriker
23	Bravo 149	1877	HT1	Pinzgauer horse/Austria		4	Noriker
24	635 Vulkan	1887	HT1	Pinzgauer horse/Austria		4	Noriker
25	Zarif Sejer	1921	HT1	Frederiksborg/Denmark		3	Knabstrupper
26	Smoky	1955	HT1	Spotted horse/Denmark		3	Knabstrupper
27	80 Atrus 55	1866	HT1	Pinzgauer horse/Austria		3	Noriker
28	126 Optimus	1890	HT1	Pinzgauer horse/Austria		3	Noriker
29	El Bedavi	1830	HT1	Arabian	1833, Babolna	34	Haflinger
30	Siglavy	1810	HT1	Arabian	1814, Lipizza	11	Lipizzan horse
31	Kuhailan Haifi	1923	HT1	Arabian, desert bred	1931, Poland	6	Shagya Arabian, Arabian, Partbred Arabian, Riding pony
32	Ibrahim	1899	HT1	Arabian/Egypt	Poland/1910 GB	3	Riding pony, Welsh-A, Arabian
33	Khalil x Saklavi Jidran	1876	HT1	Arabian/Egypt	1910, GB, 1929 BRD	3	Arabian, AngloArabian, Shagya Arabian
34	Kohailan Adjuze	1876	HT1	Arabian, desert bred	1885 Babolna	2	Shagya Arabian
35	Kuhailan Zaid	1924	HT1	Arabian, desert bred	1931, Babolna	2	Shagya Arabian
36	Siglavy Bagdady	1895	HT1	Arabian, desert bred	1902, Babolna	2	Shagya Arabian
37	Dahoman	1846	HT1	Arabia, Syria	1852, Babolna	1	Shagya Arabian
38	Hadban	1891	HT1	Arabian, Egypt	1897, Babolna	1	Shagya Arabian
39	Kbran	1864	HT1	Arabian	1901, Poland	1	Austrian Warmblood
40	Mahmoud Mirza	1851	HT1	Arabian, Iraq	1866, Babolna	1	Shagya Arabian
41	Barq	1840	HT1	Arabian/Egypt	1880, GB	1	Riding pony
42	Traveller	1890	HT2	Quarter horse/U.S.		3	Quarter horse
43	Connemara Boy	1921	HT2	Connemara pony/Ireland		1	Connemara pony
44	Mahoma I	1900	HT2	Paso Fino/Colombia		1	Paso Fino
45	Nesapolitano	1790	HT2	Neapolitan Horse/Italy		15	Lipizzan horse
46	Pluto	1765	HT2	Frederiksborg/Denmark		12	Lipizzan horse
47	Favory	1779	HT2	Kladruber/Czech Republic		10	Lipizzan horse
48	Maestoso	1773	HT2	Kladruber/Czech Republic		9	Lipizzan horse
49	Pepoli	1767	HT2	Orig. Italien/Italy		4	Kladruby
50	Tulipan	1680	HT2	Neapolitan Horse/Italy		3	Lipizzan horse
51	Trinket 1803	1885	HT2	German classic pony/Germany		2	German Shetland pony
52	Incitato	1802	HT2	Siebenbürger stallion/Hungary		2	Lipizzan horse
53	Gosel	1898	HT2	Hucul/Romania		6	Hucul
54	Hroby	1894	HT2	Hucul/Romania		3	Hucul
55	Ousor	1929	HT2	Hucul/Romania		2	Hucul
56	Prislop	1932	HT2	Hucul/Romania		2	Hucul
57	Goral	1924	HT2	Hucul/Romania		1	Hucul
58	Polan	1929	HT2	Hucul/Romania		1	Hucul
59	Gazlan	1840	HT2	Arabian, desert bred	1852, Lipizza	4	Shagya Arabian
60	O Bajan	1881	HT2	Arabian	1886, Babolna	3	Shagya Arabian, Austrian Warmblood
61	Shagya	1830	HT2	Arabian	1836, Babolna	3	Shagya Arabian
62	Batractar	1913	HT2	Arabian, desert bred	1817, Weil	2	Riding pony
63	Gazal Poolok	1990	HT2	Arabian	1895, BRD	2	Abkal Theke
64	Dahman Amir	1887	HT2	Arabian	1890, Stamsajówka	1	Partbred Arabian
65	Latif	1903	HT2	Arabian	1909, Pompadour	1	Partbred Arabian
66	Mersuch	1898	HT2	Arabian	1902, Babolna	1	Partbred Arabian
67	Soujoin	1894	HT2	Arabian	1895, Weil	1	Shagya Arabian
68	Byerley Turk	1684	HT2	Arabian	1689, England	6	Warmblood, English Thoroughbred, Hannoveran, Oldenburg, Quarter Horse, Sachsen Anhaltiner
69	Godolphin Arabian	1724	HT2	Arabian or Turkoman	1729, England	5	Paint, Austrian Warmblood, Warmblood, Hungarian Warmblood
70	Darley Arabian, exluding descendants of Whalebone	1849	HT2	Arabian, Syria	1704, England	44	Standardbred, Oldenburger, Trakehner, Warmblood, Dutch Warmblood, Welsh-A, Welsh-B
71	Darley Arabian, via Whalebone	1807	HT3	Arabian, Syria		59	Bavarian Warmblood, Austrian Warmblood, English Thoroughbred, Hannoveran, Holstein, Oldenburg, Quarterhorse, Partbred Arabian, Riding pork, Hirvelander Horse
72	Hrímfur frá Geitaskarði	1915	HT4	Icelandic horse/Iceland		5	Icelandic horse
73	Hörvir frá Vélmundarstöðum	1946	HT4	Icelandic horse/Iceland		1	Icelandic horse
74	Iseifs-Gráni frá Geitaskarði	1910	HT4	Icelandic horse/Iceland		1	Icelandic horse
75	Njal	1891	HT6	Norwegian Fjord/Norway		15	Norwegian Fjord horse
76	Jack	1817	HT6	Shetlandpony/U.K.		14	Shetland pony
77	John Rein	1882	HT6	Shetlandpony/U.K.		7	Shetland pony, Tigerpony
78	Prince of Thule	1872	HT8	Shetlandpony/U.K.		1	Shetland pony

Stallions with descendants in our dataset are listed, giving their origin, HT and their distribution in extant horse breeds (as estimated from our dataset).
doi:10.1371/journal.pone.0060015.t001

doi:10.1371/journal.pone.0060015.t001

At least 202 (48.3%) of the modern horses in our dataset obtained their Y chromosomes from Original Arabian founders. Among the HT2 ancestors are many influential stallions, including the famous founders of the English Thoroughbred: Darley Arabian, Byerley Turk and Godolphin Arabian.

The English Thoroughbred, which is best known for its use in horse racing, has a complete studbook since 1791 and all registered males can be traced back to one of three popular founders. Among them, the paternal line of Darley Arabian currently represents almost all male Thoroughbreds [47]. Pedigree analysis revealed that all HT3-carrying males can be traced to the single Thoroughbred "Whalebone", born 1807" a son of "Pot8os". Hence, the mutation leading from HT2 to HT3 must have occurred either in the germline of the famous racehorse "Eclipse" or in that of his son "Waxy" or grandson "Pot8os" (Fig. 4). The frequency of HT3 rose to 96.5% in the English Thoroughbred and to 41% in modern sport horse breeds in our dataset within 15–20 generations (Fig. 3A). Among 418 long pedigrees, we observed no obvious pedigree errors in the English Thoroughbred and Standardbreds, whereas 17 (4.06%) errors were found in Shetland ponies and Lipizzan, Warmblood and Welsh horses. This observation does not permit us to check the correctness of the studbook, which must be undertaken with maternal lines using mtDNA [48]–[51]. The low initial variation in the founders leads to a low resolution, making pedigree errors in the male line difficult to detect.

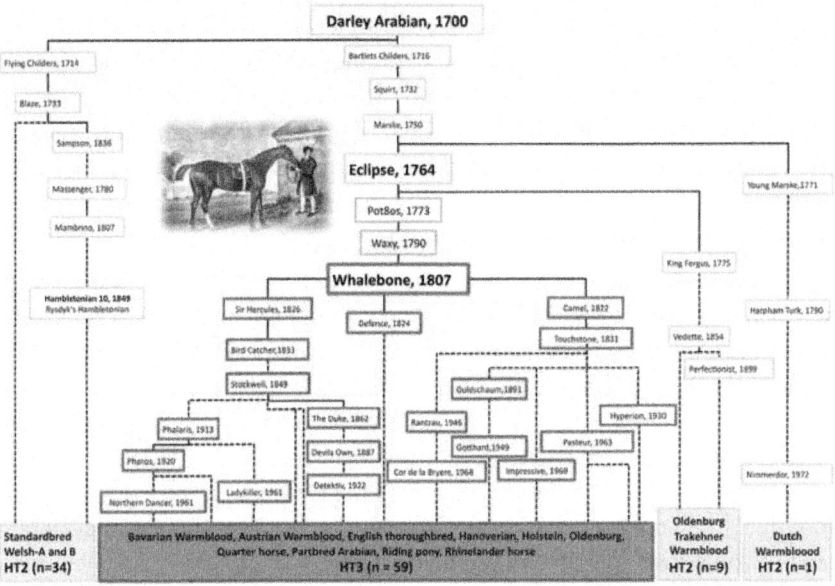

Figure 4: Pedigree of Darley Arabians progeny depicting the origin of HT3 from HT2.

Breeds of analysed males are listed on the bottom and the haplotypes of their ancestors are reconstructed (HT2-yellow, HT3-red, unknown-grey). Selected famous stallions are shown by name; dotted lines connect relatives where at least one ancestor is omitted. No descendants from "Pot8os" and "Waxy" were available apart from "Whalebone, 1807". The mutation leading to HT3 must have occurred either in the germline of stallion "Eclipse" [54] or in his son "Pot8os" or in his grandson "Waxy" and rose to very high frequency in the English Thoroughbred and many sport horse breeds through the progeny of the stallion "Whalebone".

doi:10.1371/journal.pone.0060015.g004

DISCUSSION

The analysis of Y-chromosomal and mtDNA markers offers an invaluable tool for the demographic characterization of populations. To date, the domestic horse was the only livestock species for which paternal lines could not be traced due to the lack of Y-chromosomal variability. We have identified five variants that presumably arose independently and that result in three major and three breed-specific HTs.

Of these five variants, only two arose by single basepair mutations. The other three events arose in a restricted region YE3 with extensive sequence similarity to a X-chromosomal region and consist of a gene conversion tract of about 300 basepairs, a single basepair deletion, which may also be caused by a gene conversion, and a deletion of about 900 bps. X to Y gene conversion strongly influences allelic diversity in specific human Y-chromosomal regions [4],[41]. In this study, we report the first observation of an X to Y gene conversion in a farm animal. The hyperpolymorphic region YE3 may be a hotspot for gene conversions and structural rearrangements on the horse Y chromosome, but a closer investigation is needed.

All domestic horse HTs are closely related. HT1, which is the ancestral haplotype as inferred by comparison with the Przewalski horse, seems to be the only haplotype to have survived through domestication to extant breeds. All other HTs arose directly or indirectly from HT1, presumably after domestication. The finding of low nucleotide diversity of the modern horse Y chromosome is consistent with previous studies, in which no variation was detected [21]–[23]. As the establishment of haplotypes depends on the individuals selected for the initial screen the significance of ascertainment bias has to be kept in mind [6], [52]. To overcome this problem, we screened microsatellite markers in a random sample over all haplotypes/regions and detected no variation. As microsatellites are highly mutable, the absence of significant microsatellite variation on the horse Y chromosome confirms the very recent origin of all haplotypes in our sample. Since we only used puredbred horses from a restricted region we note, that the global horse population may harbour more y-chromosomal variability.

The low diversity of the Y chromosome contrasts with the high diversity of mtDNA haplotypes observed in modern horses [15]–[19]. The difference is likely caused by the low effective population size of the horse Y chromosome due to a strong variation in male reproductive success. This may be due to the polygynous breeding patterns in wild horses and to a stronger bottleneck in male horses during domestication and might be further exacerbated by the intensive breeding practices in this species [20], [21].

The samples used in our study derive from purebred modern, mainly European, breeds, which have undergone intensive selection for particular traits during the past two centuries [30]. The refinement was mainly achieved through the disproportionate use of selected popular stallions and their descendants that were crossed to local mares. With the use of pedigree information, available in studbooks and open access databases, we inferred the impact of the upgrading process on the male horse population. We find that only a limited number of founders contribute to the extant horse haplotypes (Fig. 3B, Table 1).

Based on the pedigree information, we traced the effects of three introgression waves (the Neapolitan, the Oriental and the English waves) on NRY markers. The importance of the Thoroughbred in the English wave is clearly seen through the spread of HT3. In the "Original Arabians", the Neapolitan horse and the central and Eastern European founders, the proportion of HT2 is about 50%. In founders from northern Europe, i.e. Iceland, Norway and the British Isles, and the Iberian Peninsula the frequency of HT2 was very low. The distribution of HT2 is consistent with the movement of stallions from the Middle East to Central and Western Europe via the Neapolitan and Oriental waves (Fig. 3C, Fig. S10). The high proportion of HT2 in Central European horses may also be derived from earlier introgression of horses from the Middle East or even from ancient colonization [20].

Only three northern European horse breeds, Icelandic horses, Shetland ponies and Norwegian Fjord horses, were either not or were hardly subjected to these introgression waves and were therefore able to maintain autochthonous Y chromosome variants. These breeds have a comparatively isolated history. Due to the specific geographical and social structures in northern areas, locally adapted breeds were presumably of a higher value than imported animals from Central Europe and the Near East. In the case of Icelandic horses, the import of horses was restricted since 930 A.D. and has been prohibited since 1909 [30]. The three breeds are not necessarily closely related to one another. Icelandic and Norwegian Fjord horses branch from the root of modern breeds, when using 46244 autosomal loci [53]. As we only detected the ubiquitous HT1, little can be deduced about the ancestry of the Barb, the Iberian breeds, or the Swedish Coldblood Trotter.

The history of the domestic horse is marked by recurrent adjustment to changing socio-economic needs. Over the course of the last two centuries, when most modern breeds were established, the horse has been undergoing a transition from a working animal also used by the military use towards a domestic animal used in leisure and sports activities. This has largely been achieved through the use of a few, very popular, sires that have been extensively shifted among breeds. Due to the breeding practices during the last 200 years, "popular sires" and their sons have fathered an inordinate amount of offspring and replaced local Y chromosomes. Unique variants can only be found in a few northern European breeds. The restricted genetic diversity of the modern horse Y chromosome reflects what has survived through the species' dynamic history.

CONCLUSIONS

We describe the first polymorphic markers on the paternally inherited part of the Y chromosome of the domestic horse. We have used the new markers to investigate the paternal gene flow between horse populations and breeds. We document the influence of popular sires, particularly a single Thoroughbred stallion line, on many extant horse breeds. Our data on male lineages complement the information on the well established maternal mtDNA lineages and enhance the genetic documentation of the history and dynamics of modern horse breeds. The new polymorphic markers and haplotypes enable horse breeding practices to be monitored and verified. Although six haplotypes do not offer a high resolution, this study provides a first backbone phylogeny for deeper population genetic studies of the Y chromosome variation in horses.

DATA DEPOSITION

BAC sequences have been deposited in Genbank (http://www.ncbi.nlm. nih.gov/genbank) under accession numbers JX565700–JX565709, sequence alignments under accession numbers JX646942– JX647045. Illumina short read sequences generated in this study are available at the Sequence Read Archive (SRA) under the Accesssion number ERP001668 (http://www.ebi. ac.uk/ena/data/view/ERP001668). Y-chromosomal SNPs have been submitted to the NCBI dbSNP database: ss#711581504, ss#711581506, ss#711581507.

ACKNOWLEDGMENTS

We thank C Guelly at the Core Facility Molekularbiologie (Meduni Graz, AT) for 454 and A Sommer at the CSF NGS Unit Vienna for Illumina sequencing; A Ertl, M and V Dobretsberger and private horse owners for sample supply; N Kreutzmann, M Schwender and D Rigler for assistance in the lab. We are grateful to P Burger, K Traxler, T Leeb, S Mueller and S Macho-Maschler for their comments on the manuscript and and we thank G Tebb for editing.

AUTHOR CONTRIBUTIONS

Conceived and designed the experiments: BW GB. Performed the experiments: BW JB. Analyzed the data: BW PS CV TD. Contributed reagents/materials/analysis tools: JB PS. Wrote the paper: BW CV TD GB.

REFERENCES

1. Xue Y, Wang Q, Long Q, Ng BL, Swerdlow H, et al. (2009) Human Y chromosome base-substitution mutation rate measured by direct sequencing in a deep-rooting pedigree. Curr Biol 19: 1453–1457. doi: 10.1016/j.cub.2009.07.032

2. Repping S, van Daalen SK, Brown LG, Korver CM, Lange J, et al. (2006) High mutation rates have driven extensive structural polymorphism among human Y chromosomes. Nat Genet 38: 463–467. doi: 10.1038/ng1754

3. Jobling MA, Samara V, Pandya A, Fretwell N, Bernasconi B, et al. (1996) Recurrent duplication and deletion polymorphisms on the long arm of the Y chromosome in normal males. Hum Mol Genet 5: 1767–1775. doi: 10.1093/hmg/5.11.1767

4. Trombetta B, Cruciani F, Underhill PA, Sellitto D, Scozzari R (2009) Footprints of X-to-Y gene conversion in recent human evolution. Mol Biol Evol 27: 714–725. doi: 10.1093/molbev/msp231

5. Ellegren H (2011) Sex-chromosome evolution: recent progress and the influence of male and female heterogamety. Nat Rev Genet 12: 157–166. doi: 10.1038/nrg2948

6. Underhill PA, Kivisild T (2007) Use of y chromosome and mitochondrial DNA population structure in tracing human migrations. Annu Rev Genet 41: 539–564. doi: 10.1146/annurev.genet.41.110306.130407

7. King T, Jobling M (2009) What's in a name? Y chromosomes, surnames and the genetic genealogy revolution. Trends Genet 25: 351–360. doi: 10.1016/j.tig.2009.06.003

8.   Ding ZL, Oskarsson M, Ardalan A, Angleby H, Dahlgren LG, Tepeli C, Kirkness E, Savolainen P, Zhang YP (2012) Origins of domestic dog in Southern East Asia is supported by analysis of Y-chromosome DNA. Heredity 108: 507–514. doi: 10.1038/hdy.2011.114

9.   Brown SK, Pedersen NC, Jafarishorijeh S, Bannasch DL, Ahrens KD, et al. (2011) Phylogenetic distinctiveness of Middle Eastern and Southeast Asian village dog Y chromosomes illuminates dog origins. PLoS One 6: e28496. doi: 10.1371/journal.pone.0028496

10.   Sundqvist A, Björnerfeldt S, Leonard J, Hailer F, Hedhammar A, et al. (2006) Unequal contribution of sexes in the origin of dog breeds. Genetics 172: 1121–1128. doi: 10.1534/genetics.105.042358

11.   Ramirez O, Ojeda A, Tomas A, Gallardo D, Huang L, et al. (2009) Integrating Y-chromosome, mitochondrial, and autosomal data to analyze the origin of pig breeds. Mol Biol Evol 26: 2061–2072. doi: 10.1093/molbev/msp118

12.   Cliffe KM, Day AE, Bagga M, Siggens K, Quilter CR, et al. (2010) Analysis of the non-recombining Y chromosome defines polymorphisms in domestic pig breeds: ancestral bases identified by comparative sequencing. Anim Genet 41: 619–629. doi: 10.1111/j.1365-2052.2010.02070.x

13.   Meadows J, Hanotte O, Drögemüller C, Calvo J, Godfrey R, et al. (2006) Globally dispersed Y chromosomal haplotypes in wild and domestic sheep. Anim Genet 37: 444–453. doi: 10.1111/j.1365-2052.2006.01496.x

14.   Edwards CJ, Ginja C, Kantanen J, Pérez-Pardal L, Tresset A, et al. (2011) Dual Origins of Dairy Cattle Farming – Evidence from a Comprehensive Survey of European Y-Chromosomal Variation. PLoS ONE 6(1): e15922. doi: 10.1371/journal.pone.0015922

15.   Vila C, Leonard J, Gotherstrom A, Marklund S, Sandberg K, et al. (2001) Widespread origins of domestic horse lineages. Science 291: 474–477. doi: 10.1126/science.291.5503.474

16.   Jansen T, Forster P, Levine M, Oelke H, Hurles M, et al. (2002) Mitochondrial DNA and the origins of the domestic horse. Proc Natl Acad Sci U S A 99: 10905–10910. doi: 10.1073/pnas.152330099

17.   Cieslak M, Pruvost M, Benecke N, Hofreiter M, Morales A, et al. (2010) Origin and history of mitochondrial DNA lineages in domestic horses. PLoS One 5: e15311. doi: 10.1371/journal.pone.0015311

18.   Lippold S, Matzke NJ, Reissmann M, Hofreiter M (2011) Whole mitochondrial genome sequencing of domestic horses reveals

incorporation of extensive wild horse diversity during domestication. BMC Evol Biol 11: 328. doi: 10.1186/1471-2148-11-328

19. Achilli A, Olivieri A, Soares P, Lancioni H, Hooshiar Kashani B, et al. (2012) Mitochondrial genomes from modern horses reveal the major haplogroups that underwent domestication. Proc Natl Acad Sci U S A 109: 2449–2454. doi: 10.1073/pnas.1111637109

20. Warmuth V, Eriksson A, Bower MA, Barker G, Barrett E, et al. (2012) Reconstructing the origin and spread of horse domestication in the Eurasian steppe. Proc Natl Acad Sci U S A 109: 8202–8206. doi: 10.1073/pnas.1111122109

21. Lindgren G, Backstroem N, Swinburne J, Hellborg L, Einarsson A, et al. (2004) Limited number of patrilines in horse domestication. Nat Genet 36: 335–336. doi: 10.1038/ng1326

22. Wallner B, Brem G, Mueller M, Achmann R (2003) Fixed nucleotide differences on the Y chromosome indicate clear divergence between Equus przewalskii and Equus caballus. Anim Genet 34: 453–456. doi: 10.1046/j.0268-9146.2003.01044.x

23. Wallner B, Piumi F, Brem G, Mueller M, Achmann R (2004) Isolation of Y chromosome-specific microsatellites in the horse and cross-species amplification in the genus Equus. J Hered 95: 158–164. doi: 10.1093/jhered/esh020

24. Lippold S, Knapp M, Kuznetsova T, Leonard JA, Benecke N, et al. (2011) Discovery of lost diversity of paternal horse lineages using ancient DNA. Nat Commun 2: 450. doi: 10.1038/ncomms1447

25. Ling Y, Ma Y, Guan W, Cheng Y, Wang Y, et al. (2010) Identification of y chromosome genetic variations in Chinese indigenous horse breeds. J Hered 101: 639–643. doi: 10.1093/jhered/esq047

26. Brem G (2011) Der Lipizzaner im Spiegel der Wissenschaft. Vienna: Austrian Academy of Sciences Press. 338 p.

27. Hamann H, Distl O (2008) Genetic variability in Hanoverian warmblood horses using pedigree analysis. J Anim Sci 86: 1503–1513. doi: 10.2527/jas.2007-0382

28. Druml T (2009) Functional Traits in Early Horse Breeds of Mongolia, India and China from the Perspective of Animal Breeding. In: B F, V S, R P, A S, editors. Horses in Asia: History, Trade and Culture. Vienna: Austrian Academy of Sciences. 9–16.

29. Grilz-Seger G and Druml T (2011) Lipizzaner Hengststämme. Graz: Vehling Medienservice und Verlag GmbH. 312 p.

30. Hendricks B (1995) International Encyplopedia of Horse Breeds. Norman: University of Oklahoma Press. 486 p.

31. Out AA, van Minderhout IJ, Goeman JJ, Ariyurek Y, Ossowski S, et al. (2009) Deep sequencing to reveal new variants in pooled DNA samples. Hum Mutat 30: 1703–1712. doi: 10.1002/humu.21122

32. Godard S, Schibler L, Oustry A, Cribiu E, Guerin G (1998) Construction of a horse BAC library and cytogenetical assignment of 20 type I and type II markers. Mamm Genome 9: 633–637. doi: 10.1007/s003359900835

33. Raudsepp T, Santani A, Wallner B, Kata S, Ren C, et al. (2004) A detailed physical map of the horse Y chromosome. Proc Natl Acad Sci U S A 101: 9321–9326. doi: 10.1073/pnas.0403011101

34. Quail MA, Kozarewa I, Smith F, Scally A, Stephens PJ, et al. (2008) A large genome center's improvements to the Illumina sequencing system. Nat Methods 5: 1005–1010. doi: 10.1038/nmeth.1270

35. Andrews S. (2010) FastQC: a quality control tool for high throughput sequence data. Available: www.bioinformatics.bbsrc.ac.uk/projects/fastqc/Accessed 2012 January.

36. Li H, Durbin R (2009) Fast and accurate short read alignment with Burrows-Wheeler transform. Bioinformatics 25: 1754–1760. doi: 10.1093/bioinformatics/btp324

37. Li H, Handsaker B, Wysoker A, Fennell T, Ruan J, et al. (2009) The Sequence Alignment/Map format and SAMtools. Bioinformatics 25: 2078–2079. doi: 10.1093/bioinformatics/btp352

38. Nei M (1987) Molecular evolutionary genetics. New York: Columbia University Press. 512 p.

39. Buggs RJ, Chamala S, Wu W, Gao L, May GD, et al. (2010) Characterization of duplicate gene evolution in the recent natural allopolyploid Tragopogon miscellus by next-generation sequencing and Sequenom iPLEX MassARRAY genotyping. Mol Ecol 19 Suppl 1132–146. doi: 10.1111/j.1365-294x.2009.04469.x

40. Paria N, Raudsepp T, Pearks Wilkerson AJ, O'Brien PC, Ferguson-Smith MA, et al. (2011) A gene catalogue of the euchromatic male-specific region of the horse y chromosome: comparison with human and other mammals. PLoS One 6: e21374. doi: 10.1371/journal.pone.0021374

41. Iwase M, Satta Y, Hirai H, Hirai Y, Takahata N (2010) Frequent gene conversion events between the X and Y homologous chromosomal regions in primates. BMC Evol Biol 10: 225. doi: 10.1186/1471-2148-10-225

42. Rosser ZH, Balaresque P, Jobling MA (2009) Gene conversion between the X chromosome and the male-specific region of the Y chromosome at a translocation hotspot. Am J Hum Genet 85: 130–134. doi: 10.1016/j. ajhg.2009.06.009

43. Conrad DF, Keebler JE, DePristo MA, Lindsay SJ, Zhang Y, et al. (2011) Variation in genome-wide mutation rates within and between human families. Nat Genet 43: 712–714. doi: 10.1038/ng.862

44. Roach JC, Glusman G, Smit AF, Huff CD, Hubley R, et al. (2010) Analysis of genetic inheritance in a family quartet by whole-genome sequencing. Science 328: 636–639. doi: 10.1126/science.1186802

45. Clutton-Brock J (1999) A Natural History of Domesticated Mammals. Cambridge, Massachusetts: Cambridge University Press. 248 p.

46. Natanaelsson C, Oskarsson M, Angleby H, Lundeberg J, Kirkness E, et al. (2006) Dog Y chromosomal DNA sequence: identification, sequencing and SNP discovery. BMC Genet 7: 45. doi: 10.1186/1471-2156-7-45

47. Cunningham E, Dooley J, Splan R, Bradley D (2001) Microsatellite diversity, pedigree relatedness and the contributions of founder lineages to thoroughbred horses. Anim Genet 32: 360–364. doi: 10.1046/j.1365-2052.2001.00785.x

48. Kavar T, Brem G, Habe F, Sölkner J, Dovc P (2002) History of Lipizzan horse maternal lines as revealed by mtDNA analysis. Genet Sel Evol 34: 635–648. doi: 10.1186/1297-9686-34-5-635

49. Bower MA, Campana MG, Whitten M, Edwards CJ, Jones H, et al. (2011) The cosmopolitan maternal heritage of the Thoroughbred racehorse breed shows a significant contribution from British and Irish native mares. Biol Lett 7: 316–320. doi: 10.1098/rsbl.2010.0800

50. Bower MA, Campana MG, Nisbet RE, Weller R, Whitten M, et al. (2012) Truth in the bones: resolving the identity of the founding elite thoroughbred racehorses. Archaeometry 54: 916–925. doi: 10.1111/j.1475-4754.2012.00666.x

51. Hill EW, Bradley DG, Al-Barody M, Ertugrul O, Splan RK, et al. (2002) History and integrity of thoroughbred dam lines revealed in equine mtDNA variation. Anim Genet 33: 287–294. doi: 10.1046/j.1365-2052.2002.00870.x

52. Lenstra JA, Groeneveld LF, Eding H, Kantanen J, Williams JL, et al. (2012) Molecular tools and analytical approaches for the characterization of farm animal genetic diversity. Anim Genet 43: 483–502. doi: 10.1111/j.1365-2052.2011.02309.x

53. McCue ME, Bannasch DL, Petersen JL, Gurr J, Bailey E, et al. (2012) A High Density SNP Array for the Domestic Horse and Extant Perissodactyla: Utility for Association Mapping, Genetic Diversity, and Phylogeny Studies. PLoS Genet 8(1): e1002451. doi: 10.1371/journal. pgen.1002451

54. *Eclipse At New Market With Groom* (British racehorses of the 18th century), by Stubbs George (1724–1806); source: Wikimedia Commons, public domain, *copyright has expired.*

CITATION

CHAPTER 1

Mao, W. and Lee, J. (2008) A Combinatorial Analysis of Genetic Data for Crohn's Disease. Journal of Biomedical Science and Engineering, 1, 52-58. doi: 10.4236/jbise.2008.11008.

CHAPTER 2

Anil V Parwani M.D., Ph.D, Joseph Geradts M.D., Eric Caspers M.D., G Johan Offerhaus M.D., Charles J Yeo M.D, John L Cameron M.D., David S Klimstra M.D., Anirban Maitra M.D., Ralph H Hruban M.D. and Pedram Argani M.D., Immunohistochemical and Genetic Analysis of Non–Small Cell and Small Cell Gallbladder Carcinoma and Their Precursor Lesions, DOI: 10.1097/01.MP.0000062656.60581.

CHAPTER 3

Kuramoto T, Nakanishi S, Ochiai M, Nakagama H, Voigt B, Serikawa T (2012) Origins of Albino and Hooded Rats: Implications from Molecular Genetic Analysis across Modern Laboratory Rat Strains. PLoS ONE 7(8): e43059. doi:10.1371/journal.pone.0043059.

CHAPTER 4

Dominik T Schneider1, Susanne Zahn1, Sonja Sievers1,2, Katayoun Alemazkour1, Guido Reifenberger3, Otmar D Wiestler4,†, Gabriele Calaminus1, Ulrich Göbel1 and Elizabeth J Perlman, Molecular genetic analysis of central nervous system germ cell tumors with comparative genomic hybridization, doi:10.1038/modpathol.3800607.

CHAPTER 5

Sue S Yom, Asif Rashid, David I Rosenthal, Danielle D Elliott, Ehab Y Hanna, Randal S Weber and Adel K El-Naggar, Genetic analysis of sinonasal adenocarcinoma phenotypes: distinct alterations of histogenetic significance, doi:10.1038/modpathol.3800315.

CHAPTER 6

Sousa-Neves R, Rosas A (2010) An Analysis of Genetic Changes during the Divergence of Drosophila Species. PLoS ONE 5(5): e10485. doi:10.1371/ journal.pone.0010485.

CHAPTER 7

Huang Q, Li G, Husseneder C, Lei C (2013) Genetic Analysis of Population Structure and Reproductive Mode of the Termite Reticulitermes chinensis Snyder. PLoS ONE 8(7): e69070. doi:10.1371/journal.pone.0069070.

CHAPTER 8

Pan Y, Zhang H, Zhang D, Li J, Xiong H, Yu J, et al. (2015) Genetic Analysis of Cold Tolerance at the Germination and Booting Stages in Rice by Association Mapping. PLoS ONE 10(3): e0120590. doi:10.1371/journal.pone.0120590.

CHAPTER 9

Yates WR, Johnson C, McKee P, Cannon-Albright LA (2013) Genetic Analysis of Low BMI Phenotype in the Utah Population Database. PLoS ONE 8(12): e80287. doi:10.1371/journal.pone.0080287.

CHAPTER 10

Dong Zhang, Wenqian Kong, Jon Robertson, Valorie H Goff, Ethan Epps, Alexandra Kerr, Gabriel Mills, Jay Cromwell, Yelena Lugin, Christine Phillips and Andrew H Paterson, Genetic analysis of inflorescence and plant height components in sorghum (Panicoidae) and comparative genetics with rice (Oryzoidae), DOI: 10.1186/s12870-015-0477-6.

CHAPTER 11

Petersen JL, Mickelson JR, Cothran EG, Andersson LS, Axelsson J, Bailey E, et al. (2013) Genetic Diversity in the Modern Horse Illustrated from Genome-Wide SNP Data. PLoS ONE 8(1): e54997. doi:10.1371/journal.pone.0054997

CHAPTER 12

Badro DA, Douaihy B, Haber M, Youhanna SC, Salloum A, Ghassibe-Sabbagh M, et al. (2013) Y-Chromosome and mtDNA Genetics Reveal Significant Contrasts in Affinities of Modern Middle Eastern Populations with European and African Populations. PLoS ONE 8(1): e54616. doi:10.1371/journal.pone.0054616.

CHAPTER 13

Wallner B, Vogl C, Shukla P, Burgstaller JP, Druml T, Brem G (2013) Identification of Genetic Variation on the Horse Y Chromosome and the Tracing of Male Founder Lineages in Modern Breeds. PLoS ONE 8(4): e60015. doi:10.1371/journal.pone.0060015.

INDEX